Environmentalism Unbound
Exploring New Pathways for Change

Robert Gottlieb

The MIT Press
Cambridge, Massachusetts
London, England

First MIT Press paperback edition, 2002
© 2001 Massachusetts Institute of Technology

This book was set in Sabon by Best-Set Typesetter Ltd., Hong Kong, printed on recycled paper, and bound in the United States of America.

Library of Congress Cataloging-in-Publication Data

Gottlieb, Robert, 1944–
 Environmentalism unbound: exploring new pathways for change /
Robert Gottlieb.
   p.   cm.—(Urban and industrial environments)
 Includes bibliographical references and index.
 ISBN 0-262-07210-6 (hc : alk. paper), 0-262-57166-8 (pb)
  1. Environmentalism—United States. 2. Environmental justice—United States.
 3. Pollution prevention—United States. I. Title. II. Series.
GE197 .G67 2001
363.7—dc21

                                                        00-060932

Environmentalism Unbound

Urban and Industrial Environments
Series editor: Robert Gottlieb, Henry R. Luce Professor of
Urban and Environmental Policy, Occidental College

For Marge, Casey, and Andie, *for making it all worthwhile*
And for Ruth, Bob, Liz, Thanasi, Cathy, and Suzanne,
for sharing the passion

Contents

Preface: Common Visions and Strategies for Change

The Pico Farmers' Market

Each Saturday, about a mile from where I live, I shop at the local farmers' market. While purchasing the fresh produce that's in season, I get a chance to chat with friends, neighbors, and the farmers who sell to me each week. Spread out over a large vacant lot on the corner of Pico and Cloverfield boulevards in the city of Santa Monica in southern California, this market, known by the locals as the Pico Farmers' Market, is a visual feast of colors and foods. It comes with the sounds and bustle of shoppers and farmers, of mariachi bands and Bolivian folksingers. An active, vibrant public space, it virtually explodes on the scene each Saturday all year round, made possible by the continuous southern California growing season. As a public space, it helps recast what is otherwise a nondescript street of gas stations and small shops. These rim a predominantly Latino and African-American low-income neighborhood of bungalow houses and mildly decaying one- and two-story apartment buildings typical of post–World War II Los Angeles; a neighborhood that had also been split in two with the construction of an urban freeway in the 1960s.

There are now fifty farmers' markets operating in the Los Angeles region. The Pico market is one of a handful of these markets situated in low-income or bridge communities. Shoppers at the Pico market come from the neighborhood where the market is located as well as from several surrounding middle-class communities in Santa Monica and Los Angeles. The market manager has informed me that during the summer as many as 10 percent of the Pico market shoppers use coupons available through the Women, Infants, and Children (WIC) program, an indicator of low-income participation. As it turns out, this bridge role has been important to the success of the market, as it has with some other markets around the country, such as Baltimore's Waverly Farmers' Market and the Heart of the City Market in San Francisco.[1] Sales have grown steadily since the market first opened in 1991. With annual revenues of $1.2 million, it rivals some of the largest farmers' markets in the state, if not the country. Some of the items sold, such as dates from the Imperial Valley, an espresso machine at the entranceway, and porta-bella mushrooms from a farmer in Camarillo, are attractive to its middle-class customer base. But many of the other items—fresh nopales, hot habanero peppers, elegant long-stem Asian green beans, baby bok choy, and much more—appeal to the market's diverse base of shoppers, many of them from the different ethnic groups that constitute multicultural Los Angeles.[2]

Leasing a stall at the market can mean significant business for one of the more than forty farmers who sell at the Pico market each week. Many of these farmers sell at other farmers' markets as well and have relied on the direct marketing of the food they grow as an essential part (50 percent or more) of their revenue stream.[3] The farmers, like the shoppers, are a diverse group. They include a Latino farmer and her family from Oxnard who sells high-quality Chandler strawberries, calico corn, and a wide array of vegetables. There is a Central American immigrant family that has been forced to shift its farming operation three times when its leased lands were subdivided for residential and commercial units. There is also a group of Latino farmworkers who have started their own farm in Orange County and have become totally dependent on the farmers' market system for their revenue. In the next stall is a Chinese farmer from Riverside who sells an exquisite array of cut flowers. Sprinkled throughout the market are several Japanese farmers who specialize in Asian vegetables. And, at the entranceway, there is a third-generation Anglo farmer from San Diego who specializes in Valencia oranges, white grapefruits, and baby avocados.

Although not usually identified as such, environmental themes are present everywhere, essentially connected to every part of the market's activities. Pesticide use is one crucial though complex subtext of the farmers' market system; a number of farmers who sell at the market do not spray or have limited pesticide use, although not everyone is certified as organic. Connection to food source provides an important environmental link, helping sustain alternative relationships between producers and consumers—relationships that have powerful land use implications, reconnecting city and countryside, and providing what an earlier generation of urban environmental advocates called "a region—to live in!" Demonstrating the growing cycle, from the earth back to the earth, is another idea that has been raised. The farmers' market manager, a local visionary and one-time bagel shop owner who also functions as a city bureaucrat, has sought to convince city officials to use a small area of the vacant land for a compost site. Food waste generated at the end of the selling day would be turned into compost, to then be used for food growing within the city at community and school gardens. A farmers' market fruit and salad bar has been established as part of the lunch program for a nearby low-income elementary school and, due to its success, the program has been expanded to other schools in Santa Monica and other school districts as well. Such a program is seen as a way to support the local farmers and to provide fresh food for the school cafeteria, to help establish healthy eating patterns for kids otherwise deeply enmeshed in the fast-food culture. Perhaps most important, the Pico market provides a community space for shopper and farmer alike; at once the occasion and location for these different types of food system activities. And embedded in these alternative activities is an implicit, if not explicit, critique of the outcomes of the dominant food production and consumption system, whether hazardous chemical use, waste generation, long-distance transport, or intensive energy use, involving global players with little or no connection or loyalty to place. Stated differently in the context of this book, the activities associated with the Pico market can be seen as helping to plant the seeds for a new kind of environmentalism that can contribute to a new type of social agenda.

Linking Environmental Justice and Pollution Prevention

Two recent currents within environmentalism that have suggested opportunities for developing a new type of movement and agenda are environmental justice and pollution prevention. Both these currents have identified strategies for wide-ranging changes in industrial organization and urban life. Neither environmental justice nor pollution prevention, however, has been able to articulate a common vision or comprehensive strategy for change, whether in the form of community and worker empowerment or in terms of urban and industrial restructuring. The difficulty in establishing connections and broadening perspectives has limited the arena for environmental action. Such difficulties have also reinforced environmentalist tendencies to focus on separate forms of pollution, undertake efforts to modify discriminatory environmental burdens, or create distinctive advocacy modes that separate "natural" environment from "human" environment issues.

As a consequence, environmental justice and pollution prevention advocacy, similar to the state of environmentalism itself, is constantly forced into a more defensive mode. Concentrating on incremental change within the limits of the current environmental policy system, the movement finds itself losing its ability to envision new social and ecological relationships. Some mainstream environmental groups, in seeking to distance themselves from a much criticized command-and-control environmental policy system, have acquired the language of the market, promoting voluntary action by industry and market-derived solutions. But while a publicly responsive or "social" market may offer opportunities for environmental change, such as through tax shifting and incentive- and disincentive-based programs, most of the talk about markets precludes a public role. These private or voluntarist initiatives, once separated from community pressures and policy interventions, become then narrowly framed. "Beyond compliance," the mantra of the voluntarists, emerges as a form of regulatory avoidance, not the proposed breakthrough into new technologies or respect for the environment in the design of products and processes or the development of livable communities.

As these changes have taken place, environmental advocacy continues to find it difficult to answer the charge that it is just a special-interest activity. Confining the movement to such a special-interest role has also influenced the course of environmental justice and pollution prevention.

The concept of environmental justice, once a promising focus for social movement activity, has been narrowly interpreted by policymakers as risk discrimination. This approach represents a limiting or bounded interpretation of the environmental justice agenda, associated with the state's ability to remedy discriminatory environmental burdens, rather than a more proactive emphasis on the restructuring of social and ecological relationships. Yet many environmental justice groups have sought to expand their activities beyond the risk burden approach, influenced by the multiple needs of their constituencies. Many have begun to focus on economic development concerns such as job creation, or community needs such as food access or transportation.

Pollution prevention, meanwhile, has largely been shaped by the language of voluntarism. Yet in its origins, pollution prevention sought to identify a new approach to industry activity and responsibility. A tension emerged between pollution prevention as a public program for industrial restructuring and a more bounded view of pollution prevention as a form of environmental management predicated on the view that industry and the market always know best. With the 1994 election of a new Republican majority in Congress hostile to environmental regulation, the debate over environmental policies was narrowed to either regulatory relief as part of the beyond compliance/voluntarism framework for industry action or the elimination of regulation altogether. Absent from this debate has been the potentially significant role that pollution prevention can play in redirecting environmental activity toward new industry/community/environment relationships.

Despite this narrowing of policy options, the imaginative, transformative powers of environmental action have not disappeared. A new framework for social change activity can be identified by reconstructing and linking environmental justice and pollution prevention through a more radical, community- and workplace-centered, or production-focused and place-based approach. How that reconstituting of environmental activity can occur remains an open-ended question, influenced by the ways in which environmental issues are identified and whether a more expansive environmental discourse takes root, and environmental actions and policies are reconstituted.

A Research Agenda: Identifying Pathways

In 1992, partly in response to the urgent concerns of my students to link their environmental research to community development needs in the wake of the Los Angeles civil disorders, I began to identify a series of research questions and shape a set of research projects. These, I hoped, could evaluate the potential for the development of a new type of environmentalism. At the time, I was working on two manuscripts. The first, *Forcing the Spring*, sought to identify a broader historical view of environmentalism, including the historical roots of the environmental justice approach. It also analyzed the distinct segments of the contemporary environmental movement in relation to that history. The second book, *Reducing Toxics*, was a product of the Pollution Prevention Education and Research Center (PPERC), which I had cofounded and which subsequently became part of the Urban and Environmental Policy Institute. *Reducing Toxics* sought to analyze the evolution of pollution prevention policy and industry decision making and identify what new forms of decision making would be necessary for pollution prevention to extend beyond its voluntarist confines.[4]

The research for *Forcing the Spring* and *Reducing Toxics* and my subsequent projects led me to the central argument of this book. By reconstituting and integrating environmental justice and pollution prevention into a common vision and set of strategies for change, I argue that environmentalism can help lay the groundwork for fundamental social and environmental change, to make industries more socially responsive and greener, and to make communities more livable. But how can that best be accomplished, given environmentalism's own complex and often disjointed history? And how does a common vision and set of strategies for change provide answers to questions that seem elusive at best, and impenetrable in terms of the prevailing modes of environmental policy, industrial organization, production choices, and community life?

Through this research agenda I had hoped to explore and evaluate some key issues regarding industry activities and community life, and to test the argument about how a new type of environmentalism could emerge. A series of more specific questions emerged:

• When an industry is identified as a polluting industry, how can environmental solutions be developed on the basis of common need? And how can a community of interests be identified in the search for those solutions?

• How do workers become engaged in identifying pollution prevention solutions, when workers, such as those in low-wage sectors, are sometimes seen as resisting product or process innovations?

• How does the deskilling of work influence environmental outcomes, and, conversely, how can the reintroduction of work skills contribute to environmental and social change?

• How can communities influence production systems to address their social and environmental impacts, including those presumed to be meeting basic human needs, such as food, clothing, or transportation?

• How do regional or sectoral considerations frame environmental problems and solutions? Can environmentalism become part of the new regionalism?

• How can environmental justice inform the process of revitalization of urban, rural, and "natural" environments and communities?

• And, given the historical and contemporary ambiguities in environmental discourse, can a more embracing language emerge that identifies and elaborates the connection or the fusion of the social and the ecological?

The research projects I've been engaged in during the past five years were designed in part to address those questions. While pursuing the research, the projects also came to represent efforts to identify and understand the barriers and opportunities for social and environmental change.

The three research areas discussed here represent different entry points for evaluating the book's basic premise. The first area involves a small industry in crisis (dry cleaning), the debate over alternative pollution prevention approaches, and whether such alternatives can help establish a new community of interests. The second area involves a set of products (janitorial cleaning products) that may be hazardous to workers, and how such hazards can and have been addressed in the context of the search for social and environmental justice. The third area explores the barriers and opportunities in constructing a community or regional food systems approach in the face of a globalizing food system that has changed the very nature of how we grow, make, and consume our food.

Each of these areas in turn provides discrete lessons about the prospects for change and the role of a renewed and less bounded environmentalism in helping to bring about such change.

Collaborators

This book is a product of many collaborations. Much of the work has been an outcome of the research and activities associated with the Urban and Environmental Policy Institute at Occidental College and prior to that of the Pollution Prevention Education and Research Center at Occidental and UCLA. My interest in environmental justice and pollution prevention was stimulated by the conversations, joint research, interdisciplinary courses, and outreach I undertook with many partners both inside and outside the institute. My work with the late Julie Roque was central to this collaboration; I continue to miss her as close friend and colleague and as someone who helped me understand what was possible, while keeping a critical perspective on the issues and movements we were exploring. John Froines, my closest colleague for more than ten years, first came up with the idea of a pollution prevention center while we watched our sons playing soccer. John is that rare academic pursuing scientific research who understands and can act upon the intricacies and complexities of policy. At Occidental, my ongoing collaboration with Peter Dreier has helped link my research on environmental issues to a broader context of urban and industrial change.

In terms of my projects, the research and activities associated with each of the areas described—garment care, janitorial work, and food systems—were made possible because of my partners and collaborators. The garment care chapters have been strongly informed by my work with Peter Sinsheimer, and earlier with Jessica Goodheart. Peter's persistence, research capacities, and insights have been invaluable in placing our center's research and the arguments associated with it at the heart of the debates regarding the barriers and potential for change in this much-beleaguered industry. Jessica was instrumental in organizing and shaping the Cleaner by Nature study that established the basis for identifying a new pathway. Jenni Cho, a former student, helped me to understand the importance and range of issues associated with the Korean cleaner community. In our study of janitorial cleaning products, Andrea Brown (who has been a partner in the development of our Center and co-authored one of the articles on this specific research) and Andrea Gardner played pivotal roles. For the food system chapters, Michelle Mascarenhas and Andy Fisher have helped build the foundation for what is proving to be a critical aspect of that common vision and strategy for change. Michelle's work with me extends in each of the arenas regarding food, schools, community health, land use, and urban life. The section on food and schools was based on a joint article we had written, while other parts of those chapters benefited from her insights. Andy, a member of the Seeds of Change group highlighted in these chapters, is now executive director of the Community Food Security Coalition, an important player in the development of a new type of environmental and social justice politics.

There are a number of other collaborators and colleagues who have helped me shape this work, explore different sources, develop the projects and research evaluations, and help frame the discussion about the ideas and strategies for change. They include: Dave Allen, Steve Ashkin, Andrea Azuma, Ken Geiser, Erika Higgins, Andrea Hricko, Liz Hill, Ohad Jehassi, Brian Johnson, Lewis MacAdams, Jo Patton, Debbie Raphael, Lucia Sanchez, Rodney Taylor, Leila Towry, Craig Tranby, Elaine Vaughan, and Jack Weinberg. I am particularly grateful for those who reviewed the manuscript or parts of the manuscript and provided valuable insights about the book's structure, research, and arguments, including Beth Braker, Gail Feenstra, Andy Fisher, Helen Ingram, Rod MacRae, Michelle Mascarenhas, Jan Mazurek, Peter Sinsheimer, Maureen Smith, and Mark Winne. Support from the Occidental community included Wendy Clifford and John de la Fontaine. Clay Morgan, my editor for this book and for the Urban and Industrial Environments series at MIT Press, helped see this project through. Clay's predecessor at MIT, Madeline Sunley, was also instrumental in helping establish this series.

With each of these collaborators, I've sought to understand the ways in which research informs action and our actions provide relevance for our research. For those who have shared in the research and explored pathways through their actions, their efforts have become a part of my own continuing exploration about a renewed environmentalism's contribution to an agenda for change.

I

Breaking Boundaries

1
Environmentalism Bounded:
Discourse and Action

Reinterpreting the Roots

Environmentalism today can be characterized as a set of skilled organizational actors, fundraising apparatuses, effective though often outgunned lobbyists, and highly motivated, even passionate groups of individuals seeking to save some environmental amenity or overcome some environmentally destructive practice or action. Environmentalism can also be considered a collection of groups that operate in a largely reactive and ad hoc basis, with only a narrow focus for their actions and identifiable goals. While the environmental movement in recent years has sought to broaden its base of participants by strengthening its capacity to address daily life issues, an undercurrent of malaise and uncertainty still prevails. One wonders: has the once great promise of environmentalism dissipated? Is this a movement that is best at reacting, but not envisioning, subject to the limits imposed by a system that effectively discourages opportunities for social change from below?

Contemporary U.S. environmentalism is considered to have emerged out of the turbulence of the urban and industrial restructuring of the latter half of the nineteenth century and first two decades of the twentieth century. The current arguments about whether Nature is socially determined and thus needs to be "reinvented" are primarily rooted in the interpretations of the great upheavals that were taking place in that period.[1] These shifts included, among other changes to landscape and social organization, the growth and restructuring of mining, grazing, agriculture, transportation of goods and people, timber production, and other forces that literally reconfigured whole territories and large land bases and watersheds within the United States. While some of these changes affected places that were identified as wilderness, they were nevertheless human inhabited and socially defined. In the process, the relationships between urban and rural as well as human places and wilderness areas were recast within the new language and logic of urban and industrial development. Groups were formed either to address the excesses of a resource-based capitalism or to try to segregate capitalist development from certain protected areas valued for their aesthetic, spiritual, or recreational associations. These groups did not use the language of what we today refer to as environmentalism. Instead, traditional environmental history identified two types of approaches: preservation of (nonhuman) wildness on the one hand, and the conservation and efficient management of (socially utilized) resources on the other. Environmental advocates and movements were thus assumed to fall into either of these camps. In one camp were the preservationists seeking to protect, appreciate, and enjoy Nature, or natural places. In the other camp were the conservationists seeking to efficiently or instrumentally use Nature as potential resources for social goals, albeit by eliminating the destructive tendencies of an unregulated market.

This prevailing historical view of the conservationist and preservationist roots of environmentalism came to be challenged by a more socially rooted environmental history that emerged in the 1970s and 1980s. These historians were strongly influenced by the new social movements that had embraced the developing language of ecology and environmentalism and that made only partial reference to the earlier conservationist and preservationist approaches.[2] Urban and industrial issues such as mode of transportation, technology choice, land use, or hazardous materials use and the problems of pollution and waste byproducts and disposal (among a complex array of problems) became significant if not dominant environmental themes. They constituted the arena in which environmental action was unfolding as well as the subject matter for historical research and analysis. Environmentalism, despite the long-standing assumptions about its location as an exclusively Nature-centered movement, appeared to have become the place where ecological and social issues were necessarily joined.

Disputes over Discourse

Despite the growing importance of urban and industrial issues in the agendas of most environmental groups, there remained during the 1970s, 1980s, and 1990s, disjointed and at times contradictory elaborations of what can be called *environmental discourse*. These elaborations were reflected in the use of language and presentation of ideas and concepts about the environment, much of which still assumed the long-standing separation of the social from the ecological.

These disputes over discourse have also been reflected in the divergent pathways and discordant messages of environmentalism in the last three decades, despite the movement's successful appeal. The significance of these disputes over environmental discourse should not be underestimated. Since discourses, as David Harvey has argued, are "manifestations of power . . . the coded ways available to us for talking about, writing about, and representing the world," they also become ways to shape actions as well as perceptions. More than many other social movements, environmentalism has become associated with compelling ideas and images—whether of Nature (the value of wilderness) or Society (the negative associations of urban pollution or hazards). These images are made manifest by language and representation. The power of environmental discourse also makes it fair game for varying interpretations and associations. Claims made by different actors, such as the chemical or oil and gas industries, about their commitment to the environment and participation as "environmentalists" often muddies the debate about nature and environment and further underlines the importance of who controls the discourse. Environmental politics should therefore be seen not simply as debates over how to act or what policies to establish. Conflicts of interpretation over the terms of environmental discourse also become debates over how to influence the language that people use in talking about the environment.[3]

Since Thoreau, environmental discourse has been most often embedded in representations of Nature as an unspoiled world separated from human activity. By way of example, go to any chain bookstore and the environmental selections are almost invariably housed among the Nature books. Once captured in language, the idea of Nature takes on its own life, further creating the impetus for actions that seek to decouple the environment/society connection. These are actions rooted in, stimulated by, and ultimately experienced largely through words and visual representations. Moreover, a discourse that distinguishes and separates the built environments of cities, rural areas, and wilderness, underlies what Harvey has called "a pervasive anti-urban bias in much ecological rhetoric."[4] Environmental discourse therefore refers not only to representation and the power of meaning, but the historical interpretations of those meanings and the battleground over action and agenda associated with such interpretations. Three key arenas—land and resources, city and countryside, and work and industry—reflect those environmental debates and the evolving ideas about the separations and intersections of environment and society.

Land and Resources

Hunters

When my wife was growing up in the Fox River Valley in northwest Illinois during the 1950s, the region was still considered largely rural. While the valley had already been developed as an agricultural region linked to the city of Elgin (the major urban center in the valley), Dundee, the small town where she lived, was considered nonurban or even semirural. The Fox River flowed directly through the town and into the surrounding countryside of farms and woodlands, and could also be identified as a "natural river" or at least establishing some semblance of open space. A fair number and variety of fish and bird life could be found in and around the river, while other types of wildlife, though dwindling, could be found in the surrounding countryside. On the banks of the Fox lived my wife and her family in a small house that had been built during the Civil War. Next door lived her Uncle Harold, who, along with his friends and his children, had developed a great passion for fishing and hunting. Some of this activity took place upstream on the Fox, or further north in Wisconsin or Michigan where open space and woodlands were even less developed and richer in wildlife.

Over the years, Uncle Harold had developed a great skill at duck carving, because duck decoys were an essential part of the hunting experience. By the mid-1960s, when my wife went off to college, Uncle Harold had become the expert craftsman, having carved, in just a few years, the entire array of ducks that inhabited the states of Illinois and Wisconsin. This passion for his craft and his parallel passion for the outdoors had been an integral part of Harold's life ever since he had explored the river as a boy. Uncle Harold, who had retired from his work at the local gas station, devoted much of his time to carving and was increasingly recognized as a major figure within this hunting subculture. For Harold, the great pleasure of carving was its association with the hunting experience, an experience that established a link to the natural world not dissimilar to Aldo Leopold's hunting experiences in Sand County in southern Wisconsin where Uncle Harold had also roamed with his son and his hunting buddies. For uncle Harold and his fellow duck carvers, the carving remained part of the connection to the "natural world," an extension rather than separation of his social universe and activities with the experience of Nature.

By the mid-1960s, the land use dynamics affecting the urban-rural relationships of the Fox River Valley began to change. A few years earlier, a chemical plant had located upstream from where my wife's family and Uncle Harold lived, and rapidly began to affect the ecology of the river. No longer would the river freeze over during the winter, while the numbers and kinds of ducks on the river also began to decline. At times, odors from the chemical plant became so strong that the river itself began to have a distinctive chemical smell. In the early 1970s, an environmental monkeywrencher named "the Fox" captured some of the discharges from the chemical plant and dumped these chemical wastes back at company headquarters, much to the delight of even some of the more conservative residents of the community.

But it was the land sprawl associated with contemporary forms of urbanization that most influenced the type of place the Fox River Valley was becoming. New housing developments and the construction of giant shopping malls underlined the rapid conversion of farmland and other open spaces through the continuing reach of Chicago's northwestern suburbs. By the late 1960s, my wife's sister, six years younger, was already defining her growing up in the Fox River Valley as a semisuburban rather than a small town or rural experience. The evolving relationship of the midsized city of Elgin to the surrounding countryside and small town environment of the valley extended this pattern of suburbanization. Elgin, during the 1950s, 1960s, and 1970s, had essentially deindustrialized, with a loss of manufacturing jobs and a declining relationship between farm activity and the nearby city infrastructure that had anchored the valley. Seeking to capitalize on the growing trend in Illinois and elsewhere to combine nostalgia with the more perverse forms of economic development, Elgin in the early 1990s began its experiment with "riverboat gambling." This would become Elgin's way to make the city more financially secure and capture a bit of the income available at the edges of the suburban growth.[5]

Forced to seek out hunting and fishing opportunities further from his home in the Fox River Valley, Uncle Harold continued his duck carving. He would work methodically in his basement workshop or sit at the edge of the river watching and observing the flight of the Canadian geese or the pintail ducks that passed by his home. For Uncle Harold, duck carving was simply and purely the craft of the hunter. That association remained for him until his death at the age of ninety-one, even as his work increasingly began to be perceived as a form of "art," a representation of an abstracted natural world rather than an expression of the connection between place, activity, and the universe of natural and social environments.

Hikers

The wind blew up from the river, fresh and mysterious, against my face. The air was alive with the faint odor of juniper. Far, far away, beyond the river, beyond the canyons, beyond countless miles of mesa, so far away that they were sometimes mountains of earth and sometimes mountains of an ancient, dried-out moon, rose a snow-covered divide that seemed to bound the universe. Between me and this dimmest outpost of the senses was not the faintest trace of the disturbances of man; nothing, in fact, except nature, immensity, and peace.

If it is not always possible to bring the people to the forests, it is sometimes possible to bring the trees to the people.[6]

—Robert Marshall

Robert Marshall, socialist, cofounder of the Wilderness Society, and U.S. Forest Service employee, was in his Forest Service capacity a strong advocate for a socially responsible "use" philosophy in managing the forests. Marshall also promoted a concept of wilderness that included both a protectionist approach (strictly opposed to development) and an emphasis on participation in and access to the wilderness experience by low-income groups and other urban residents. These potentially contradictory perspectives were unified for Marshall in his view of the need to link city and countryside, or urban and rural places, a link bounded by the *experience* of Nature, as well as the management of Nature and its resources. For this inveterate hiker, unlike many of his prowilderness contemporaries of the 1920s and 1930s, the concept of seeking to protect an "untamed wilderness" became related to the need for "open-air recreation for low-income groups."[7]

His was particularly the hiker's vision. For Marshall, the ability to hike off the beaten track, in places unexplored, where Nature emerges as a source of inspiration for all those who encounter it, represented a source of fulfillment. "It was particularly elating to stride along the world above the 10,000-foot contour where an energy unknown at ordinary elevations seemed to be liberated. One felt like keeping on forever and forever," Marshall wrote of a 1938 forest wilderness trip in the Wind River Mountains near Mount Baldy in California. Marshall's hiking became a form of self-discovery, and his essays, letters, and books were indeed rooted in those traditions of Nature writing as self-discovery. "Off the trail you are on your own," he characterized the New Deal Forest Service perspective on wilderness or semiwilderness Forest Service areas. "Goodbye to the endless repetition which machinery imposes; goodbye to a life where the same events are certain to be repeated, to the world where you know exactly what lies down the street, or around the corner. Here is a new world, clean and shining; a new life!"[8]

Self-discovery for Marshall was active and participatory. His descriptions of the white and Eskimo people living in the Koyukuk wilderness region in Alaska speak of the "tone, vitality, and color to the entire functioning of life" associated with daily life in the region. The experience of the hiker captures, at least in its limited episodic form, that "tone, vitality, and color." Hiking is action, exploration, a form of encounter and discovery. It becomes an experience of recreating self, a form of "outdoor culture," as regional planner and Marshall's Wilderness Society colleague Benton MacKaye put it.[9]

For Marshall as well as for MacKaye, hiking also included, as an essential mission of wilderness advocacy, bringing Nature into the daily lives of people in the form of open or green space, connecting such spaces in urban as well as what MacKaye called "primeval" or wilderness settings. Marshall's own ambivalence about "the mechanistic civilization of urban life," his celebration of the "grandeur of primeval woodlands," and his promotion of wilderness exploration as a source for genuine adventure and self-discovery that he wrote about in the 1920s and early 1930s, reemerged as a more explicitly social vision during the New Deal. The experience of Nature for Marshall became relational, establishing a continuum of places and relationships and opportunities for both the hiker and the observer. Though Marshall's name eventually came to be associated with the grandeur of the wilderness areas of the West (e.g., the Bob Marshall wilderness area in Montana), he became increasingly focused during the 1930s on less spectacular yet crucial wilderness and semiwilderness areas more accessible to urban residents. Like MacKaye, who envisioned an Appalachian Trail as both a hiker's universe and as part of a broader urban-rural reconnection and reconfiguration, Marshall focused on places like the Adirondacks, the Appalachians, and other less spectacular eastern and midwestern mountains, hills, and grasslands. These, for Marshall, were also places in which one could become reinvigorated, through hiking and other outdoor experiences, to make Nature more directly a part of daily life.[10]

Nature Consumers

While hunting and hiking established direct human connection to the natural world, those connections would become influenced and redefined by the forces of the market through much of the twentieth century. This redefinition was significantly shaped by a rapidly growing recreation industry that sought to associate natural places and wildness with scenic resources to be consumed as objects of recreation.

The connection to environment as a form of consumption reinforced and extended this view of the separate spheres of environment and society. Samuel Hays's important insight that the post–World War II environmental movement increasingly became defined by its consumption orientation clearly emerges through the meanings ascribed to such words as "nature" and "environment." Defense of the environment becomes a defense of scenic resources. This was perhaps best captured with the publication of the large-sized, coffee-table picture books of the 1950s and 1960s with their spectacular representations of areas like Dinosaur National Monument, which straddled the Utah, Wyoming, and Colorado borders. These representations, in turn, became important fundraising tools and were central elements of the discourse-centered strategies of prominent groups like the Sierra Club and the Wilderness Society. Almost inevitably, given the constituency that embraced such environmental amenities (seen as recreational opportunities), environmental discourse came to be defined primarily in middle-class terms, a part of the suburban (or not specifically urban) quest for green and open spaces. Nature also became a place to be consumed, a cultural artifact defined by what John Urry has called the "tourist gaze," where the familiar (daily life in the urban environment) becomes separated from the faraway or exotic (Nature or wildness as tourist places).[11]

For groups like the Sierra Club and the Wilderness Society, the wilderness experience emerged as a particular type of recreational activity associated most prominently with the appreciation of a spectacular type of Nature. This monumentalist and recreation-based view of Nature, also linked to the idea of national identity in the nineteenth and early twentieth centuries, emerges fully blown in the 1950s and 1960s with the significant expansion of recreational opportunities and their affiliated industries. It was in this period that the relationship of consumer to product and its influence on environmental discourse became particularly pronounced. Environment became, through images, language, and ideas, primarily associated with the "great unspoiled world of nature" that groups like the Sierra Club so celebrated. The environmental experience then becomes representational (a commodity to be valued) rather than relational (connected to experience). "Wilderness, after all, is the most precious of our scenic possessions," wrote David Brower (then the Sierra Club's executive director and soon to emerge as a kind of preservationist icon) in the 1967 edition of the club's *Wilderness Handbook*.[12]

Perhaps more than any other environmental figure of the 1950s and 1960s, Brower helped establish the representation of wilderness (Nature to be consumed) as the centerpiece of contemporary environmental discourse. Brower's genius, following John Muir's footsteps, was his ability to create powerful imagery and language in connection with scenic places, or scenic resources as Brower and others defined these places. The pivotal environmental battles of the post–World War II period (the battle to save Dinosaur National Monument in the 1950s, the fights to stop facilities at Grand Canyon and Diablo Canyon in the 1960s, or the Storm King/Hudson River battles in the early 1970s) were framed as part of this Nature-centered discourse. This occurred even as other arguments (e.g., the economic inefficiency of a particular development) were also employed.

Even when more elaborated scientific discourses emerged (such as conservation biology or ecosystem protection) associated with the passage of administrative strategies or legislation (e.g., the Endangered Species Act), the concept of Nature as a visually and aesthetically appealing recreational resource set the parameters of policy for agencies like the National Park Service. These agencies set out to preserve the parks as "scenic pleasuring grounds," as Richard Sellars put it, primarily through policies that emphasized recreational amenities, such as the construction of better roads and even the use of pesticides to control for such pests as bark beetles. The environment ultimately became a commodity in the development of a market-driven recreational tourism. This consumption of Nature, whether in the form of national parks or other types of recreational resources, also served to sever the connections that nearby residents and local communities had established over time, thereby robbing such places of any social context. Nature could be protected, as long as it could also be consumed.[13]

Resource Managers

In 1933, Robert Marshall published *The People's Forests*, a plea for public ownership of the forests. Marshall's approach combined a social and environmental justice perspective (the forests belong to the people, the need to protect rural communities) with a more traditional conservationist or wise-use perspective (efficiency in the public interest). In this and other writings, Marshall sought to reinvigorate the social or "public interest" goals of conservationism that had largely disappeared by the 1930s.[14]

The social justice perspective that Marshall championed can be located in early conservationist discourse, even as it sought to occupy a middle ground between industry and labor, or capitalism and socialism. But that social agenda was constantly undermined by an expansive, resource-based capitalism. While preservationists increasingly associated with the recreation and tourism interests such as the railroads and the park concessionaires, the conservationist or resource management approach became strongly associated with the drive for increased production. This objective was directly linked to the role of the resource and extractive industries whose concept of "management" was synonymous with more extensive development. As early as the late 1910s and early 1920s, the U.S. Forest Service, the crown jewel of Gifford Pinchot's resource-based social progressivism, had evolved into a management arm of the timber industry. That process of industry capture extended to the Bureau of Reclamation, the Bureau of Land Management, and other resource management agencies, including those established during the New Deal for their job creation and rural development functions.[15]

In key environmental/development conflicts, such as the battle over control of the power supply provided by the Hetch-Hetchy project, the public interest argument about the need for public ownership became increasingly marginalized. Even where public ownership prevailed, as in the case of Los Angeles with its municipally owned water and power utility, the management of such resources was associated with expansionary urban real estate and development interests. The submerging of the public role due to the triumph of development on behalf of private interests was particularly noteworthy in the West, where some of the sharpest resource battles were to take place. By the 1960s, this public-private, development-oriented, resource management consensus had emerged full blown. As one example, the public and private resource managers constructed the "Grand Plan," so named by a Los Angeles Department of Water and Power official. The plan involved a linked development scenario for California and the Southwest, based on a string of Colorado River Plateau coal-fired power plants, water projects, and coastal-based nuclear power plants. Such collaborations among both regional interests and the public and private utilities signaled the elimination or marginalization of any public-interest objective, including environmental goals, within the resource management/conservationist perspective.[16]

This approach toward resource management had not gone unchallenged. Around the time that *The People's Forests* had been published, Marshall, Pinchot, and others had sought to regroup progressive conservationist forces and identify a renewed public interest role in resource management. But even the limited social planning initiatives associated with the New Deal, such as the Tennessee Valley Authority and the strategies for rural development through public and private partnerships, were themselves captured by the growth-minded urban and industrial interests who were so effectively reshaping the city and countryside relationships. Moreover, these shifts in the conservationist or resource management mission were further reinforced by the continuing role of science and expertise in resource management decision making and outcomes first elaborated in the Progressive Era. Resource managers increasingly relied on science and expertise (often purchased) to resolve potential conflicts or competing interests. Once a problem was identified (for example the depleting stocks of salmon or dwindling forests due to particular types of development) then scientists and experts could be employed by the resource managers to most effectively manage the problem, while not challenging the goal and nature of the development objectives. For example, California Governor Pat Brown, who presided over some of the most crucial water and resource development initiatives of the postwar period, was a key advocate of using science and expertise to provide scientific and technical justification for some of the massive water and transportation and other infrastructure projects he promoted, projects that also reshaped the relationship between the developed (and resource short) and less-developed (and resource-rich) regions.[17]

Much of this expertise-driven interpretation of the resource manager approach could be located in progressivism and subsequently in the New Deal. However, one could also derive from some of the New Deal initiatives that never came to fruition another type of interpretation about resources. These approaches explored resources as "materials" and situated the issues of nature protection versus development in the context of "place." This argument about appropriate materials use and a public role to counter the pressures of the market paralleled the interest of some New Deal regional planning advocates to consider the public role and the concern about place in addressing the nature-development-resource management relationships. By focusing on place, particularly at its most intimate levels (home, neighborhood, community, the daily life connections to Nature), the regionalists saw opportunities for countering the anti-Nature, unbalanced, market-driven forms of development. But while these perspectives floated at the edges of the land and resources debates in the 1920s and 1930s, they would only reappear in environmentalist discourse more than five decades later when they reentered the arguments about pollution prevention, materials use, and the environmental justice focus on "livable communities."[18]

Through the 1980s and 1990s, the nature-protection/consumption and resource-management approaches that had first emerged during the Progressive Era increasingly came to be contrasted with a promarket approach opposing any form of government or public role. The promarket position had also ebbed and flowed in the development of policies, but eventually achieved a peak moment with the triumph of the Republican "Contract with America" and the 1994 Congressional elections. That election also served to highlight the limits of the traditional nature protection or resource management discourse. For environmentalism to break out of its nature consumption or resource management boundaries, it also had to be associated with livable places. Like the hiker's inspiration, the duck carver's craft, or the trees and garden in the city, these were places that needed to be experienced and not just valued.

City and Countryside

The River in the City
Can Nature survive the city? The recent history of the Los Angeles River provides a cautionary but hopeful perspective. The L.A. River's now-straightened path winds through the heart of the city, encased in concrete for most of its length, and re-engineered to become the largest

urban flood control project of the Army Corps of Engineers. It has had, after twenty-five years of nearly continuous construction and 3 million barrels of concrete poured into its fifty-two-mile pathway, a thorough makeover. But with three soft bottom stretches and a new source of reclaimed water, the river has also reinvented itself as a home for over 200 species of birds, and poplars and elms and a wild array of vegetation that have also begun to coexist with urban life. With these changes, the struggle over the river's identity has become a type of discourse battle, influencing not only the river's course, but whether the land bordering it, including many of Los Angeles's diverse neighborhoods, can also reemerge as more livable community places.[19]

For most of the twentieth century, the idea of a free-flowing or "natural" urban river had been considered an oxymoron. For more than 100 years, urban rivers had come to be defined, like the city itself, as an "anti-environment," a form of battered and even disappearing Nature. Whether the Chicago River, which had its course reversed to send contamination downstream, or Cleveland's "burning river" (the Cuyahoga River), renowned for the fires that had been sparked by the discharges from the steel and chemical plants along its edge, urban rivers appeared to be "rivers" in name only. The L.A. River, more concrete than actual water and bounded by graffiti- and pocked-marked walls, had become perhaps best recognized as the backdrop for countless Hollywood scenes. Whether the car chase in *Repo Man*, the alien's motorcycle pursuit in *Terminator 2*, or the gigantic, irradiated ants crawling out of the river's storm drains in the 1950s science fiction film *Them*, the river has served as a convenient setting for the dark side of the urban environment.[20]

For most of the year, the Los Angeles River is nothing more than a semidry trickle. Then suddenly, even with just a modest rainfall, its concrete barriers are barely able to contain rapidly surging waters, a danger further magnified because the surrounding development and channelization have caused the loss of the absorptive capacity of the riverbeds and surrounding lands. Since 1991, the river upstream has also begun to flow more continuously, thanks to discharges from an upstream water treatment plant. Whether flood control channel or "effluent-dominated waterway," as its various engineer-managers also like to call it, the river has come to represent an urban land use and water supply management tool for the myriad of agencies that help define it. But for the residents living at its edge, the river has also become, with its toxins and assorted wastes that gather along its pathway, an urban eyesore and often menacing concrete environment identifying unsafe places. Perhaps more than any other place in the city, the Los Angeles River describes the region's environmental limits—the lack of livable spaces in modern-day Los Angeles.[21]

But can one consider L.A. River a "river," albeit an urban river, in more than name? On one side of the discourse battle reside environmental advocates for whom the Los Angeles River offers the opportunity to invent a new kind of urban environmental landscape. This reenvisioning of streams, bird havens, and reconstructed habitat becomes a form of urban environmental reconstruction, a plea to identify and save some aspect of nature in the city. While environmentalists succeeded in placing the L.A. River onto the policy agenda as an environmental asset, other *community* advocates have begun to focus on the river as a place to revitalize neighborhoods. This reenvisioning includes bike and walk paths, parkways, greenways, or new forms of community economic development that seek to turn the river into a community asset rather than hostile environment. But does the history of the river and its role in the development of Los Angeles belie the claims of its environmental and community advocates?

The river, in its pre-Anglo days, had been a modestly meandering river, rich with marshes, wildlife, grapevines, and wild roses, "a very lush and pleasing spot, in every respect," as its first Spanish chronicler, Father Juan Crespi, noted in his diary. The local Indians called the river "pwinukipar" or "full of water," as much for the swell of the winter storms as for the streams and changing patterns of river flows that crossed the Basin. The river also identified a regional watershed, of plant and animal habitat, of water-consumptive vegetation and trees such as willows and cottonwoods, and of the watercress, waterfern, and duckweed that had become part of the life that nestled by the river. During massive surges of flood waters, the river also provided an agricultural boon by depositing fresh layers of silt that increased the fertility of the land.[22]

With the first massive land booms of the 1880s, the process of recon-figuring the river and establishing land speculation as the dominant mode of urban settlement in Los Angeles became intertwined. The river began to serve as a source of water supply, encouraging development as well as being a potential barrier to that development due to its erratic and unpredictable nature, establishing a possible limit to what could be built and where development was to be directed.[23] Its role as water supplier quickly outstripped by the pace of development and the rapid extension of the urban boundaries, the river lost any significance other than as barrier for the land syndicates and other promoters of expansion. No longer reliant on the river, Los Angeles, like its eastern counterparts such as New York, Boston, and Philadelphia, which had procured distant water sources as a condition for expansion, continued to stretch the urban (or, more appropriately, suburban) boundary line on the basis of imported water. This imported water, first from the Owens River water-shed at the base of the Sierras and subsequently from the Colorado River, and then finally from northern California's Sacramento River system, helped create an expansionary spiral that became synonymous with the notion of *sprawl* in the twentieth century. In just fifteen years, Los Angeles expanded three times its size after the Owens aqueduct came on line in 1913, while the Colorado River Aqueduct, completed in 1941, promised to enable expansion to reach every direction in the basin. When the northern California water flowed over the Tehachapis in 1972, south-ern California now stretched from the desert to the sea. Southern Cali-fornia had become a region noted for its cycle of imported water fueling urban expansion that fueled the search for more distant imported water.[24]

In the 1930s, the Army Corps of Engineers, armed with new federal flood control legislation aimed at creating public works jobs, had decided to "declare war on the Los Angeles River" after another series of massive floods wreaked havoc to the developed areas. The Corps proposed to tame the river through a series of construction programs estimated to cost $200 million. This included a massive leveed channel from the river's headwaters to the ocean, the development of three flood control basins, and various channel improvements including those along the river's trib-utaries and washes that emptied into the river. "Rebuilding this fractious stream presents engineering novelty," one commentator exclaimed, and the Army Corps embraced its role of redesigning a river's course and its very function.[25] But the Corps intervention did not happen without opposition, including objections from the self-help cooperative groups and progressive political figures such as supervisor John Anson Ford. The Corps' actions, they argued, threatened "water and soil conservation," while the concrete channel would "speed the delivery of soil-saturated waters into the bottomless sea." Critics also argued that the transformation of the river into a concrete bed—this "killer [now] encased in a concrete strait jacket," as one water agency publication triumphantly viewed the river—would make possible further unrestricted development along the now channelized floodplain.[26]

The Corps' reconstruction of the river as flood control channel during the 1940s and 1950s cemented the image of Los Angeles as the antienvironment. The river was straightened, the last of the willows were removed, and fortresslike walls were erected. Los Angeles now had its own "water freeway," as one Army Corps official characterized the river, capable of transporting the unwanted floodwaters and various urban debris from near its origins at the Chatsworth Hills to its mouth at San Pedro Bay. The "water freeway" was an apt expression for this cement embankment containing the flow of water within it, an artifact of urban resource management that easily lent itself to these darker images of the Los Angeles urban environment.[27]

In 1985, a poet and longtime radical activist, Lewis MacAdams, decided to undertake an L.A. River performance art piece, to demonstrate that this "latter-day industrial hell," as he later wrote, could become the object of a forty-year art project identifying the symbols and opportunities for urban environmental transformation. Entering near the confluence of the Arroyo Seco and the river near downtown, MacAdams and his performance artists cut a hole in the fence along the river to explore its concrete channel riverbed. "We asked the river if we could speak for it in the human realm," MacAdams and his fellow performance artists declared. Hearing no response, they decided a new kind of urban environmental advocacy group was needed, which they called the Friends of the Los Angeles River (FoLAR).[28]

More than any other urban environmental group at the time, FoLAR's advocacy was as much about "discourse" as about policy or even activity. To reinvent the river as a river, to create a living environment in the city, particularly for a place like Los Angeles, which had the lowest per capita park space in the country, was first and foremost a battle of definition. Part of FoLAR's task was to engage the resource managers, landowners, and policymakers who had become wedded to the flood control and sewage effluent management functions. However, FoLAR also needed to convince the residents along the river that this could become a living river.[29]

For Los Angeles, with its disjointed set of suburban tracts and inner-city neighborhoods, and diverse, complex, and multiracial communities, the L.A. River could also represent, as some urban environmental and community advocates have now come to believe, a link between the social and the ecological. The green agenda of more open space could also become an urban agenda of community-based initiatives. Its objectives could include schools, housing, community gardens, parks, and biking and hiking trails, along with new jobs through new enterprises and environmentally preferable forms of development—all of which could be developed in tandem with the remaking of the river landscape. These connected agendas could also draw on the growing research indicating that river restoration—creating some places where the river or stream banks might once again meander and habitat might flourish—could in fact represent a more appropriate and effective form of flood protection in those areas. Uprooting the concrete and relandscaping and redesigning along the edge of the river had become the basis of a new politics and a new set of policies for managing the river and building more livable places.

Urban and Suburban: The Escape from the City

The association of the city as an anti-environment has its roots in the degraded environments of the industrial city of the middle and late nineteenth centuries. This was reflected in the sharp income disparities and increasing levels of poverty, corrupt urban political machines, growing shortages of affordable housing, heavily polluted waterways, and mounds of garbage and horse manure that piled high in residential neighborhoods. The environmental effects associated with the industrial city were most striking in its working-class quarters. There one could find the "dark, colorless, acrid, evil-smelling" environment of the city, with its "open sewer rivers" and "great mounds of ash, slag, rubbish, and rusty iron," as Lewis Mumford characterized this period of urbanization. With its residences and resources subordinated to the "sordid convenience of the manufacturer," the industrial city became the darkened city, with its "black clouds of smoke that rolled out of the factory chimneys."[30]

The rise of the industrial city in the United States, however, reversed the urban growth patterns of its European counterparts. Nineteenth-century European working-class communities were most often situated at the edge of the urban core, while upper-class residents located in the central city neighborhoods (which were also most desirable in terms of their environmental amenities). But the inner core areas of U.S. cities often became the most degraded. For example, the transformation of Pittsburgh into "steel town" in the late nineteenth and early twentieth centuries coincided with a pattern of urban development that placed the most environmentally contaminated neighborhoods near the mills located along the riverbeds and adjoining flatlands. These patterns were repeated in most urban-industrial centers of the late nineteenth century, where there continued to be a need of close proximity to daily work. This industrial component of the city complemented and reinforced the changing urban and working-class demographics (increasing immigration from southern and central Europe in the Eastern and mid-Atlantic cities, northern Europeans in the Midwest, and Mexicans, Asians, and dispersed inmigrations from throughout the United States in the western cities). These changes, partly fueled by a rapidly expanding industrial capitalism's continuing search for cheap labor sources, also influenced the patterns of settlement and the decline of infrastructure in the central areas of the industrial city. The conditions of urban life in the core—the extraordinarily overcrowded tenements, the increased waste generation, the spread of massive disease epidemics, heavily polluted landscapes and riverways—established the industrial city as a place out of ecological and social balance in terms of resources, watersheds, waste flows, and livability.[31]

At the same time that the industrial city witnessed this environmental decline, a combination of real estate, commercial and financial interests, transportation and sanitary engineers, and progressive reformers helped lay the groundwork for the middle-class outmigration from the central city into the vacant and lightly settled land at the periphery. This migration was aided by the increased capacity of the streetcars to travel longer distances (financed and subsidized by municipalities that had been largely captured by these expansionary interests). It also was facilitated by breakthroughs in sanitary engineering that allowed new developments to be established with an adequate sanitary and waste management infrastructure, making physically possible the middle-class migration to this "wholly new residential environment—the modern suburb," as Sam Bass Warner put it. This escape from the industrial city by the middle class and more affluent residents was also in part the search for the "rural ideal" with its newly recaptured environmental amenities of pleasing natural surroundings and open spaces. Even the community garden and school garden movements of the first decade of the twentieth century promoted the garden concept as a way to resocialize youth to the value of the land and thereby make agriculture-based rural living more attractive (and relieve the congested conditions of the city). As a consequence, a tension emerged within the urban environmentalism of the late nineteenth and early twentieth centuries. Should the environmental conditions of daily life be defined by restructuring and redesigning the industrial city itself? Or should the search for green and open spaces be defined as the need to escape from the city?[32]

These contrasting positions—restructure the urban environment or be removed from the environmental hazards of daily life by escaping from the city—became a dominant theme among urban environmental advocates and the urban and regional planning movements. Life inside the factory and the tenement (with their unhealthy air, poor sanitation, and lack of basic amenities) increasingly was contrasted with the need for the "clean air" of the outdoors and for greener spaces. The "parks and playground" advocacy of the settlement house reformers such as Jane Addams was partly constructed around the need for public space and open space, as well as healthy outdoors activity in the unhealthful and environmentally degraded urban ghettos of a rapidly urbanizing and industrializing America. This form of environmental advocacy was at once an urban and industrial reform movement. Urban reformers such as Florence Kelley led campaigns against the environmental hazards of the working-class quarters in the inner city. These included tenement housing conditions, as well as the suffocating conditions of work in factories or the piece work/sweatshop conditions in homes. This critique of urban daily life was also embedded in the advocacy of open space and green space as creating breathing space for more livable neighborhoods, contesting and transforming the blackened and dense anti-environment of the working-class quarters.[33]

However, this movement for open space also came to be associated with anti-urban perspectives as well. While the social reform–oriented urban environmentalists emphasized the need to mitigate conditions in the immigrant and working-class neighborhoods, a wing of the planning and open space movements began to emphasize an escape from the city through the construction of new communities in outlying suburban or exurban areas. This desire to escape from urban congestion and to recapture a connection to Nature further situated wilderness advocacy as a form of nostalgia.

In time, many suburban communities would make explicit this identification with "natural" surroundings, by adopting such names as "Park," "Forest," "Meadows," "Oak," "Grove," and "Hills." In fact, when Levittown was first conceived to help launch the next great wave of suburbanization after World War II, the Levitt family sought to associate their new suburb of Levittown as an island of Nature. Indeed, the development was to be called "Island Trees," though the name was dropped since there were no trees (except those to be planted) in the development area.[34] This arcadian discourse (or what Evan Eisenberg identified as the quest for the "middle realm where wildness and civilization are mingled in just the right proportions") linked such values as "simplicity," "harmony," and individual "autonomy" with physical nature. This in turn contrasted with the association of the "urban" with disorder, conflict, and diverse values (also associated with immigrants, ethnic and racial difference, and ultimately class distinction). The anti-poor, anti-immigrant themes dominated much of this "escape from the city" discourse of the 1910s and 1920s, extending from John

Muir's dispassionate attack on the "low lands" to Henry Ford's attacks on the "pestiferous growth" of the city and its inhabitants.[35] The suburbs, conceived of as part countryside, part urban, but "truly residential areas," as environmental historian Samuel Hays put it, also became associated with open space and Nature. The suburbs became the preferred and more attractive natural environment; and it was "the resulting experience of contrast between the city and the land beyond it," as Hays concluded in his interpretation of this bias in environmentalism, where much of the post–World War II environmental movement could be found.[36]

Automobile Country

If you keep taking my name in vain, I'll make rush hour longer.
—sign off a freeway, southeast of downtown Los Angeles

During the 1920s, Benton MacKaye, Lewis Mumford, and other "urban-scale" advocates associated with an ecologically oriented regional planning movement had hoped that the advent of the automobile would create the basis for a new type of transportation and land-use strategy. Such a strategy, embedded in what MacKaye, in an article in *The New Republic*, called the "townless highway" concept, would provide for new public spaces in urban areas while contributing to new ways to reconnect city and countryside. Transportation systems could be designed to separate the road from the pedestrian, while residences would not need to come into contact with motorways. For MacKaye, the environmental and cultural benefits associated with decoupling roads and vehicles from community life would be transparent. The townless highway strengthened the capacity to create a regional framework for resource use and the provision of goods and services. It would link urban communities of scale to rural areas and country life (including farming communities as "the open wayside environment") as well as to the indigenous or primeval areas with their "proper access to wild places."[37]

By 1930, however, it had become apparent that any expectations about the automobile's contribution to what MacKaye called the "stable and settled and balanced and cultivated way of life" were misplaced, given the automobile's actual role in urban and regional restructuring. The automobile and the highway, as MacKaye bemoaned, had emerged as the instrument for the construction of a "single roadside slum" that was "insidiously wiping out all these precious assets [primeval, rural, and cosmopolitan/urban] together." MacKaye despaired that the automobile had in fact contributed to a new pattern of unplanned migration outside the cities and along highway routes, labeling this construction of the "motor slum in the open country," "as massive a piece of defilement as the worst of the old-fashioned urban and industrial slums."[38]

Through much of the twentieth century, the United States became automobile country. Long recognized for its role in the development of mass forms of production and powerful cultural influences, the automobile dominated the urban landscape for more than half a century. Far more than its streetcar predecessor, the automobile extended the process of growth along the edge of the city. For suburbanization proponents, the automobile made possible the escape from the city, providing, as James Flink put it, the means "to afford a simple solution to some of the more formidable social problems of American life associated with the emergence of an urban-industrial society." Dispersion was the key. The highway engineer's role, one official wrote in a 1923 *American City* magazine commentary, was to "help scatter the dwelling houses and residence facilities of cities of more than 25,000 population far and wide out into the outlying open country in some appreciable way to relieve congestion in American cities."[39]

This process elaborated earlier transit-oriented shifts but also identified a new type of land-use pattern, the automobile suburb. "In highways," the president of the Studebaker Corporation would declare in 1940, "lies a new national frontier for the pessimist who thinks frontiers have disappeared." That same year, an Urban Land Institute publication identified that nearly one-third of privately owned lots in cities were now vacant, due to migration to the suburbs, spurred in part by road building. The automobile suburb extended and restructured the streetcar suburb both in the overall pattern of settlement (less dense); length and direction of the journey to work; deconcentration of employment; and new forms of low-density, residential architecture. Conceived as an environmentally attractive residential link with Nature and green space, the automobile suburb created instead an environmental disconnect between population, resources, and ecological boundaries within and among regions.[40]

The shift from urban to suburban, which significantly accelerated after World War II, was in part an expression of the centrality of the automobile in American life. Already by the mid-1950s, when debates over construction of the interstate highway system were taking place, nearly three-fourths of the American public already owned an automobile, with increasing numbers of two-car households becoming the norm rather than the exception. The automobile had provided the means by which a dispersal of jobs and services (for example, supermarkets) could be facilitated as part of this inexorable march away from the urban core and toward the suburbs. It influenced, at a cultural and economic level, the transformation of the countryside, both the increasing destabilization of rural, farm-centered communities and the ease by which farm products via the rise of truck transport could become industrial raw materials in a dispersed and globalizing food system.

The passage of the Federal Aid Highway Act of 1956, which provided the funding mechanism as well as the strategic goal of establishing the national interstate highway system, was the capstone of this process. It set up the framework for funding the construction of the interstate highway system by linking it to a small increase in the tax on gasoline and diesel fuel. The federal government then provided a 90 percent match for funding through this tax, with the funds exclusively dedicated to highway building. At the same time, engineers and highway lobbyists controlled the funds and made decisions about whether, how, and where each highway would be built, becoming the ultimate arbiter of land use and planning for both urban and rural communities. The new highway system thereby placed the automobile as the centerpiece of both urban life and the city/countryside/suburb relationships.[41]

Key to the passage of the original interstate highway legislation had been its linked urban orientation and military justification. Proponents of the highway program (known as the National System of Interstate and Defense Highways) used the hothouse rhetoric of the Cold War to justify an extraordinary expenditure of public funds in an era when the public role in stimulating new infrastructure development had become more problematic. In an effort to increase its support among urban members in Congress, the Bureau of Public Roads designated 2,175 miles of the system within city limits. The construction of this urban highway system brought MacKaye's "motor slum of the open country" into the city with even more powerful social and environmental impacts. Following the passage of the Federal Aid Highway Act, a highway building frenzy impacted every major urban center in the country. Cities and neighborhoods were reconfigured, neighborhoods were obliterated or physically divided, and the new spatial economic or class divide became the freeway rather than the railroad tracks.[42]

This massive urban reconfiguration through highway building and automobile dominance also extended to relationships of community to workplace (by the late 1990s, 91 percent of commuters in the United States were traveling to work by car). It significantly influenced access to core community needs such as shopping, while other core services such as health care and schools became increasingly dependent on the automobile in relation to location and transportation. It also became an important factor in the decline of social interaction (by the late 1990s, 92 percent of commuters' automobiles had only one occupant and the length of time for the average commute had increased to nearly one hour each day). Now both a country of automobiles and an automobile country, this "rapture-of-the-freeway," as Joan Didion once described Los Angeles, seemed to constitute the primary connection to place, with its nonstop journey from home to work and back again.[43]

Regional Visions and Metropolitan Realities

Two years before his "Townless Highway" essay, MacKaye wrote *The New Exploration*, a slim volume that laid out the regional planning movement's concept of regionalism as a central, unifying framework for establishing what he called "the right kind of environment." "Environment is to the would-be cultured man what air is to the animal—it is the breath of life. So far as outward matters go, environment is the basic ingredient of living, as air is to existence," MacKaye proclaimed in this critical exposition of some of the key arguments of the regional planning movement of the 1920s.[44]

For MacKaye and other advocates of the regional idea, the rapid urbanization of the late nineteenth and early twentieth centuries had not only created environmental stresses associated with manufacturing activities in the urban core but had also led to the shifts in land use and transportation that he would elaborate on in his "Townless Highway" article. These land-use changes in turn had also promoted new forms of urban expansion, or what MacKaye called the "super-urban" or "the overcity—the outer layers of the tide which overwhelms the city." The regional idea, on the other hand, sought to establish more direct linkages between city and countryside that allowed for open space and self-sustaining primeval and rural environments. It also provided what MacKaye called the "cosmopolitan approach" to urban life, with its appropriate use of natural and cultural resources.[45]

Three of the key figures of the regional planning movement of the 1920s and 1930s—MacKaye, Mumford, and Catherine Bauer—each articulated an important albeit distinctive contribution to this regional or "cosmopolitan" vision. They sought to extend the traditional planning focus on urban form to address issues of urban life. Their intent was to identify a new type of urban ecology focused on the total urban environment. They did so by promoting such concepts as ecological balance similar to what today might be called regional "watersheds," or other resource-based systems. They became associated with a new type of discourse—"not merely the concern of a profession [but] a mode of thinking and method of procedure," as Mumford described the advocacy of regionalism. They also became fully engaged in what Bauer described as promoting the need for a decent living environment inside the urban core as well as in the critical city/countryside relationship. But these arguments on behalf of a new type of urban environmental discourse failed to influence policy or help shape new social movements even in the early years of the New Deal when the spirit of social experimentation was greatest. By the late 1930s, the idea of an urban environmentalism began to recede from view, to only reemerge in different forms in the late 1960s and 1970s, without any connection to this earlier regional politics.[46]

MacKaye provided the baseline conceptual framework for a reconsideration of the city/countryside and industrial–natural world relationships. Planning, for MacKaye, was exploration, visualization, and revelation—exploring the "industrial wilderness" as well as the rural interior. A new kind of regional reorganization between city and countryside needed to be established through "natural reservoirs," as MacKaye called them—mountain crestlines, escarpment canyons, and wetlands, to establish "levees" to inhibit metropolitan flow. For MacKaye, environment ("part geography and part folks") applied "not only to the mountain scene where perhaps we spend our Sunday, but to the home and town where of need we spend our work days. And it applies to the motor wayside in between."[47]

This New England Yankee, who studied forestry at Harvard, was one of the first of the bright young men to serve in Gifford Pinchot's Forest Service.[48] Throughout his career as forester, regional planner, and wilderness advocate, MacKaye focused on the question of how "to adapt man's ways to nature" and how to preserve not just natural or material resources but also human resources and values. His embrace of the concepts of regionalism and regional planning were centered on the notion that the human-nature relationships could be reoriented through a renewed city/countryside relationship. Led by MacKaye, the regionalists were most concerned with the megacity—the ever-expanding metropolitization of both agricultural land and MacKaye's "primeval" or undeveloped lands. Both MacKaye and Mumford were particularly intrigued by the opportunities provided by older technologies such as hydroelectric and solar energies (replacing fossil fuel systems such as coal) and by emerging technologies, including the automobile. The regionalists felt these technologies could help establish natural regional patterns of land use and resource flows, where science and technology could be wedded to new urban and rural patterns and the preservation and appreciation of wild places. MacKaye's best-known planning and exploration activity—the development of the Appalachian Trail—was perceived by him to be a wilderness-centered and community-based reconstruction of city and countryside relationships. In this context, the trail became not simply recreation but, as MacKaye liked to put it, "re-creation."[49]

Where MacKaye had both radical and conservationist/roots with an abiding passion for the natural world and livable communities, Mumford was the historian and public intellectual seeking a broader, more encompassing vision of how to construct a more sustainable form of regional development and urban life. Mumford was an intellectual promoter and historical analyst of the new type of urban environment. This "regional city," as Mumford's regional planning ally Henry Stein characterized it, would be grounded in the new environmental technologies, with parklands and other green spaces surrounding and ultimately controlling the spread of the urban core. These would be cities of scale rather than areas of metastasizing sprawl, where infrastructure development such as flood protection, water resource and energy development, and sewage and waste treatment could be consistent with Mumford's "natural region." Drawing on the decentralized anarchist vision of Peter Kropotkin and the garden city concepts of Ebenezer Howard, Mumford's city of scale (elaborated in a special 1926 issue of *The Survey* under the heading "Regions—To Live In") was at once a throwback to the image of the New England town community and a modernist appreciation of how science and technology could reorganize city and countryside. Modern technology, Mumford and other regionalists anticipated, could provide the necessary impetus for the "basic communism which is latent in the emerging economic order." But that vision was short-lived. By the late 1940s and 1950s, the megalopolis appeared truimphant, and the promise of science and technology—and most notably the role of the automobile—had become the instrument of unchecked power, "unregulated overexpansion," and "megalopolitan civilization."[50]

Distinct from the intellectual constructs of Mumford and MacKaye, Catherine Bauer became not just the planner and visionary, but the organizer and mobilizer of the livable—and democratically constructed—urban community. Her concept of modern housing, elaborated in her 1934 book by the same name, identified a framework for residential and urban development. Drawing on European experiences, Bauer spoke of construction for use rather than for profit. Housing, for Bauer, needed to be established outside the speculative market and integrated into a broader, more comprehensive community or neighborhood planning process. Residences would be located adjacent to parks, schools, and other community facilities and businesses. Housing—and by extension the urban neighborhood—would constitute "some sort of visible organic form," part of the social and environmental goal of improving the quality of urban life, beginning with land, air, and sun. At the same time, such a process would act as a counter to the tendency toward development of the periphery and erosion of the urban core. "Government-assisted community housing," Bauer argued, could provide "a real tool for regional planning and industrial re-centralization."[51]

What more directly distinguished Bauer from her regionalist counterparts was her emphasis on the need for civic action—creating a militant grassroots housing movement consisting of those most directly interested, such as families who needed housing and workers who needed jobs. Such a movement, drawing on trade unions and community organizations, insured that any planning initiative or political advocacy would be grounded in a democratic discourse of participation and community renewal. Bauer's advocacy was linked in part to the interest of labor groups in cooperative and nonprofit housing and community development models. It also helped identify a policy- and project-based framework for the housing and community planning aspects of the regional idea during the mid-1930s that could become the basis for "improving human environment in a modern industrial society."[52]

But the key regional-planning-based New Deal initiatives, despite interest and participation by various community, labor, and regional activists and planners, ultimately shifted urban policy and resource and social-planning initiatives in a different direction. In the area of housing, for example, New Deal programs favored both "slum clearance" and public housing as housing of last resort, as well as the middle-class-oriented programs of the Federal Housing Administration (FHA), which emphasized construction of subsidized housing in low-density environments at the urban edge. At the same time, Catherine Bauer's vision of democratic public housing that generated community and environmental values had given way by the postwar period to the large, highly standardized public housing behemoths. These projects, whose density, as Bauer put it, made them "seem much more institutional, like veterans' hospitals or old-fashioned orphan asylums," projected a type of "charity stigma," with each project proclaiming, visually, that it served the lowest income group. These projects were also environmental black holes, recreating the "anti-environment" of the industrial city in its post–World War II central city form.[53]

During the 1940s and into the post–World War II period, the massive resource development projects, such as the valley authorities like TVA and Bonneville, also shifted from cooperative ventures and local agriculture and community support programs, to industrial programs that concentrated capital, land ownership, and certain types of agriculture, energy, and industry activity. These programs not only undercut democratic and cooperative initiatives but precipitated the march toward the suburbs and the exurbs and their stresses on both urban and rural or primeval environments. The massive resource management agencies that had been established, ranging from TVA to Southern California's Metropolitan Water District and the various entities associated with Robert Moses's efforts to reshape the New York region, became the dominant de facto urban planning agencies. In the process, they had aided in the process of dispersing regional growth and creating "fragmented metropolises," to use Robert Fogelson's powerful metaphor for Los Angeles.[54] By the 1950s, this concept of suburbanization as fragmentation could be applied to the older industrial cities of the East and Midwest and to the dispersed sunbelt cities of the Southwest and western United States. The dominant theme in each of these cases was the escape from the city concept, with its powerful racial, class, and environmental implications.

The triumph of the suburban "pastoral" mode also became the flip side of the decline of the "anti-environment" urban core. "They [Americans] dislike the city's variety and concentration, its tension, its hustle and bustle," William Whyte would write in his 1957 report on urban America, whose introduction he titled, "Are Cities Un-American?" While the period between 1900 and 1950 could be characterized as the first significant trend toward suburban and outlying developments, the period that followed represented the most far-reaching shift in population, land use, and social and environmental conditions reconfiguring city and countryside. This process was intimately linked to the triumph of the automobile and the escape from the city. Moreover, the declining central city was now joined by the economically and environmentally declining older suburbs at the inner ring of the expansion. The fragmentation of the region, with its substantial environmental and economic dislocations experienced during the first half of the twentieth century, had indeed become the fragmented—and exploding—metropolis at the turn of the twenty-first century. The environmental deterioration of the central cities combined with loss of higher-paying manufacturing jobs, breakdown of infrastructure, and the prevailing crises concerning such basic needs as food, housing, transportation, and education, now became complemented by the additional and often quite extraordinary environmental and equity burdens associated with the expanding metropolis. The search for open space, so central to the discourse of the escape from the city, had become synonymous with the environmental degradation of spaces, both urban spaces and now those exurban places often located at the edge of the expanded metropolis.[55]

Work, Industry, and Environment

The Taylorist Idea

If the industrial city came to be conceived as the "anti-environment," then the workplace (from Blake's "dark satanic mills" in the early stages of industrialization to the evolving forms of industrial activity in the Progressive Era) could be considered the industrial equivalent of the anti-environment. For Progressive Era urban reformers such as Josephine Goldmark, the workplace, with its "bad air, bad light, overcrowding, dirt, and unsanitary conditions," required significant environmental reform. For some Progressive Era workplace reformers, the focus on the workplace environment was extended to include control of workplace conditions and related production decisions. Workplace reform advocates like Florence Kelley and Alice Hamilton argued that the control of the conditions of work by workers—loyalty to "the producer, not the product," as Alice Hamilton put it—provided the starting point for environmental reform. But for these urban and industrial reformers who promoted the cause of efficiency and more capable resource management, their loyalty was toward a production system based on increasing worker productivity rather than one where the worker became the beneficiary and instrument of environmental and industrial change.[56]

Among the advocates of production efficiency, Frederick Winslow Taylor, the founder, promoter, and ideologue of scientific management in industry and the workplace, can be considered the industrial equivalent of the production-oriented, resource-managing conservationists. Though not associated with Progressive party activities, Taylor's elevation into a major ideological figure was in part accomplished by progressive/ conservationist arguments. The Taylorist system itself became far more visible due to Louis Brandeis's famous conservationist-oriented "efficiency of production" argument during the railroad rate case held before the Interstate Commerce Commission (ICC) in 1911. The railroads (allied with the railroad unions) wanted the ICC to increase their freight rates in the Northeast, while shipping interests opposed such an increase. This interest-group debate, however, became transformed when Brandeis shifted the discussion to questions of efficiency in production. Brandeis embraced Taylor's concept of scientific management as a way to demonstrate that the railroads were not able to account for their costs, nor were they able to maximize their production potential. For Brandeis, the capacity of a company to adhere to this "gospel of efficiency," a term that so effectively characterized the discourse at the time, became central to his argument. The scientific management approach, Brandeis argued, differed from ordinary systems of production "much as production by machinery differs from production by hand." The Eastern Rate case also enabled Taylor and his concept of scientific management to assume center stage in the development of these approaches during the later years of the Progressive Era.[57]

Taylor's arguments can be broadly characterized as the division of "planning" from "doing" in the workplace. Such a division would be accomplished through tools like job standardization or the setting of precise work tasks, and an emphasis on efficiency and worker productivity to the exclusion of all other factors. For Taylor, this restructuring of the workplace involved "the deliberate gathering in on the part of those of management's side of all the great mass of traditional knowledge, which in the past has been in the heads of the workmen, and in the physical skill and knack of the workman, which he has acquired through years of experience, and then recording it, tabulating it, and, in many cases, finally reducing it to laws, rules, and even to mathematical formulae . . . by the scientific managers."[58]

Such a restructuring also represented an undermining of the role of craft unions in the late nineteenth and early twentieth centuries and their association of work with knowledge. Taylor was particularly hostile to trade unions as interfering in this process of transferring knowledge from the workplace to the engineer/manager. Taylor's famous comments about the need to "deskill" work ("In our scheme, we do not ask for the initiative of our men. We do not want any initiative. All we want them to do is to obey the orders we give them, do what we say, and do it quick.") were a direct counterpart to his comments about the need for management to reinvent itself as a form of expertise. This expertise was to be based on the understanding that industrial activities, in the form of work tasks to be performed within the framework of scientific management, were, at their core, "monotonous, tiring, and uninteresting." For Taylor, those who did the work "could not derive or fully understand its science." "Labor's sole duty [in the Taylorist workplace] is to obey their scientific managers, and not to supply initiative, i.e. brains," wrote one reform-oriented socialist. On the other hand, Taylor saw the great contribution of the expert as identifying how each worker could achieve a "state of maximum efficiency," while at the same time insuring worker acceptance of the "unpleasant, disagreeable things" they were now obliged to do.[59]

Whether through wage increases associated with piece rate work, more rigorous industrial discipline at the factory level through the use of the stopwatch or time/motion studies, or the strategies of "de-skilling," it is easy to see why the Taylorist and "Fordist" (mass-production forms) concepts were deemed synonymous. For while Fordism can be best understood as a production strategy (e.g., greater automation, assembly-line procedures), Taylorism fit more directly into a productionist discourse. The Taylorist call for production efficiency became the industrial equivalent of the conservationist appeal to prevent "our forests vanishing [and] our water-powers going to waste," as Taylor put it, with workers identified as resources to be managed, and products seen as continually reproduced and made abundant.[60]

In the United States, scientific management, prior to World War I, was perceived by the craft-union-dominated American Federation of Labor as a direct challenge to its "equity right" in addressing the conditions in which its members worked. It also contrasted with the ideas of reformers like Jane Addams who wanted to insure that scientific management and efficient industrial organization did not undermine what Thorsten Veblen had called the "instinct of workmanship." This crucial value,

Addams believed, was central to a worker's sense of well-being. While scientific and technical advances in machinery aided the Taylorist cause of greater productivity, Addams argued that it was imperative in the future that advances in production lay "not in the direction of improvement in machinery, but in the recovery and education of the workman."[61]

However, already by the post–World War I period, the AFL leadership and several of its progressive allies began to accept and ultimately embrace some of the concepts of scientific management and efficiency as part of a broader shift toward a union-management cooperation approach. The key to the Taylorist/AFL marriage was labor's acceptance of the goal of maximizing production. By 1928, one labor analyst lauded the labor movement's new emphasis on production as opposed to the "old rough-and-ready trade unionism [which] battled over the division of the product." "The new, suave, discreet unionism," this analyst concluded, "talks the language of the efficiency engineer and busies itself about ways and means of increasing output." The new "gospel of production" thus emerged by the 1920s as the dominant discourse about work and industry, complementing and extending the gospel of efficiency.[62]

Along those lines, Herbert Hoover, that quintessential engineer/capitalist championing productionist and efficiency values, would argue, in an address to the Federated American Engineering Societies, that the only real objective in industrial activity was "maximum production." His argument also linked the goal of production to an ever-expanding form of consumption. "There is no limit to consumption," Hoover said in his talk, which was reprinted in the *Bulletin of the Taylor Society*, "except the total capacity to produce, provided the surplus of productive power is constantly shifted to new articles from those that have reached the saturation point of demand." In the words of the progressive muckraker Ida Tarbell, the scientific management or productionist idea embedded in Taylorism also meant that the "world could take all we could make [and] that the power of consumption was limitless."[63]

This idea of unfettered production leading to limitless consumption also evolved during the 1920s into an unabashed celebration of the capitalist goal of "sales" rather than promoting the utility of goods or products. With "sales" or market-driven productionist goals and values linked to efficiency as well as the de-skilling of work, the new "modernist" or "Fordist" industrial order generated far-reaching structural environmental outcomes. These in turn changed the very nature of the work process and the relationships between industries and the communities where they were located. In this way, the Taylorist discourse effectively shifted the concerns of the urban environmental reformers about the anti-environment of the workplace and the need for livable places to the goals of contributing to a system of ever-expanding production and unlimited consumption.[64]

The Hazards of Work, Problems for the Environment

By the 1920s, the triumph and celebration of the gospel of production could not completely hide the fact that the new productionism carried with it the burden of occupational and environmental hazards. Changes in production methods were identifying new kinds of workplace environmental problems. These stemmed from reorganization of the workplace, use of new technologies, processes, and products, scale of operations or intense use of equipment or labor, and development of entirely new hazardous industries. It also included the restructuring and further expansion of what Alice Hamilton called the "dangerous trades," such as lead-based industries. Some of the scandals of the Progressive Era, including the radium poisoning of workers producing watch dials and the mercury poisoning of hat makers, anticipated these new kinds of occupational environmental hazards. At the same time, advocacy groups such as the Workers' Health Bureau and some trade union locals were also beginning to focus on the role of hazardous new technologies as a central issue of workplace conditions, the organization of work, and workers' rights. The link of work, industry, and environment was becoming more visible and acute, potentially undermining the Taylorist view of production efficiency.[65]

Perhaps the most striking example in the 1920s of these new tensions involved the debate over the introduction of leaded gasoline. Prior to the 1920s, in the early years of automobile design, the issue of inefficient performance through engine knock ("the nightmare of all automotive engineers," as industry researcher Charles Kettering put it) emerged as a central challenge for this new industry. Auto industry researchers, led by Kettering, became convinced that the source of the efficiency problem was associated with the source of the fuel. Through his work at the Dayton Laboratories and subsequently at General Motors and the Ethyl Corporation, Kettering and his assistant Thomas Midgley identified the introduction of the lead additive, tetraethyl lead, as one possible solution for reducing the knock, improving the efficiency of the engine, and significantly increasing the power of the automobile. This lead additive represented a technical and marketing breakthrough for the automobile industry, which, Kettering would later describe as "one of the greatest steps in the world."[66]

The technical changes associated with reducing knock were part of the major shifts in automotive technology, production methods, and marketing during the 1920s that helped establish the automobile as the centerpiece of the new gospel of production and the rise of what G.M.'s Alfred Sloan called the "mass-class market." The reduction of the knock, while associated with higher octane ratings and greater gasoline efficiency, was also a key component of the new production and marketing strategies for the automobile industry. Leaded gasoline laid the groundwork for touting the automobile's "comfort, convenience, power, and style," as Sloan would later characterize its contribution to the revolution in automotive production in the 1920s. Power plus efficiency ultimately meant both greater sales and a crucial new language of consumer identity. For both the chemical and automobile industries, the introduction of leaded gasoline came to represent a financial and marketing windfall.[67]

A potential barrier to this financial and productionist bonanza were the recognizable hazards of the manufacture and use of tetraethyl lead. Knowledge of lead toxicity dated back at least two thousand years to the Roman period. The toxicity of tetraethyl lead compounds had been a specific focus of research for more than forty years before the introduction of leaded gasoline. It could "fairly well be predicted" that this new fuel additive would likely be hazardous, as Alice Hamilton, one of leaded gasoline's key critics, put it. Given this cautionary knowledge, why and how did such an extraordinary shift in work, industry, and environment take place? The answer is located in the nature of the discovery itself: if tetraethyl lead could so dramatically improve automobile performance, then such a feature simply overwhelmed whatever caution seemed warranted. It was the direct counterpoint to the precautionary principle: introduce first, evaluate (if forced to) later. Indeed, viewing their research as central to productionist outcomes, Kettering and Midgely constructed arguments that continually sought to redefine and thereby eliminate any environmental consideration from their research agenda.[68]

Given the temper of the times, the introduction of leaded gasoline proceeded quickly and inexorably. Several facilities were established, as the rush to production became a rush to corner the market of this lucrative new product. Just a few months after the lead additive was identified as a solution to the knock, the first leaded gasoline went on sale in Dayton, Ohio, in February 1923. By 1924, the first moderately priced car that could accommodate the use of leaded gasoline, the "Chrysler Six," was also in production. Producing tetraethyl lead had emerged as a "war order priority," as Irénée du Pont put it.[69]

Expanding production as a war-order priority complemented the changes associated with work and production that were occurring in this period. With the emphasis on output, little if any attention was paid to worker knowledge and care. Workplace hazards, however, were experienced from the outset of production, creating problems of insomnia, headaches, loss of appetite, delirium, and delusions resembling alcoholic intoxication. While these various cognitive and neurological impacts were highlighted in the press, several deaths at tetraethyl lead production facilities shifted the discussion from whether there were hazards in producing the additive to whether the product should be produced at all. Labor groups, such as the Dayton, Ohio, Central Labor Council, which issued a "danger" flier on tetraethyl lead, and the Workers' Health Bureau, which called for the banning of the lead additive, further shifted the debate to one of worker knowledge and rights regarding this new occupational exposure. In several cities and states, including New York, Pittsburgh, and New Jersey, tetraethyl lead use was banned by health agencies. These actions were influenced by media coverage and by worker and public response. "Whatever may be responsible for the poisonous nature of this fluid," one newspaper account characterized the mood at the time, "it has left a fearful impression upon the public mind."[70]

For a brief moment, the web of work, industry, and environment relationships were made more transparent. Faced with the prospects of government intervention and a possible product ban, the tetraethyl lead producers and their research collaborators constructed two lines of defense. They first argued that the workers themselves were responsible for their own health problems ("the heedlessness of the workers in failing to follow instructions," as Midgley put it). At the same time, tetraethyl lead was identified as "an industrial hazard as distinguished from a menace to public health," as one chemical trade publication put it. This position contrasted with Hamilton's concerns regarding what she characterized as the "possible danger to the public from small quantities of lead dust," a concern that had influenced actions to ban the product. The position distinguishing occupational hazards (which were now apparent) and environmental hazards (which leaded gasoline advocates defined as remote due to low concentrations and limited exposures) was also elaborated in an industry-funded report by the U.S. Bureau of Mines.[71]

These arguments and reports, however, did not allay the public fears. Fearing more enforced bans, the leaded gasoline manufacturers voluntarily suspended production and the U.S. surgeon general convened a conference and established a committee to analyze the occupational and environmental issues associated with its production. Some occupational health and environmental advocates such as Hamilton saw the surgeon general's intervention as a precedent for creating outside review of an industry decision to introduce a new hazardous product or process. However, the report of the surgeon general's committee allowed industry to resume production, albeit with greater attention to limiting occupational exposures. The surgeon general's committee had also raised concerns regarding increased use of leaded gasoline creating the potential for a serious environmental hazard and the need for future monitoring. Despite these concerns, the industry interpreted the outcome as carte blanche for a resumed—and expanded—shift in production and use.[72]

Both in the United States and England (where leaded gasoline production had also been suspended and then resumed based on the

U.S. surgeon general's report), production levels increased quickly and dramatically. Just three years after the surgeon general's report, "war order production" for leaded gasoline had once again become the industry norm, with sixty refiners, who accounted for two-thirds of all the gasoline then on the market, producing leaded gasoline. For nearly the next half-century, leaded gasoline became a fixture in fuel production and automobile use, reaching as much as 300,000 tons or the equivalent of 25 percent of the total lead used in the United States by the late 1960s. By the early 1970s, at its peak use, virtually all gasoline contained an average lead content of nearly 2.4 grams per gallon, an extraordinary contribution to the amount of lead in the ambient environment.[73]

The introduction of leaded gasoline in the 1920s can be considered representative of a period when occupational and environmental exposures proliferated, and the labor movement increasingly removed itself from issues concerning the nature of work and work processes. It also indicated how the Progressive Era tradition of an active role for government had become fully subsumed under the continuing and powerful logic of productionism. The new types of supplier/user relationships particularly characteristic of the chemical and petroleum industries heightened those changes. New products and technologies were influenced by the drive to create new markets for a range of consumer and industrial products, which in turn created new patterns of occupational and environmental exposure. These changes also paralleled an erosion of worker and public rights about industry decision making, a by-product of the Taylorist idea and the Fordist organization of production. These changes, however, also provided a transition from "productionism" as a Progressive Era notion of "scientific management" in the workplace, to the triumph of "Sloanism," a variant of the gospel of production that emphasized rapid product turnover. Through these strategies, constantly evolving products ("keeping the customers reasonably dissatisfied with what they have," as Kettering put it) would become central to the goal of expanded markets or the Taylorist idea of producton efficiency leading to "limitless consumption," independent of any environmental or occupational consequence.[74]

During the 1930s and 1940s, the rise of a mass-production-focused unionism challenged some of the productionist strategies that had so negatively impacted the conditions of work. The new industrial unions, under the banner of the CIO, sought to identify worker rights on the shop floor, including most significantly the right to join unions. Yet the growth of industrial unionism, while crucial in fighting for the worker's right to organize, failed to address and ultimately stem the Taylorist conceit that reducing or eliminating worker knowledge and participation in the organization and outcomes of work were central to a new social order. The union focus on wages, job security, and retirement benefits, while critical to the notion of a decent job and a living wage, also meant a more exclusive focus on the social welfare of its members. Absent was any focus on the social nature of the production system, including its occupational and environmental outcomes.

The focus on occupational hazards as an arena for worker action only reemerged significantly in the 1950s, 1960s, and 1970s. This included grassroots activity (miners and textile workers) and community or environmental challenges to which some unions (the Oil Chemical and Atomic Workers Union and, to a lesser extent, steel and auto unions) felt the need to respond. But by the 1970s the division between jobs and environment was even more dominant. The post-Fordist strategies of flexible production and worker participation in work process issues didn't overcome that division, since worker participation strategies did not extend to the worker's social role within a production system. Worker participation ultimately became another productionist strategy.[75]

Even as post-Fordist strategies in labor/management cooperation and the new initiatives to "tap the know-how of ordinary workers" were advanced, pressures generated by globalization and downsizing also stimulated resistance and even sabotage of the team concepts designed for greater productivity. For example, General Motors's Saturn plant in Tennessee, "the poster boy of the right way to do it," as the executive director of the USC Center for Effective Organizations put it, experienced disillusion and occasional hostility from its workforce as reskilling and team concepts gave way to production pressures. The Saturn plant's adoption of the Japanese team concept, moreover, did not significantly extend into the areas of environmental and occupational health impacts associated with various production choices. Indeed, whatever environmental changes had been instituted, whether in terms of solvent substitutions or emission and discharge issues from chemical use, they were largely influenced by regulatory concerns, particularly fears about potential Superfund liability, since General Motors had become a responsible party at a substantial number of its production sites. Despite its advocates' claims of worker productivity, flexible production meant neither a more knowledgeable and reskilled workforce nor a new framework for addressing the work/environment relationship.[76]

At the same time, the mainstream environmentalism that had emerged by the 1970s functioned on the basis of the division between work, product, and environment, whether in terms of policy or the advocacy of consumer, occupational health, and environmental movements. This additional separation of the spheres of daily life paralleled the division of city and countryside (with the urban core identified as the anti-environment) and the erosion of the regional vision of balanced, ecological communities in the wake of the auto-induced fragmented metropolitan realities. By the end of the century, the environmental cause had become more cri-de-coeur than agenda for action, a still unfulfilled search for some alternative way to define our social and ecological universe.

Beyond the Gospel of Consumption

The divide in environmental discourse presents a divide over focus as well as interpretation of the proper arena and content of environmental action. The separation of the social from the ecological, which still persists in much of environmental discourse, splits Nature, as Raymond Williams has argued, into unrelated parts. But such a split also extends to a divide in human activity itself, between those seeing themselves as producers and those who see themselves as consumers. "The consumer wants only the intended product," Williams writes, "all other products and by-products he must get away from, if he can. But get away—it really can't be overlooked—to treat leftover nature in much the same spirit: to consume it as scenery, landscape, image, fresh air."[77]

This consumerist focus not only serves to separate nature from human activities, or the social from the ecological, but also establishes a focus on environmental change as a form of consumer action. This may be reflected through nature-related activities (the consumption use of wildlife, the appreciation of scenic resources), or the idea that constraints on human consumption (whether population control or consumption of resources, including scenic resources) represent the central environmental goal. What is eliminated from this scenario is the producer role—the importance of production in the complex of relationships between the physical world and human activities. It is the products of our contemporary urban and industrial society—stemming from why, how, and what gets produced, as well as why, how, and what gets to be consumed—that describe the state of the environment. "The pollution of industrial society," Williams argues, "is to be found not only in the water and in the air [and on the land] but in the slums, the traffic jams, and not these only as physical objects but as ourselves in them and in relation to them." "In this actual world," Williams concludes, "there is then not much point in counterposing or restating the great abstractions of Man and Nature. We have mixed our labour with the earth, our forces with its forces too deeply to be able to draw back and separate either out. Except that if we mentally draw back, if we go on with the singular abstractions, we are spared the effort of looking, in any active way, at the whole complex of social and natural relationships which is at once our product and our activity."[78]

Also eliminated is the connection to *place* and the idea that "Nature" also resides in place, whether in city, suburb, rural area, or designated wilderness area. Environment as representation—and consumption activity—freezes the moment of connection, or appreciation, much as a still-life portrait of a landscape is valued. If the concept of the landscape is representation of Nature defined as scenes observed "from the outside looking in," as Mindy Thompson Fullilove and Robert Fullilove have written, then the concept of place, by contrast, "influences us because we are inside a moment in space and time." In terms of the ways we connect or disconnect the social and the ecological, the differences are crucial in situating environmental discourse—and environmental action. On the one hand, we have the language that situates Nature as "out there," a "kind of abiding metaphysical absolute against which we can judge messy, contingent culture," as Michael Pollan has put it. In contrast, Pollan argues, using the garden as his central metaphor, we need to place "Man *in* Nature," recognizing that in doing so, we would need to establish "a more complicated and supple sense of how we fit into nature." The contrasting and at times conflicting notions of environmental discourse and action often reflect this situational perspective—whether one is part of (inside the experience), or separated from (outside looking in) the environment.[79]

Environmentalism as a form of consumption and a language that separates the social and the ecological stands in contrast to an environmentalism of daily life within an evolving social order. Since the 1970s, new forms of environmental discourse and action—both community and production related—have sought to shift the terrain for this complex set of social movements. The possibility of breaking free from a bounded environmentalism to become a broader, more socially inclusive movement capable of challenging the very structure and logic of a capitalist social order has become available once again. No longer limited by its consumption orientation, this environmentalism of daily life offers what Hilary Wainwright has characterized as a "totalizing vision."[80] To put forth a totalizing vision requires not just a change in discourse but the ability to identify new strategies for action. Such a shift creates a more effective challenge to an implacable system whose genius has long been its capacity to undermine or capture oppositional ideas and eventually incorporate them into the dominant paradigm of urban expansion, limitless consumption, and of loyalty to the product rather than the producer.

2

Livable Regions and Cleaner Production: Linking Environmental Justice and Pollution Prevention

Overcoming Divides

Among environmental justice colleagues, Love Canal housewife turned antitoxics organizer Lois Gibbs liked to tell the story about how, in her travels, she encountered the problem of split constituencies, which so often undermined efforts to build a new type of environmental and social justice coalition. At meetings with community residents concerned about nearby hazardous industries, Gibbs would appeal for worker and community solidarity to build such a coalition, but often without much response. She'd then ask members of the audience, often mostly women, how many of their spouses worked in those plants. A large number would raise their hands. She then encountered a similar response when she spoke with workplace groups. Almost invariably, the workers would be concerned that community protests were going to lead to job loss if, as the workers thought possible, their plant was shut down due to protests about the hazards to the community. Community/worker solidarity seemed even more problematic in this situation. At meetings with workers, Gibbs would ask the mostly male audiences whether they lived in the communities whose residents were mobilizing against the facilities where they worked. Again, a large number of hands would be raised.[1]

Gibbs had made her point, but the lessons for her audiences were nevertheless still elusive. The problem for Gibbs, who was primarily community focused, or for her workplace counterparts, such as organizers with the Oil, Chemical and Atomic Workers' Union, was how to overcome a divide that had its roots in language and the separate spheres of workplace and community organizing. The answer for Gibbs and other community and workplace organizers is still difficult to find. But while Gibbs's story is cautionary, it also points to the places where a new environmentalism, in search of a common language of daily life experiences and concerns, needs to be located.

Narrowing Choices

During the 1980s and 1990s, environmental justice and pollution prevention came to represent two of the most promising approaches for expanding the agenda of the environmental movement. Environmental justice provided an opportunity to broaden environmentalism's base by situating environmental issues in relation to race and poverty and by identifying with the language of social justice. This included concerns about both *equity* and *quality of life*—growing disparities in income distribution (locally, nationally, and globally), community health concerns, transportation problems for the poor, the livability of cities and rural areas, and more.

Similarly, pollution prevention allowed the environmental movement to establish a wider appeal. Like environmental justice, pollution prevention could reach new constituencies, including workers subject to industry claims about the environmental movement's "elitism" and its presumed disregard for jobs. Pollution prevention seemed better positioned to change the nature of the jobs-versus-environment debate, in part by identifying a common ground between communities and workers. It appeared capable of doing so by directly challenging the way industry made decisions that had significant environmental impacts.

During the 1980s, when the environmental justice and pollution prevention approaches emerged as significant tendencies within environmentalism, the environmental movement seemed to be growing. At this time mainstream environmentalism, led by large, staff-based, expert-oriented, national- and global-oriented groups, sought to counter the antigovernment, antiregulatory approaches that pervaded the debates in Washington, particularly during the Reagan presidency. More than most other social movements at the time, environmentalism seemed capable of assuming the mantle of the public interest. Environmentalists advocated for clean air, clean water, protecting the ozone layer, shifting away from the fossil fuel economy with its potential for global warming, and even for addressing population growth (in the face of the Reaganite hostility toward population control as a prochoice position). These positions did not appear to contradict, but could instead reinforce, the new environmental justice and pollution prevention arguments about clean jobs and green industries and healthy and livable communities, including low-income communities. A broader, more inclusive environmentalism seemed possible. This possibility was also fueled by attacks against already existing environmental policies and programs (which led to substantial increases in paper membership through mail-order campaigns of groups like the Sierra Club and the Environmental Defense Fund). It facilitated the rise of community-based environmental justice groups and even efforts to link unions and pollution prevention advocates through such campaigns as "Superfund for Workers," all of which gained momentum due to the counterproductive rhetoric of the antienvironmental Reaganites like James Watt.[2]

But while each of the different segments of the environmental movement expanded their reach through the 1980s, their limits also became increasingly apparent. For mainstream environmentalists, membership and much of the actual organizational growth remained on paper and not based on constituent activity. Earth Day 1990, the media-oriented extravaganza that offered little by way of consolidating and expanding organizational reach or shifting discourses, underlined this tension for mainstream environmentalism. Earth Day 1990 did provide an opportunity to facilitate increased fundraising and mass mailing–related membership growth for the mainstream environmental groups. But little if any new policy initiatives, let alone organizing activity, came to be associated with this hyped event. While Earth Day 1970 contributed to the rise of the mainstream environmental organizations and the legislative, administrative, and legal infrastructure of an environmental policy system, Earth Day 1990 can be seen as closing out, or at least significantly interrupting the two-decade growth era of mainstream environmental organizations. It also began to reveal an identity crisis regarding the structure and outcomes of the environmental policy system that had taken shape in the 1970s and 1980s.[3]

The election of Bill Clinton and Al Gore in 1992 was interpreted by some in the movement as a vindication and triumph of the mainstream environmental approach. Similar to the Carter years during the late 1970s, which established a revolving door between the administration and mainstream environmental groups for environmental policy advocates and policymakers, the first two years of the new Clinton administration witnessed a wholesale convergence of the activist/policymaker roles. A new generation of revolving-door activist and policymaker positions suggested the possibility of new experimentation in policy. A "new paradigm in environmental governance" seemed possible, as one of those revolving-door figures put it. At the same time, the embrace of the new Clinton/Gore administration indicated the inattention to movement-building activity, even as a significant degree of satisfaction was being expressed in terms of improved access to the levers of power in Washington.[4]

The mood of change was in the air for the more radical and grassroots wings of environmentalism during the first years of the Clinton administration. The EPA declared that environmental justice and pollution prevention would be the cornerstone of how the agency presented itself to the public, and even new administrative units and specific funding streams for both approaches (as well as a linked environmental justice/pollution prevention grants program) became available. Al Gore's own presentation of environmental issues, *Earth in the Balance,* became a well-read text, and environmentalists were pleased to note that one of Gore's more striking proposals, a tax on carbon, was being put forth by the Clinton team as one of its first environmental initiatives. Though wary of signs of membership loss and funding opportunities since their peak in the period immediately preceding Earth Day 1990, the mainstream environmentalists felt more than ever capable of turning ideas into policies. Even a shift in priorities by their long-time antagonists in industry toward what some were calling the "greening of industry" or "sustainable business development" seemed possible, if not likely.[5]

Those contented feelings didn't last long. The carbon tax idea was nearly dead on arrival and signaled that the new paradigms in governance did not include the polluter pays principle. While the Clinton EPA issued broad pronouncements about environmental justice and pollution prevention through executive orders and modest new funding programs, the mood in Congress was decidedly antiregulatory. With the Republican takeover after the 1994 elections, policy choices for environmental advocates appeared to narrow to different conceptions of more flexible regulation that often translated into a reduced public role. Environmentalists feared that some of the most extraordinary antiregulatory initiatives would now dominate Congressional debate. The discussions about new types of regulatory approaches to influence the direction of industry and land-use decision making that pollution prevention and environmental justice had once promised were simply eliminated as a factor in policy deliberations.[6]

In face of such narrowing choices and a wide range of anti-environmental initiatives, the opportunities presented by environmental justice and pollution prevention emerged as perhaps the only route for a renewed environmentalism. But similar to the constraints faced by mainstream environmentalists, where debilitating debates such as the Sierra Club referendum on whether to adopt an aggressive anti-immigration stance threatened to further unravel environmentalism's appeal, environmental justice and pollution prevention advocates and ideas faced difficult choices. Environmentalism would either remain bounded as an interest group framework for action and policymaking, or become a reinvigorated set of ideas and call to action. Environmental justice and pollution prevention could point the way toward this renewal or further identify the limits of how this crucial twentieth-century movement had become yet more segmented in the new millennium.

Environmental Justice: Place Matters

A Summit

It had seemed, in the period around Earth Day 1990, that an environmental justice movement with a common focus and vision was not only possible but in reach. The movement's growing appeal was its ability to link issues of race, environmental discrimination, equity, and social and environmental change. Yet, just two decades earlier, during the first Earth-Day period, African-American leaders like Gary, Indiana, Mayor Richard Hatcher, Whitney Young of the Urban League, and George Wiley of the National Welfare Rights Organization had argued that the emerging environmental movement represented a diversion from the civil rights agenda and potential competition for scarce government resources. "The nation's concern with environment," Hatcher was quoted in *Time* magazine a few months after Earth Day 1970, "has done what George Wallace was unable to do: distract the nation from the human problems of black and brown Americans." In another widely cited statement, Young declared that the "war on pollution is one that should be waged after the war on poverty is won."[7]

For the civil rights movement of the late 1960s and early 1970s, environmentalism simply ignored the needs of the poor. Unlike the middle-class environmentalists, the poor were "part of a different revolution" whose focus was on justice, as Norman Faramelli of the Boston Industrial Mission put it. Even more disturbing for some critics were the potential disproportionate costs to be assumed by the poor to pay for environmental improvements, with related prospects of job loss and the undermining of economic progress for blacks and other nonminority poor. This latter problem—the tension between jobs and the environment—was particularly charged since "the industries in which blacks have a foothold and a potential for economic improvement are precisely those [autos, chemicals, fabricated metals, primary metals, etc.] where we find the greatest ecological hazards," as one African-American political scientist argued at the time.[8]

Yet it was those very problems of community and industrial hazards as they affected the poor and low-income communities of color—of social and environmental and economic justice—that became the focus of the emerging environmental justice movement of the 1980s and 1990s. This, however, was not a new focus. The idea of merging the concerns of justice and the environment, whether in communities, workplaces, or more broadly "regions to live in," had multiple roots in earlier urban and industrial movements. Even during the heyday of the civil rights movement in the 1960s, such issues as lead paint in housing or "slum rats" were seen by some advocates as examples of inner-city pollution that required a justice-oriented environmental approach. Similarly, the hazards from leaded gasoline were primarily identified at that time as an equity- or justice-related concern due to the proximity of highways to low-income neighborhoods and communities of color where the greatest concentrations of lead in the ambient environment could be found. By the 1970s, the distinction between an environment- and justice-oriented agenda had also come to divide such groups as the Sierra Club and the Conservation Foundation, exacerbated by the emergence of potentially divisive issues as population control. At the same time, new urban- and justice-oriented initiatives among environmental activists like the Urban Environment Conference (which focused on such issues as housing, transportation, and urban land use), sought to counter these divisions and strengthen the links between movements.[9]

Despite these and earlier efforts, neither environmental justice nor an urban environmentalism were assumed during the 1970s to be distinctive environmental movements. This occurred, in part, because the issues associated with the environment/justice link were generally located elsewhere in the language of social movements. At the same time, social justice activists accepted the prevailing view that environmental discourse was primarily, if not exclusively, nature centered and wilderness oriented (despite the growing environmental focus on industrial pollution and urban-oriented issues, even among the mainstream groups).[10]

Despite the rich and diverse history that had previously bridged social and ecological themes, it wasn't until the early 1980s that a more explicit connection was made that was better able to challenge the prevailing view about what constituted an environmental issue or environmental group. The key to this change was the growing prominence, by the early 1980s, of an assortment of antiwaste and antitoxics local groups involved in community mobilizations and confrontations. The event most frequently cited as identifying a new type of civil rights–oriented environmentalism was the 1982 protest against a proposed PCB hazardous waste facility to be sited in a poor and predominantly African-American rural county in North Carolina. The community protest became prominent in part because of its depiction of "environmental discrimination"—situating the decision about siting environmentally hazardous facilities or pursuing negative or locally undesirable land uses as having a discriminatory intent. This argument, elaborated more prominently in a 1987 United Church of Christ study, particularly cor-related race and location in facility siting.[11] Consequently, the associa-tion of race and environment came to be embedded in the language of discrimination or "distributive justice," despite the wide range of other historical and even contemporary environment/justice associations.[12]

The wastes and hazard issues were crucial in distinguishing environ-mental justice from mainstream environmentalism, particularly in the early and mid-1980s when the environmental justice groups first began to coalesce. To begin with, the antitoxics groups argued against the dom-inant environmental policy approach that sought to identify the most effective end-of-pipe, waste treatment strategies available. While main-stream environmental groups often remained critical of these same policy approaches *in terms of their implementation* the groups nevertheless tended to accept the end-of-pipe, or pollution-control policy framework itself. Legislation such as the 1976 Resource Conservation and Recov-ery Act, which distinguished between better management and treatment as opposed to prevention of wastes, had been developed with at least the tacit if not explicit buy-in of the mainstream environmental groups. While these groups also maintained an interest in prevention, as repre-sented in other legislation such as the Toxic Substances Control Act, which provided for pretesting of chemicals before they entered the market, they were most absorbed by the issues of permitting, rule making, and enforcement of policies. Functioning as both defenders of the environmental laws and critics of their implementation, the main-stream groups were constantly deciding whether to challenge, modify, or accept the latest interpretation of the mitigation and treatment approach. However, it was some of those very same treatment technologies that were being employed in low-income communities that had become the focus of protest for the antiwaste and antitoxics groups.

Community opposition in South Central Los Angeles during the mid-1980s to a proposed solid waste incinerator illustrated the suspicion about discriminatory intent and the potential divide that could occur between mainstream and environmental justice groups. In the early

1980s, the Los Angeles Bureau of Sanitation (BOS), the local agency responsible for managing the city's growing solid-waste stream, began to promote mass burn incineration as an alternative to expanding the city's existing landfill sites adjacent to middle-class homeowner neighborhoods. Recognizing that environmental-based residential opposition to any waste disposal or treatment facility needed to be minimized or deflected, the BOS decided to locate the first of the three proposed 1,600-ton-per-day incinerators in an African-American and Latino neighborhood in South Central Los Angeles. The sites for the two other planned incinerators would be closer to middle-class neighborhoods in the west side of Los Angeles and the suburban San Gabriel Valley to the east. By locating the first of these incinerators in a low-income community of color, the BOS planners made two assumptions. They judged there would be only minimal community opposition to the incinerator, due to the appeal of additional jobs and the establishment of a community betterment fund. Then, once the South Central incinerator was built, environmental opposition from local residents in the other two areas could be challenged on the basis of "racism," since the middle-class environmental-oriented residents, it was assumed, would not oppose the South Central site. However, significant community opposition did develop in South Central, based on substantive environmental concerns, mistrust of the arguments about presumed benefits, and procedural concerns about the lack of community input. Alliances also formed between South Central community groups, some mainstream environmentalists, and neighborhood groups from other communities. These alliances in turn strengthened the community opposition to the South Central incinerator and led to the call for alternative approaches to incineration and landfilling, such as waste reduction and recycling.[13]

While the plans to build the South Central incinerator were proceeding, the state Waste Management Board, which was also interested in a transition from landfills to incinerator technology, commissioned a report by a political consultant on how best to facilitate the siting of such facilities. This report to the Waste Management Board, subsequently often cited by environmental justice groups, also assumed minimal environmental opposition to incineration technologies in low-income communities or communities of color. The report recommended that the agencies involved in siting try to engage mainstream groups like the Sierra Club, since they might be supportive or at least neutral in assessing the value of the treatment technology. While this report was only advisory and speculative, it immediately resonated among environmental justice groups as illustrative of the discriminatory intent around facility siting decisions and their concern that some mainstream environmental groups might be on the opposite side of the issue. In the battle over the South Central incinerator and similar siting-related confrontations, several mainstream environmental groups, such as the action-oriented Greenpeace, supported the antitoxics, antiwaste facility community-based groups. But for many of the mainstream environmentalists, there still remained clear differences in style and language as well as focus between their approach and that of the more confrontational environmental justice groups.[14]

As the conflicts over siting and appropriate waste-management strategies intensified during the mid- and late 1980s, the community groups involved in these issues increasingly sought to develop a common identity and focus. In 1985 and 1986, they coalesced around a campaign to reauthorize and keep intact key liability and funding mechanisms for the primary clean-up legislation, the Comprehensive Emergency Response Compensation and Liability Act, also known as Superfund. The campaign by the community groups, which they called "Superdrive for Superfund," further demonstrated the visibility of the community mobilizations and the heightened media attention about potentially catastrophic toxic releases such as the one that occurred in Bhopal, India, in 1984. As a result, the Superfund Amendments and Reauthorization Act (SARA) kept much of the original approach intact, while adding new community right-to-know provisions about chemical releases into the environment.[15]

With the passage of SARA, the antitoxics groups had emerged as important new players within the environmental arena. Despite a single-issue focus (most often formed around efforts to block the siting of a hazardous facility), the antitoxics groups increasingly began to identify broader production issues and outcomes as a central concern. This was perhaps best captured in the shift of the popular slogan "not in my backyard" to "not in anybody's backyard," or the more graphic slogan "plug up the toilet" to shift the focus to a change in production practices. At one level, the antitoxics movements of the 1980s began to associate the concerns of community with broader production issues linked to the concepts of toxics use reduction and pollution prevention. However, an additional and ultimately more significant focus on where as well as what was being sited emerged for several of the antitoxics groups, particularly those located in communities of color. This focus on discrimination, captured in the arguments about patterns of hazardous facility siting, emerged as the primary ingredient characterizing the environmental justice position.[16]

Despite the success around the Superfund campaigns and the various local community mobilizations, by the late 1980s some emerging fault lines among the community groups were beginning to appear. While nearly all the antitoxics groups, many led by women, were able to effectively mobilize local residents, distinctions between groups began to surface. The racial demographics of the different communities clearly played a role. One of the leading national antitoxics groups, the National Toxics Campaign (a group that had been instrumental in promoting pollution prevention and toxics use reduction) fell apart, due partly to tensions over race. Particularly for several of the antitoxics and anti-hazardous facility groups that had mobilized in communities of color, their struggles came to be defined primarily on the basis of the issue of discriminatory intent. For these groups (as distinct from several of the other antitoxics groups located in predominantly white low-income or middle-income communities) their struggle was seen as a fight against *environmental racism*, establishing a direct civil rights–environmental association. Blocking the siting of a hazardous facility also became a struggle to stop one more unwanted burden for a community of color.[17]

By 1990, a number of these civil rights–focused, community-based environmental groups decided to proceed on three related fronts. They challenged the mainstream environmentalists (as well as some of the community-based coalitions such as the National Toxics Campaign and Lois Gibbs's Citizen's Clearinghouse for Hazardous Wastes) for the absence of people of color in leadership and key staff positions. They simultaneously challenged the EPA for its failure to address issues of discriminatory intent and to open its decision-making process to include the impacted communities. Also, as part of their own growth process, a number of the groups issued a call to convene a gathering, a People of Color Environmental Leadership Summit, to coalesce the local community groups of color and establish a common framework and mission statement.[18]

As part of the challenge to the mainstream groups, letters were sent to the leaders of the Group of Ten, a semiformal association formed immediately after the 1980 election of Ronald Reagan, which consisted of the "chief executive officers" of the largest mainstream environmental groups. The letters raised the issues of race in relation to leadership, staffing, and agenda. Responses to these challenges varied widely. Several of the mainstream environmental organizations decided to create new staff positions or programs that suggested a new focus on environmental justice. In the case of two of the organizations, the Natural Resources Defense Council, and, subsequently, the Environmental Defense Fund (now Environmental Defense), new offices were opened in Los Angeles to focus on local environmental justice concerns, and new staff were hired in Washington to establish environmental justice activities. These initiatives, however, were also justified by their capacity to raise new sources of funding or, as in the case of EDF, because of a large, highway-related court settlement ($1.5 million) that called for the development of an environmental justice program. Despite the new focus of these offices, the mainstream groups, in keeping with their overall structure, continued to be largely concerned with national legislation and policy such as the Clean Air Act. Thus, while these offices and the staff positions established by other mainstream groups like the Sierra Club broadened the policy focus of the mainstream groups, there still remained a significant gap in terms of specific constituent or community-based organizing activity.[19]

Among the community coalitions, responses also varied. Some groups were not able to resolve the differences among their constituent groups or to successfully identify a multiracial form of organizing or coalition building. Other groups sympathetic to or embracing an environmental justice perspective successfully made the transition to support the arguments around discriminatory intent and the link to a civil rights dis-

course, while simultaneously seeking to explore new pollution prevention and community-based social and environmental agendas. A few
groups, like the San Francisco-based Urban Habitat, succeeded in developing multi-ethnic and cross-issue agendas. But the continuing focus on
discriminatory intent in facility siting, rather than these new agendas and
coalitions, most directly shaped the language and early definitions of
environmental justice.[20]

The dialogue initiated with the EPA, which at first bordered on confrontation, also had mixed results. In 1990, a letter from a group of environmental justice academics and activists to the EPA took the agency to
task for failing to address community concerns, including the question
of discriminatory intent in facility siting. As a consequence, the agency
established an Environmental Equity Workgroup, which sought to shift
the argument about the need for a new orientation in policymaking and
its social and environmental impacts, to the agency's preferred focus on
environmental equity. The key to the EPA definition of environmental
equity was its focus on whether and how a disproportionate share of
negative environmental consequences resulted from industrial, municipal, and commercial operations as well as from federal policies or
programs.[21]

Identifying such burdens, according to the EPA's approach, was best
accomplished through scientific measurement and risk analysis. "Risk is
central to equity," EPA administrator William Reilly argued, while risk
analysis made possible a distinction between risks, as well as who was
subject to the most burdensome risks (whether as a factor of income,
race, location, etc.). Comparative risk analysis, the buzzword of note
during the early 1990s, became the way to create a more equal playing
field of risk burdens, a kind of "equity of burden" approach.[22] By tying
the issue of equity to risk, policymakers could also avoid the potentially
controversial arena of intervening around the sources of the risk rather
than mitigating the risks for the most burdened constituencies. It allowed
the agency to distinguish between the notion of a scientific assessment
that was both "measurable and quantifiable" and a program that would
require broader socioeconomic factors be taken into account in evaluating issues of justice and discrimination. It also reinforced the arguments
of Reilly and other agency officials that policy needed to be based on

the "scientific understanding of risk" rather than have policymaking influenced by what EPA characterized as "public risk perceptions."[23]

While pursuing its new "equity of burdens" approach, the EPA continued to maintain its focus as a mitigation- and management-oriented agency. Through much of its previous two-decade history, the EPA had focused its policies on how to mitigate environmental problems after they had been created. Despite early rhetoric about a "systems" approach, the mitigation framework was end-of-pipe, pollutant by pollutant, and media-or-species specific (whether an air, water, land, endangered species, or resource problem). To attack the problems at their source (whether industry, sector, global, region, or community based), while attractive rhetorically, was constrained by the legal, political, and administrative biases concerning government regulation, which limited the degree and nature of public intervention in industry and even land-use decisions. At the same time, issues of race and discrimination were also considered separate if not irrelevant and even counter to the environmental policy domain.[24] The EPA's Office of Civil Rights, in the 1970s and 1980s, played only a minimal role within the agency itself. In responding to a critical report by the U.S. Civil Rights Commission, the EPA commented that it was a pollution abatement agency, to be distinguished from an agency principally concerned with community development with its greater attention to issues of social equity and place. EPA officials argued that any decisions about where to locate facilities, such as sewage treatment plants as mandated by the Clean Water Act, were only evaluated in terms of this pollution abatement focus, while others would have to consider community impacts.[25]

By the early 1990s, both the EPA and the mainstream environmental groups had begun to recognize that issues of discriminatory intent were going to require some response. Letters to the agency as well as to mainstream environmental groups were generating attention, including articles in the press, which in turn led to a flurry of meetings among the accused parties designed to mollify the criticisms about these discriminatory patterns. Buoyed by this new attention, the groups and individuals that assembled at the People of Color Environmental Leadership Summit felt they were in reach of constructing a new type of civil rights–oriented environmental movement.[26]

The summit, however, turned out to be an event not simply focused on the arguments about environmental discrimination. The fourteen-point mission statement adopted by summit delegates served as a broader statement of values and objectives for environmental justice. These included concerns about land use, ethical considerations, workplace issues, and democratic decision making. Despite what had initially appeared to be a narrow focus on facility siting, it was clear that the question of discrimination itself could become an entry point in developing a new approach to community development or what others were calling "sustainability" in the context of an argument about "justice." But could this direction be further elaborated if the focus remained on discrimination, and environmental justice remained more exclusively a type of civil rights discourse?[27]

Extending the Civil Rights Link: The Example of Title VI
The People of Color Environmental Leadership Summit offered potentially divergent strategies for those who participated and saw themselves in the forefront of a new type of movement. A civil rights/discrimination-oriented focus appeared compelling. This included issues of discriminatory intent in environmental policy, particularly around siting, the absence of community input in environmental decision making, and the absence of people-of-color staff and leadership and environmental justice agendas among the mainstream environmental organizations. While a veritable cottage industry of reports, studies, and journal articles debated the cause or even the existence of disproportionate environmental burdens, the power of the civil rights argument resonated, and took industry, policymakers, and the mainstream environmental groups by surprise.[28]

In response, the EPA and other government agencies introduced a flurry of new initiatives in the early 1990s that focused largely on the question of environmental discrimination. The most significant of these initiatives, President Clinton's February 1994 Executive Order 12898, committed the federal government to a civil rights, or antidiscriminatory review of any environmental policy decisions that resulted in "disproportionately high and adverse human health or environmental effects . . . on minority . . . and low-income populations." The executive order was released after an intense process of debate and dialogue between the Clinton administration's EPA administrator Carol Browner and several of the groups and individuals that had been instrumental in the People of Color Summit. While the executive order most directly symbolized the Clinton administration's focus on discrimination in environmental decision making, it also masked the difficulty in reorienting the nature and structure of environmental policy through legislation or new administrative authority.[29]

For the environmental justice groups, the dominant focus on distribution and discrimination issues, particularly the "equity of burden" approach, represented an opportunity and a limiting factor in constructing a new approach. On the one hand, environmental discrimination allowed these groups to force policymakers like the EPA and mainstream environmentalists to focus on the problems experienced by communities of color that had largely been absent from how environmental policy decisions were made or how environmental agendas by mainstream groups were determined. This focus on discrimination was assumed to be the centerpiece for an environment/civil rights association. However, the "equity of burden" approach suggested that if the particular risk were minimized or the risk burdens shared, then the problem of "environmental discrimination" would be resolved. In contrast, the People of Color Summit groups were seeking to identify a broader view of the nature of the problem and what appropriately constituted an environmental focus. Instead of the "equity of burden" concept, they identified the concept of justice as key to linking a community-based environmental advocacy to a civil rights–oriented discourse.[30]

This concept had several reference points. In the area of risk assessment, the environmental justice advocates argued that risks for specific populations needed to be distinguished from risks for general populations, the method more commonly used to assess risks. Specific populations, whether in terms of age (for example, children) or location (for example, low-income communities or communities of color where certain risks could be compounded by other additional or cumulative risks), were likely to be more vulnerable populations. By focusing on the evaluation of who was at risk and where the risks were occurring, the justice (cumulative burdens) and civil rights (disproportionate risks) dimensions of an issue could be joined. By focusing on the community concerns of vulnerable populations, environmental justice could expand the conventional definition of what constituted an environmental issue, and help develop a more "aggressive overall social justice agenda," as Deeohn Ferris and David Hahn-Baker put it.[31]

As a tactical strategy, perhaps the most striking example of the civil rights association with environmental justice was the use of Title VI of the 1964 Civil Rights Act, which barred discrimination associated with federally funded programs and activities. It differed from the Equal Protection Clause in the Fourteenth Amendment (in not requiring the need to show intent to discriminate), as well as other titles in the Civil Rights Act (in its ability to pursue administrative as well as judicial enforcement). Historically used as a tool for federal intervention where intentional discrimination could be legally identified in situations involving federal contracts, Title VI had not been applied during the 1960s, 1970s, and 1980s to the environmental area or specifically to EPA in relation to its permitting or rule-making activities. Only a handful of Title VI cases related to education and health issues had made their way through the courts during this period, although one unsuccessful 1984 case associated with the siting of a highway did have clear environmental implications.[32]

The use of Title VI for environmental purposes was first explored for its application to communities subject to a range of environmental burdens. The People of Color Summit helped highlight some of those struggles over siting and contributed to the development of a series of Title VI administrative petitions. Less than six months after the first of these actions was filed, Executive Order 12898 was published, which identified a possible federal government commitment to the use and enforcement of Title VI. The executive order, with its focus on discrimination, provided a Title VI rationale for such legal strategies. Most of the suits were designed as a legal tactic to influence particular siting disputes. But they also suggested a political calculation that by introducing discriminatory intent and environmental justice considerations, such actions could become an effective tool to increase potential damages and raise awareness about the issues involved.[33]

While providing visibility, the Title VI actions, which grew to more than seventy-five separate cases in a little more than five years, appeared to be of limited value as a legal strategy. In one of the more visible and politically charged cases, a coalition of environmental justice activists filed suit against the state of Pennsylvania regarding a decision to issue a permit to a hazardous waste facility in a largely African-American community in Delaware County. The issue of "disproportionate burden" was transparent in this case. In just ten years, the Pennsylvania Department of Environmental Protection had granted permits for five hazardous waste facilities in Chester. These facilities processed a total of 2.1 million tons of waste, while the rest of the county had two small facilities (hospitals located in a predominantly white area) with a 1.4 thousand ton capacity. Chester, an economically depressed community with high unemployment and large numbers of residents on some form of assistance, also had several other solid waste and sewage treatment facilities. This situated the community as a kind of environmental sacrifice zone for waste treatment and disposal. Despite the argument about generating more jobs, community groups mobilized against these facilities, using the language of justice (the need for better jobs and a cleaner environment) to elaborate their argument about discrimination. The initial petition was denied by the district court on the grounds that, although a concentration of waste facilities could be identified, the community groups were not able to prove discriminatory intent. After the court of appeals reversed the district court, the U.S. Supreme Court, in an August 1998 ruling, dismissed the petition again on the grounds that it was moot.[34]

Even when a Title VI action proved successful, as in a California case involving the siting of a hazardous waste facility in Kettleman City (a rural, predominantly Latino community in the Central Valley), the basis for the Title VI review in legal terms remained narrow. In the Kettleman case, a court suit successfully challenged a state agency's failure around public participation in the siting decision, specifically the failure to translate the document into Spanish. But the EPA also had the option of addressing a range of other environmental burdens for Kettleman City residents, including groundwater contamination from pesticide runoff as well as air quality concerns due to pesticide sprays. As in the Chester case, a strong argument could be made that Kettleman City provided a clear-cut case of the "disproportionate burden" for an environmentally stressed community.[35]

Part of the failure to act by the EPA was embedded in the structure of the agency itself. This problem was underlined by the lack of legislative action to implement the executive order and other administrative actions, such as the establishment of an environmental justice staff position and a small grants program. EPA operations were primarily dictated by its legislative mandates (e.g., the Clean Water Act or Clean Air Act), which in turn provided budgets and administrative structure (for its "program offices" such as the Office of Water, or Office of Air and Radiation) for implementation. Without legislation, there were no budgets; without budgets, EPA agency actions remained marginal to the structure of environmental policymaking. Moreover, the separation into program offices limited opportunities to address environmental issues outside the media-specific or problem-specific boxes that had been created. Even where the EPA had the ability to intervene in relation to federal contracts, the agency remained reluctant to proceed. The issue of contracts was particularly noteworthy since the federal government had at one point established a Title VI form as part of contract language, but that had eventually been dropped. The EPA continued to be reluctant to reintroduce this language due to harsh feedback from the states. Ultimately, most Title VI administrative petitions to the EPA were denied as a matter of course, despite the potential power of the agency to intervene through application of its contract power. For the EPA, environmental justice remained primarily an issue of procedural violation rather than a means, whether through legal or administrative action, to identify the related set of environmental and social concerns at the community scale.[36]

Title VI, however, despite its failures or limited successes in the courts or the lack of administrative actions within the EPA itself, still proved to be an important new tool for the environmental justice movement, as several of its advocates suggested. Title VI actions became one of the most visible arenas where the issues of community stresses—environmental and social—could be highlighted, as in the Chester case. Title VI also functioned as a staging ground for political action, where the legal or administrative suit was seen as an extension of a broader campaign rather than simply the procedural focus or question of implementation of laws and policies that it referred to.[37] What Title VI did not address—either as legal tool or campaign device—was the agenda for change, where debates over community stresses could shift to the question of new forms of community development. It was in these broader arenas of community life that the reorienting of environmentalism—and environmental justice—would most likely occur, if it were to occur at all.

New Openings

Like Title VI, Executive Order 12898, and other discrimination-focused policy initiatives, the civil rights–environmental justice link remained ambiguous or at least incomplete. Nevertheless, the community-based struggles about particular environmental hazards could "provide a window into the processes that produce the distributive outcomes," as Sheila Foster argued. Community-based campaigns that challenged a particular siting decision or negative land use had the potential to question or at least begin to address the production choice itself that had led to the problematic environmental outcome. The "plug up the toilet" term popularized by the antitoxics groups quickly translated to "prevent the pollution in the first place." But such a shift in language did not easily translate into identifiable strategies or campaigns. Nor did the argument about environmental discrimination necessarily translate into arguments about social justice. Concerns about jobs, inadequate health care, transit dependencies, or inadequate and substandard housing, as well as issues of community economic development were often divorced from the language about environmental concerns. Some environmental justice groups addressed the dilemma by identifying themselves as community development or social justice groups, albeit with an additional environmental agenda. The difficulty for the new environmental justice groups was particularly acute when there appeared to be a conflict between goals, such as jobs and environment, or, as some critics argued, between environmental protection and economic well-being.[38]

This presumed conflict was of particular concern in relation to the restoration of contaminated industrial sites in inner-city communities. Such sites, known as "brownfields," were defined as places that had, according to the EPA, "actual or perceived contamination and an active potential for redevelopment or reuse."[39] Literally hundreds of thousands of such sites could be found in nearly every major urban center that had a history of inner-city-based industrial activity. The General Accounting

Office at one point had estimated in the late 1980s that anywhere from 130,000 to as many as 425,000 such sites could require some form of cleanup, with costs estimated in the hundreds of billions of dollars to meet various federal and state cleanup requirements. Some public officials, such as Cleveland's Mayor Mike White, identified contamination and cleanup requirements as the "number one issue facing urban redevelopment." Yet brownfields were also typically identified as places less contaminated than the higher profile sites listed under the Superfund law's National Priorities List, and therefore represented sites where standards might be relaxed compared to the more stringent cleanup standards ordinarily triggered by Superfund. In a high profile interpretation of this "lesser contamination" standard, the EPA's 1995 Guidance for regulators called for a "creative reinterpretation" of the Superfund provisions by seeking to remove as many as 24,000 to 27,000 (out of 38,000) sites from potential Superfund status. However, some of the same issues that had been sparked by Superfund—abandoned properties, the type of cleanup required, continuing liability considerations, debates over appropriate strategies for how to reclaim the land—were invoked in relation to the brownfield sites as well.[40]

Though the brownfields issue rapidly emerged as a new kind of environmental justice issue in the early 1990s, it nevertheless referenced long-standing problems of land contamination and inner-city abandonment that had plagued urban life since the rise of the industrial city in the late nineteenth century. Those problems were reinforced in the post–World War II period due to the reconfiguration of the central cities associated with the continuing trends of suburban development, the rise of the interstate highway system, housing policies that reinforced urban blight, and the loss of the central city's manufacturing base. The presence of derelict buildings, vacant and contaminated lots, and abandoned factories became particularly pronounced during the 1950s, and especially during the 1960s, when the outbreak of inner-city urban riots forced government agencies to reevaluate earlier urban policies. Most of the neighborhood initiatives that sought to tackle these assorted land-use problems were community rather than regulatory driven. Urban planner and environmental justice advocate Carl Anthony, for example, recalled an early 1960s effort by a Harlem-based civil rights project to reclaim two vacant lots that had been partitioned by dilapidated fences and "rat-chewed" garbage. The goal of the civil rights group, Anthony recalled, "was to create a demonstration project to show what neighborhoods could do with a little help from their friends." After informing neighborhood residents about potential rehabilitation strategies, a community mobilization around the lots resulted in a reclaimed community site with play spaces, sitting areas, and a common area for picnics, as well as a community office in a vacant apartment in a nearby tenement building. These actions were not called "brownfield rehabilitation," Anthony concluded, but captured instead a "vocabulary of rehabilitation [that] included words like freedom, justice, hope, and self-reliance." In contrast, Anthony argued, the contemporary discourse around brownfields referenced a commercial and legal language that tended to be dominated by developers and corporations.[41]

The language and policies associated with "brownfields" were in fact not about community initiative and land reclamation, but referred more to the reluctance of lenders and developers to purchase or utilize inner-city "brownfield" sites for fear of liability and/or reluctance to address cleanup issues. The term "brownlining," or reluctance to invest in the inner city due to contamination issues, began to be used as a variant of "redlining." At the same time, existing environmental policies had long favored "greenfield" sites, those new developments, including industrial parks, that were located in suburban and exurban areas. The first brownfields policy initiatives established by the EPA, introduced in 1995 soon after the Republican takeover of Congress and the failure to pass legislation to reform Superfund, focused on relaxation of standards to encourage investors interested in purchasing brownfields properties but concerned about potential liability triggers or cleanup costs. Agency administrator Carol Browner promised the new Congress that the EPA would be "faster, fairer, and more efficient" in implementing Superfund provisions with respect to brownfield sites. While some investors responded positively and identified inner-city redevelopment opportunities as potentially lucrative investments, a shift toward more inner-city development only occurred on a limited, piecemeal basis, despite the arguments about possible new profit centers. By identifying environmental policies such as Superfund as a cause of the lack of inner-city investment, brownfields advocacy also came to be seen as promoting further environmental deregulation and criticism of "eco-risk exaggeration" rather than environmental policy restructuring to incorporate sustainable community development strategies.[42]

The brownfields issue, however, eventually became attractive to environmental justice groups like Carl Anthony's Urban Habitat organization because of its potential for community-based development. Before the entry of the environmental justice advocates, the key advocacy groups involved in dealing with old industrial sites in urban corridors—community economic development and mainstream environmental groups—not only approached the issue of what to do with such sites differently but often had conflicting goals. That division—community groups interested in any type of job-creating opportunity and mainstream environmental groups opposed to any relaxation of standards—reinforced the earlier history of division between such groups. The environmental justice groups, however, saw the very existence of brownfield sites as a core urban environmental problem, as well as an opportunity to look at environmental justice in a regional context, as Anthony argued.[43]

While eager to articulate a vision of community-guided—as well as regional—development, many of the environmental justice groups also remained suspicious of development strategies that proposed cleanup of inner-city, low-income sites to "industrial standards." Such an approach suggested what a number of community organizers saw as a license to pollute or to simply continue to extend the pollution of low-income communities that they had been forced to live with for decades.[44] Moreover, even when redevelopment was assumed to be successful, the lack of community input and more exclusive focus on the marketability of properties rather than their broader community purpose could create new burdens. Such approaches, similar to the urban renewal programs of the 1950s and 1960s, could force out people and businesses that had "long endured the deteriorated social and environmental conditions of the community in which the brownfields lie," as brownfields activist Lenny Siegel put it. In contrast, environmental justice groups sought to establish a framework to not only promote a more sustainable and community-oriented redevelopment, but to institutionalize a monitoring and community-centered decision-making process as well. One such example was the concept of a "community impact statement" to compile and analyze "the environmental load on a community independent of any particular project proposal," as Siegel defined it. Brownfields advocacy could in this context also be seen as expanding the environmental justice/civil rights discourse to include the issues of community and economic development as well as a focus on place that had been so instrumental in the early development of the 1980s antitoxics groups.[45]

The emerging interest of the environmental justice groups in impacts from highway construction and the inevitable link between land use and transportation also offered community groups the ability to challenge public policies that had environmental, social, and economic implications for poor communities as well as for cities and regions. The highway issue was an important example of this neighborhood-regional environmental justice/transportation link. Since the development of urban highways and especially following the passage of the National Highways Act and the development of the national interstate highway system, urban core communities in working-class residential and low-income neighborhoods have been subject to the impacts from highway building projects. During the 1950s and 1960s, some of the more visible neighborhood-based opposition to highway building came from middle-class residential or commercial districts whose areas were included in the highway plans. The battlegrounds in New Orleans, the west side of Manhattan, or the Embarcadero District in San Francisco during this period identified the opposition as "preservation" oriented. But the "justice" argument with respect to urban freeway construction also emerged in this period. However, most of the conflicts over highway building plans that erupted in low-income communities failed to generate the same kind of press and policy interest. These opposition campaigns focused on the additional community burdens on low-income neighborhoods represented by freeway construction and siting, linking the issue of environmental burdens to the broader question of community development and identity. The construction of freeways through inner-city communities that brought with it elevated exposures from lead emissions, particulate matter and other criteria pollutants, and toxic air contaminants, combined with the often devastating land-use impacts from neighborhoods cut in two, made freeways an enormous social and environmental burden rather than "transportation asset." As an advocate with the National Coalition for the Transportation Crisis put it, the urban freeway system had become "a tool of both social injustice and environmental destruction."[46]

This type of environmental justice position not only raised the civil rights-oriented arguments about "discriminatory burdens" but it also led to arguments about the need to remake neighborhoods more livable in economic and social as well as environmental terms. By the late 1980s and 1990s, regional and national coalitions began to take up this cause of "livable communities," a term that eventually worked its way into the national policy discourse regarding issues of "sprawl."[47] For groups such as Urban Habitat and the Surface Transportation Policy Project (STPP, a coalition based in Washington D.C. that addressed national transportation policy), the "livable communities" argument became the occasion for establishing an "environment plus" as well as a "transportation plus" approach. From the environmental justice side, Urban Habitat has argued that the environmental justice position needs to include a land-use and community and economic development framework through such strategies as transit-oriented development. From the transportation side, the STPP, which has mobilized around the destructive social and environmental consequences of auto-centered transportation, has argued that while such consequences are ubiquitous in terms of regional impacts, they also have powerful equity/justice implications in terms of urban core concerns.[48]

These same patterns repeat themselves with other critical daily life concerns. The question of community food needs, for example, has become a concern of social justice groups focused on the "discriminatory" patterns of a food system that has increasingly marginalized inner-city neighborhoods. Along these lines, some environmental justice groups see the widespread existence of vacant lots as both burden and opportunity. On the one hand, such lots can "demoralize nearby residents, cast a pall over renewal efforts, undermine property values, and inhibit the influx of capital needed for revitalization," as one activist put it. At the same time, a vacant lot could also become "a potential opportunity and significant resource" if such land could be reclaimed, such as for urban agriculture and recreational use with community gardens. Other groups have linked more mainstream environmental causes such as air quality to particular justice-related issues faced by low-income communities. For example, groups like New York's West Harlem Environmental Action or Los Angeles's Bus Riders Union have advocated that buses shift to cleaner fuels to replace the commonly used diesel fuel. Diesel, a significant occupational as well as environmental health hazard, is also a significant community hazard for residents living near bus yards, schoolchildren exposed to diesel emissions from school buses, or in neighborhoods where there is heavy bus use. Thus, the broader air quality focus that mainstream groups like the NRDC have used in their efforts to reduce or eliminate the use of diesel can be reframed by the environmental justice groups by focusing on such community concerns as transit needs, children's health, and access to community services. The diesel issue in this way can also provide an environmental articulation of a broader community- or place-based focus. Ultimately, this type of environmental justice argument, in its most elaborated form, asserts that *place matters*.[49]

Regions—To Live In, Seventy-Five Years Later

The development of environmental justice as a form of community renewal and regional restructuring, a cause taken up by community groups like West Harlem Environmental Action and Urban Habitat, has also helped inform the development of this new type of place-based, regional politics. Nearly seventy-five years after the publication of "Regions—To Live In," the regional planning movement's discussion of the need for a more sustainable regional restructuring, a new environmental justice and social justice focus on regions has begun to re-emerge. The new regional arguments are about the misallocations of physical as well as economic resources, land-use problems as well as economic development problems. It also suggests new types of political alliances and political reconfigurations, such as suggested by Myron Orfield's argument about an inner-city, older/working-class suburban alliance in its conflicts with the wealthier outlying suburbs or exurban communities. This "idea of a new urbanism—which takes the metropolitan region as the basic unit of political analysis and governance—is a notion with broad potential appeal," University of Wisconsin professor and political activist Joel Rogers has also argued. The new urbanism can be associated with reconstructing urban environments—for example, in the advocacy for urban agriculture, for green spaces in the city, for urban-regional foodsheds, and for new transportation and land-use strategies.[50]

The dilemma for the new regionalism, similar to the kind of difficulties experienced by the environmental justice groups in the debates over brownfields redevelopment, is the question of redevelopment for what and by whom. The urban or regional strategies that environmental justice has so effectively highlighted as necessary to a broader environmental politics are still limited by the difficulties in addressing the need for jobs and the forms of economic development associated with the regional approach. If communities and regions are to be more livable and development strategies more sustainable, how does that translate with respect to issues of production? These include questions concerning the organization of the workplace and the nature of work. They include the design and use of the materials and products that also shape the nature of our communities and how we live. And if environmental justice provides a new pathway to a politics of place, then it becomes crucial to determine whether a parallel politics of production can also emerge. Such a politics needs to be able to address what we produce as well as how we produce. For environmental justice to succeed, it also has to become a movement concerned with jobs, industry, and the environment. For if place matters, production does as well.

Pollution Prevention and Cleaner Production: Identifying a Public Role

The Paradigm, or a Revolution Betrayed?

When Ken Geiser, the director of the Massachusetts Toxics Reduction Institute (TURI), took the podium at the Greening of Industry conference at the University of California in Santa Barbara in November 1997, most of the 200 or so participants did not anticipate that Geiser would evoke memories of the 1960s. Geiser spoke of his earlier embrace of a democratic socialist vision of a more just society, not dissimilar to Barry Commoner's argument about the need for social governance of the production system. Geiser pointed out that while a democratic socialist perspective had become more problematic by the 1990s, there was still a compelling need to recapture a sense of vision and possibility that 1960s social movements had once offered. Was socially and environmentally responsible industrial activity (the theme of this conference as well as for an increasing number of publications, forums, and other similar gatherings) an achievable goal, Geiser challenged his audience? Had a revolution been initiated, but not completed?[51]

The outlines of such a vision of urban and industrial restructuring had first taken form in the mid- and late 1980s, associated with such concepts as toxics use reduction and pollution prevention. Geiser himself had been one of the participants in efforts to establish a policy framework to implement such approaches. The Massachusetts legislation, the Massachusetts Toxics Use Reduction Act, that had authorized and provided the funding for TURI, was itself an early example of the promise of this new type of policy approach.[52] By 1990, with the passage of the federal Pollution Prevention Act, it appeared that a significant new thrust in environmental and industrial policymaking might be possible. But did this new thrust, or new paradigm, as some liked to call it, actually represent a change in direction in environmental governance and industry decision making? Or was it an opportunity deferred, mired in the labyrinth of the regulatory system and the inability or unwillingness to link regulation to core decisions about what industry was producing, how it was being produced, and why it was produced in the first place?

The new pollution prevention approach primarily defined itself in contrast to the previous two decades of pollution control policymaking that had carefully sought to limit the focus of policymaking. Pollution control distinguished between the government role that focused on how best to address the outcomes of production and industry-specific decisions about what and how to produce. This "end-of-pipe" concept of environmental management was also associated with legislation and regulation that sought to mitigate an environmental problem at the end point in each production system. This could be accomplished by identifying what constituted an acceptable environmental or human health outcome (identifying the acceptable levels of risk), as well as by facilitating an end point technology that could mitigate those outcomes (identifying, and, at times, forcing the development of a best-available control technology). While there had been other conceptual approaches introduced during this period, most notably the idea of market approaches to accomplish similar goals, nearly all of these had been associated with this pollution control regulatory framework.

The concept of prevention (influencing industry choices at the front end rather than controlling or mitigating production outcomes) first emerged significantly in the late 1970s and early 1980s. The prevention concept had various sources and different kinds of advocates. For some industry groups or individual companies like 3M, prevention was the equivalent of regulatory avoidance, by anticipating the possibility of regulation before it was introduced and thereby limiting its potential reach. 3M was also at the vanguard in constructing an industry position that accepted the logic of regulation in a pollution control scenario but argued vociferously that only industry knew best when it came to decision making about specific production choices. This applied particularly to the argument about pollution prevention changes in the development of products or industry processes. Pollution prevention as a conceptual alternative to pollution control regulation, in the forms promoted by 3M and other "industry-knows-best" advocates, served as a kind of proxy for promoting business efficiency. It was also seen as an extension of a less intrusive pollution control approach insofar as this business-friendly pollution prevention anticipated and effectively sought to avoid the need for some of the costly end-of-pipe controls.[53]

In the early 1980s, pollution prevention also emerged as a community-focused or environmental justice concern, since a number of low-income communities had become the sites for end-of-pipe management strategies (for example, incineration of wastes), which were themselves creating new kinds of environmental burdens. The "plug up the toilet" war cry of the early antitoxics environmental justice–oriented groups was as much a challenge to policymakers as industry. Extending their argument further into the domain of production decision making, the first significant efforts toward identifying a more holistic prevention or toxics use reduction approach were developed by antitoxics groups like the National Toxics Campaign and the Citizen's Clearinghouse for Hazardous Wastes. The arguments of the antitoxics groups paralleled and eventually helped shape the initial policy-oriented discussions about "source reduction" and "hazardous waste reduction" introduced by the Office of Technology Assessment (OTA) and picked up by state and local policymakers. When the Massachusetts legislation was passed in 1989 and parallel legislation was adopted that same year by the state of Oregon, it appeared that a major policy shift was about to occur.[54]

At the time of the passage of the Pollution Prevention Act in 1990, pollution prevention had become, for some, the generic, all-encompassing concept that represented this new paradigm. While the 1990 legislation provided no specific regulatory powers to accomplish pollution prevention outcomes, it nevertheless established a definition more in keeping with the broad view of the antitoxics groups. The language of the act situated the tasks and goals of prevention, defined as source reduction, within the design and operations of facilities prior to recycling, treatment, or disposal, rather than, as some industry groups wanted, located within the framework of managing wastes, as with off-site recycling. For its more radical advocates, pollution prevention promised a revolution in how industry operated and how process, technology, workplace, and product decisions were to be made.[55]

But did such a change occur in the decade following the passage of the Pollution Prevention Act? In a 1998 article published in the *Pollution Prevention Review* that mirrored Geiser's argument of an unfinished revolution, Joel Hirschhorn, the author of the key 1986 OTA study on hazardous waste reduction, saw policymakers, primarily at the EPA, redirecting this new paradigm away from its revolutionary implications. The "pollution prevention vision," Hirschhorn commented, "was broad, encompassing changes in manufacturing technologies and practices, chemicals and other raw materials, and even products and packaging. It also covered resource and energy conservation." However, Hirschhorn argued, a narrower view had also emerged during the 1990s that had eaten away at pollution prevention's visionary aspect by agency "incrementalists" who feared "a systematic assault against chemicals." Incrementalism, even at its best, involved only modest changes to the dominant end-of-pipe paradigm. Talk about sustainability and the green-

ing of industry that increasingly displaced the focus on pollution prevention was also less policy focused and often masked a reluctance to
bring about more fundamental change. The revolution, Hirschhorn dramatically concluded, had been betrayed.[56]

Similarly, leading pollution prevention researcher Warren Muir, in a
commentary included in the EPA's 1997 progress report on pollution prevention, concluded that the introduction of pollution prevention policy
had not had any noticeable impact on aggregate U.S. toxic chemical
waste generation and industrial practices. Muir countered arguments by
the EPA and other industry-knows-best advocates that sought to highlight the voluntary contributions by industry in reducing their toxic
releases to the environment. He pointed out that the volume of waste
generation continued to rise at the same time that various institutional
barriers within companies continued to limit its adoption. "We've
learned how to talk pollution prevention, but are a long way away from
putting it into action nationally," Muir warned. And the handful of
studies that had monitored industry's willingness to invest the capital and
restructure its decision making to accomplish pollution prevention goals,
indicated, in the context of voluntarism, only a weak commitment by
industry, if any at all.[57]

These arguments were challenged by regulators and industry officials
who spoke of a growing "greening of industry" phenomena occurring
largely outside of regulatory influences. The role of government in
encouraging industry greening was assumed to reside in the area of technical assistance and incentive-based programs, but not, to use the dominant antiregulatory metaphor, as part of a "command and control"
system. EPA pollution prevention official John Cross, for example,
warned that the romantic and interventionist interpretation of pollution
prevention as a form of social governance of industrial systems was little
more than "a stalking horse for regulatory imperialism." Adopting a perspective increasingly assumed by the EPA and even by some environmentalists, Cross argued that pollution prevention was at best only a
stage toward greater sustainability and industry greening. For the EPA,
other concepts, such as industrial ecology and design for the environment, had become more compelling approaches associated with industry activity, further reducing the need to legislate and regulate. Ultimately,

the lesson for the agencies was that the new paradigms in environmental governance referred primarily, if not exclusively, to how industry conducted itself, while regulation and policy remained the domain of the legislatively mandated end-of-pipe requirements. Thus the real power at the EPA, based on staff and budgetary resources, were the program offices. Units such as the Office of Air and Radiation, the Office of Water, or the Office of Solid Waste, had the key responsibilities for standard setting, enforcement, and overall implementation for the myriad of pollution control laws and regulations in each medium (air, water, land). While addressing environmental problems that had already been created, through legislative mandates that provided both regulatory powers and resources, pollution control permeated the regulatory system. Pollution prevention voluntarism and other greening of industry strategies, meanwhile, occupied the ideological center of the agency without any substantive role to implement its new perspective.[58]

This division between pollution control as the appropriate arena for regulatory activity and pollution prevention as an area for industry activity outside of regulatory action was underlined by the nature of the government's own pollution prevention activities. These included a series of programs that reinforced the voluntary focus. Such nonregulatory programs as "Green Lights" (encouraging energy savings through lighting efficiencies), the 33/50 program (voluntary efforts by companies to reduce their toxic releases by targeted amounts), and Climate Wise (voluntary pledges by industry to reduce greenhouse gas emissions), were touted by EPA officials as representing "new models for government/industry interaction." "One of the clearest indicators of corporate responsiveness to the need for reducing chemical releases and preventing pollution," the EPA argued, "has been a company's participation in the EPA's voluntary programs."[59]

But what motivated industry to act? Since the passage of the Pollution Prevention Act and its clear signal about the voluntarist path, EPA programs had firmly associated the agency's pollution prevention approach with the "industry knows best" perspective, identifying regulatory flexibility as well as technical assistance as the primary role for government. By the end of the 1990s, the place of pollution prevention within the environmental policy system had become something of a conundrum.

Heralded by the EPA as potentially "the most effective method for reducing risks to human health and the environment," pollution prevention had become a concept without a home, a set of policies without direction or even a set of tangible goals to be accomplished. A stepchild of the environmental policy system, pollution prevention remained in regulatory limbo, a revolution still seeking to find its champions and unable to change the existing rules of the game.[60]

The Ambiguities of Pollution Control

The rise of the contemporary system of environmental management, including the focus on end-point mitigation, had its roots in the ambiguities of the politics and policymaking of the late 1960s and early 1970s. It was in this period that a radical environmental perspective took shape, questioning the basis of industry decision making and criticizing what Iris Young called "the prerogative of private companies to produce whatever they wish however they wish."[61] In some ways, the emergence of this environmental perspective came as a shock to both industry leaders and government officials. The rapid-fire passage of legislation between 1970 and 1976, including the Clean Air Act, Occupational Safety and Health Act, Clean Water Act, Resource Conservation and Recovery Act, and Toxic Substances Control Act, was due in part to the phenomenal explosion of popular interest in environmental action and change. It also revealed an expanding interest in industry activities as a legitimate arena for government action not seen since the New Deal and the Progressive Era.

Much of the legislation, however, was ambiguous in terms of the intent and focus of regulation and strategies for implementation. On the one hand, in responding to the rise of environmental advocacy, Congress identified broad goals in the legislation, such as pollution-free waterways and a system for reviewing chemicals prior to their introduction. Some legislation, such as the Clean Air Act and the Occupational Safety and Health Act, provided potential opportunities to establish broad and even more systemic approaches to achieve certain goals. This could be found, for example, in the language in the Clean Air Act about mobile sources of air emissions that appeared to allow regulators to explore potential transportation and land-use management approaches rather than just slapping on tailpipe controls. Even a specific end-point control, such as the introduction of the catalytic converter to control certain kinds of pollutant emissions, had an inadvertent pollution prevention outcome in helping to precipitate the eventual elimination of leaded gasoline.[62]

This wave of legislation led to concerns among Congress, the president, and regulators alike, that without demonstrating immediate results, environmental advocacy could become part of the radical critique and direct-action politics that had been spawned by the social movements of the late 1960s. Events like Earth Day 1970 revealed ambiguities about the future direction of environmentalism that cut across issue, policy, constituency, and movement-building areas. While the mainstream and policy-oriented groups that formed in this period, such as the Natural Resources Defense Council and the Environmental Defense Fund, identified a more limited horizon for intervention and action (the courts, legislation, and regulatory action), even these groups had developed their own unpredictable if not quasi-radical culture. While seeking specific and manageable policy outcomes, the mainstream groups were also associated with the need for far-reaching change. Confrontational tactics, as suggested by the EDF slogan "sue the bastards," were frequently employed. During the early 1970s, the distinctions between the local and national environmental groups were also less apparent in terms of the broad view about the need for change than in the focus and tactics of environmental opposition.[63]

Faced with this unprecedented challenge concerning the government's role regarding industry decision making and core social problem areas such as land use, transportation, and even food production, Congress, the newly formed Environmental Protection Agency, and even President Nixon embraced the language of systems change and forceful intervention. "The environment must be perceived as a single, integrated system," Nixon declared in his message to Congress identifying the reorganization plan that led to the formation of the EPA. Congressional mandates identified sweeping antipollution targets medium by medium, and called for potentially enormous powers for regulators to enforce those mandates. EPA's first administrator, William Ruckelshaus, sought to demonstrate that visible enforcement would be the agency's initial signature activity, pressuring steel plants as his first target to develop smokestack emission controls in order to demonstrate compliance with regulations and tangible improvement in the quality of the air.[64]

The concept of a systemic approach, or integrated pollution control, was never directly pursued by the EPA, OSHA, or any other administrative unit that addressed environmental questions. The EPA especially developed the pollutant-by-pollutant, single-medium approach. This was partly due to the requirements of the 1970–1976 legislation that created the need for the program offices and a media-specific division of responsibility. It was also partly due to the long-standing bias in favor of policy "incrementalism" that shifted the focus to specific controls from broader industry or land-use activities. Within those limits, the outcomes of this policy framework were also mixed. As J. Clarence Davies and Jan Mazurek noted in their summary review of more than twenty-five years of pollution control policy, certain improvements in the pollutant load from industry activity have been significant if not dramatic, such as the elimination of leaded gasoline, changes in water quality due to improved sewage treatment and point source controls, and reductions in the level of some of the criteria air pollutants. Gregg Easterbrook's polemical celebration of these successful environmental end-point outcomes, *A Moment on the Earth*, also became the occasion to challenge the continuing unease within all segments of the environmental movement that change remained far too limited. But, as Davies and Mazurek also pointed out, the limits of the system could be identified by its failures within the logic of its own approach. This could be seen in terms of the lack of controls on toxic air contaminants, the failure to regulate pesticides and other nonpoint source pollutants, or the difficulty in addressing persistent toxics, such as heavy metals, without moving toward a pollution prevention system.[65]

For industry, the pollution control system also generated ambiguous responses. At first, certain sectors, such as steel and oil refining, had been strongly hostile to the development and implementation of the pollution control approach. But already by the mid-1970s, most industries had begun to accept this form of intervention as providing some certainty in outcomes (e.g., the need for end-point controls), while continuing to try to narrow the scope of the control measures. The antiregulatory arguments about pollution control that began to take shape in the late 1970s were part of a broader corporate counteroffensive about government intervention as a whole, including resource policy, occupational safety and health programs and standards, and consumer product issues. By the early 1980s, a new environmental battleground had formed: the anti-regulators, including those within the regulatory agencies, versus the pollution control and environmental policy system advocates, led by those in Congress who forced through legislation that extended rather than reduced regulatory mandates and end-point controls. Industry groups sought to maneuver through these political minefields by constantly seeking to limit the nature of regulation, accepting the need for certain controls where they were applied uniformly and with some consistency, but also encouraging the development of a more flexible, industry-centered regulatory approach. For industry, regulation with flexibility was the basis for encouraging a more voluntary or industry-driven process of environmental change.

The antitoxics/toxics use reduction and environmental justice movements that emerged in this period also expressed discontent about the limits of pollution control, while strongly defending the value and importance of interventionist approaches, including cleanup or end-point mandates. One area of concern was the failure of enforcement even in some basic areas associated with end-point mitigation.[66] But at a political level, the antitoxics groups primarily sought to broaden the framework and targeted outcomes for policy by defining toxics use reduction as requiring an integrated approach to environmental, community, occupational, and product-based hazards and exposures. The problem of a pollutant-by-pollutant, medium-by-medium approach, the toxics use reduction advocates argued, was not just that it could lead to a transfer of environmental problems, but it often discounted occupational and product-based impacts. By focusing on workplaces and products, as well as environmental and community exposures, the toxics use reduction strategy sought to bring together communities, workers, consumers, and local environmentalists in a new type of coalition, which had been an early goal of Ralph Nader and other late 1960s anticorporate activists.

The antitoxics and environmental justice groups also emphasized a public or community and worker role in decision making, beginning with, but not limited to, expanding right-to-know opportunities. At the same time, the environmental justice focus on exposures and vulnerable communities and populations led to concerns about the priorities, strategies, and methods of cleanup and mitigation. The reauthorization and effective implementation of Superfund, the most visible and highly contested legislative and regulatory measure within the pollution control universe of environmental policy, remained an important goal of the community activists during the mid-1980s. The Superfund debates also revealed the environmental justice desire to address community hazards and needs more broadly (to include housing, jobs, transportation, and land-use issues as examples), and the concern, as articulated in the brownfields discussions, for community renewal and revitalization. But the Superfund debates, which in some ways became the fulcrum for the discussions about reorienting or redefining environmental policy, given the legislation's costly and limited outcomes, also triggered the search for third-way solutions. By the early 1990s, pollution control had become an embattled system of environmental management and industry response. And although the antitoxics and environmental justice groups had helped stimulate the discussion about a new framework for policy, including a prevention or front-end approach, the policy debates through the 1990s shifted again. This time the focus was no longer on the role of the regulators and the strategies for public intervention, but on the ways to explore and identify how industry could presumably accomplish what government could not.

The Third Way

Despite its much-maligned status, the pollution control system had, by the early 1990s, achieved one often-disregarded, yet critical and significant outcome. Pollution control regulatory programs were "almost unrivaled in capturing the attention of core business decision makers," as two EPA analysts concluded from their study of industry motivation for pollution prevention. A good indication of this was the rapid growth of the environmental and health and safety compliance staff within industry during the 1980s and 1990s. Their activities were particularly aimed at avoiding some of the more costly outcomes of regulatory action, such as involvement in a Superfund site. This attention-focusing aspect of the pollution control system underlined the key industry concerns associated with the concept of command-and-control, the term of approbation that had come to characterize the extent and intrusiveness of regulation, particularly for their impact on industry decision making (even at the end point).[67]

The fierce and continuing battles of the 1980s over the degree of regulation began to shift during the early 1990s to a discussion of how and ultimately whether to regulate. The focus on comparative risk provided one direction (establishing priorities for government action on the basis of a scientific review of risks), though many of the environmental justice groups remained suspicious that comparative risk represented an indirect effort to divert attention from and ultimately reduce activity around hazardous waste sites. Superfund especially had become the *bête noir* of the critics of a more extensive regulatory approach requiring a change in industry activity. In the early and mid-1990s, a series of efforts were initiated to identify and document what a "third-way" approach would look like. A popular new term associated with this approach, "civic environmentalism," identified the importance of both local and statewide (rather than federal) activity as well as initiatives influenced by "civil society" actors, including both community groups and industry, while deemphasizing the federal role.[68]

The most influential of these efforts was associated with a series of gatherings that took place during the mid-1990s in Aspen, Colorado. The Aspen process involved a series of "dialogue" meetings at the Aspen Institute. It included companies and industry groups (3M, Monsanto, Intel, Union Carbide, At&T), government agencies and elected officials, mainstream environmental groups (NRDC, EDF, Sierra Club), and a handful of environmental justice advocates. The charge of the meetings was to construct the third-way framework for environmental management between regulatory conflict and a deregulatory path that relied entirely on market forces, as advocated by such groups as the Political Economy Research Center in Montana. After three years of meetings, a document, called "The Alternative Path," was issued by the convenors of the Aspen process. The document called for establishing a "cleaner, cheaper way to protect and enhance the environment," as its subtitle put it, and for redefining or reinventing the government role, particularly the federal government's role, in regulation. The environmental policy system, the document argued, was changing from "prescriptive, technology-based requirements to an incentive performance-based contract" based on more flexible approaches to regulation. But the Aspen process never fully resolved how to identify a programmatic shift for some of the more intractable issues such as Superfund reform, nor did it reach consensus on specific legislative or regulatory approaches. Its most important outcome was its influence over the policy discourse itself.[69]

With the new Clinton administration, the third-way concepts resonated in terms of the "reinventing government" conceptual framework—and this shift in discourse about the role of government—that the Clintonites were seeking to promote. EPA head Carol Browner and other environmental officials put forth the slogan "cleaner, cheaper, smarter" for a range of new programmatic initiatives that took center stage in the early and mid-1990s. These included the Project XL ("excellence and leadership") program, the Common Sense Initiative, and Design for the Environment. Each of these programs incorporated key elements of the third-way approach: stakeholder collaborations; regulatory flexibility; and the development of specific outcomes, often defined as pollution prevention goals, but primarily focused on reduced waste streams. Most important, the third-way programs operated on the assumption that core changes in industry activity were best accomplished by utilizing industry's own knowledge and skill base, while providing input from stakeholders and maintaining or reducing the level of regulation so that opportunities for innovation would not be undermined.[70]

While these programs were introduced with much fanfare, the third-way, government-initiated process produced few tangible results. Each of the third-way programs involved particular industry sectors such as the electronics industry (Project XL), computers and electronics, petroleum refining, and automobile manufacturing (the Common Sense Initiative), or the printing and dry-cleaning industries (Design for the Environment). Stakeholder representatives included various industry players (upstream and downstream, as well as subcontractors), local, state, and federal regulators, community and environmental justice groups, mainstream environmental groups, and labor. While these

EPA-initiated processes were more inclusive, there was still an enormous imbalance in terms of the technical capability of the nonindustry groups for addressing specific industrial process and technology issues. However, even when the nonindustry groups were able to identify possible alternative technologies, as in the Design for the Environment dry-cleaning sessions, the weight of the discussions were still oriented toward what industry would be willing or able to accomplish within its existing process and technology approaches. Nonindustry stakeholder recommendations for action thus became at best tangential and not associated with any designated outcome.

As the various initiatives evolved into more of a discussion than decision-related forum and the regulatory flexibility measures suggested primarily a reduced level of regulatory activity, a number of the community and environmental groups began to criticize the third-way approaches as unbalanced. Some of the environmental groups challenged specific outcomes, such as the Project XL process involving an Intel facility in Arizona. More broadly, the community and environmental groups as well as some of their technical support allies, such as the Toxics Use Reduction Institute, recognized the lack of a coherent alternative perspective on the issues of regulation, technology change, and community, labor, and environmental participation in a stakeholder process. As a result, a series of gatherings took place among these groups seeking to identify a new type of framework for analysis and action. These gatherings sought to bring together environmental justice, mainstream environmental, and pollution prevention advocates, using the concept of clean production to link a variety of new approaches above and beyond their earlier advocacy around toxics use reduction and pollution prevention.[71]

Part of the dilemma for the community and environmental groups was the continuing shift in the policy discourse away from a regulatory framework. By the late 1990s, a number of the third-way programs, such as the Common Sense Initiative, had either collapsed or given way to the antiregulatory perspective that swept through Washington. These included arguments regarding unfunded mandates (no regulation without federal funds to pay for it), takings (the primacy of property rights), and benefit-cost analysis (no regulation without meeting a benefit-cost test). The 104th Congress, fresh from the triumph of a large number of Republicans in the 1994 election who had championed the cause of environmental deregulation, introduced a series of bills on these issues that essentially challenged much of the environmental policy system that had evolved since the 1970s.[72]

While none of the more onerous bills was signed into law, the counteroffensive launched during the 104th Congress significantly shifted the nature of the policy debate. By the late 1990s, the code word in policy-making was voluntarism. Aside from EPA's own voluntary programs, several of which constituted the basis for the agency's pollution prevention approach, an unwillingness to regulate further, even within the confines of end-point regulation, dominated most of the policy discussions. The new paradigm turned out to be no paradigm at all, at least in terms of a role for policy.

The Search for Alternatives: The Next Generation?
With minimal outcomes from many of the third-way approaches and the ascendance of the voluntarist interpretation of pollution prevention, a series of other conceptual approaches began to receive attention among academics, policymakers, and industry as well as community and environmental activists. This "next generation" of alternatives, as introduced through a series of academic symposia, publications, and industry-based gatherings, sought to capture and extend the arguments first raised by the pollution prevention advocates about the limits and failures of the end-point mitigation and management systems. Some of these alternatives, such as industrial ecology, had long-standing reference to core environmental concerns such as resource and materials use. Others built on ideas and programs first introduced in Europe or elaborated concepts about design or process or decision-making changes that had other nonenvironmental roots in industry practices. All of the next generation alternatives served as a counterpart to government intervention. Three of these alternative approaches—industrial ecology, extended producer responsibility, and design for environment—provide a useful perspective on the direction of the next generation search.[73]

Industrial Ecology Among these approaches, industrial ecology emerged as potentially the most encompassing, by identifying production systems in throughput terms. It was also clearly rooted in an earlier environmental focus on waste issues, although by the 1990s, the waste problem began to be viewed more as a question of material flows and system efficiencies. These analytic arguments about materials extraction, use, and reuse, constituted the initial elements of the industrial ecology approach, a term first significantly promoted at a 1991 National Academy of Science colloquium.[74]

Industrial ecology has focused on both technological and engineering questions. In relation to the engineering of products and processes, industrial ecology advocates have argued that waste materials should be regarded as raw materials—"useful sources of materials and energy for other industrial processes and products," rather than being "automatically sent for disposal." The enormous inefficiencies associated with the failure to reuse or recycle materials placed stresses on both the materials base (depletion of nonrenewable raw materials) and the energy required for waste disposal and reliance on virgin raw materials. Energy use in fact emerges as a kind of industrial ecology proxy—the greater the amount of energy per unit of production, the less efficient the industrial system, and vice versa.[75]

Industrial ecology advocates have assumed a kind of technological optimism, a celebration of new technology opportunities that are linked in turn to efficiency and sustainability goals. These opportunities provide what leading industrial ecology analyst Jesse Ausubel has characterized as the "Copernican turn"; that is, the ability to incorporate "Environment" and "Nature" into Economy. Concepts such as "dematerialization" (the reduction by weight of the materials used for any given economic function) and "decarbonization" (the shift away from carbon-generating fuels and energy sources) have become industrial ecology building blocks, constituting in turn what some have called an industrial ecology "science of sustainability."[76]

Key industrial ecology advocates, such as Braden Allenby, a vice president for AT&T, have also extended the industrial ecology concept to include the development of new technologies and products as part of the business cycle. Countering the environmental justice celebration of locality and difference, Allenby has argued that "the real threats to a habitable and sustainable world in the next two centuries [may] arise from the continuing social turmoil associated with the relatively inflexible cultural and ethnic differences among people." Industrial ecology therefore tries to shift the environmentalist focus toward the global and away from the local or what are considered community-based parochial concerns, while environmental change is most directly associated with the technological advances of the high-tech industry sectors, such as biotechnology, communications, and electronics.[77]

The industrial ecology focus on material and energy flows and the new technologies of the high-tech sectors contrast with the toxics use reduction arguments that have focused on the workplace and product-related as well as environmental consequences of industry decisions. An interesting set of examples described in one of the first industrial ecology texts underlines those differences. Valerie Thomas and Thomas Spiro, in exploring the materials use issues associated with lead and cadmium, identify lead as a leading candidate for an industrial ecology approach due to its high recycling rates. While acknowledging that there are significant environmental and human health hazards associated with lead exposures, Thomas and Spiro distinguish between the emissions associated with lead paint and leaded gasoline (which, they argue, have been eliminated or controlled) and lead-acid batteries, which have emerged as a leading use of lead. But from a toxics-reduction perspective, lead-battery recycling facilities, located in both U.S. inner-city communities and third-world countries (e.g., maquiladoras in Tijuana) have represented a significant environmental and health burden for the workforce at those facilities as well as for the surrounding communities. In another example from the same text, an evaluation of the industrial ecology approach in the manufacturing of consumer products identified three examples of progress in achieving an industrial ecology approach. These include the Kodak company's recycling strategies for "single-use cameras," the recycling activities associated with plastic (PET) beverage containers, and the recycling of vehicle components and materials among automobile manufacturers. Yet, from a toxics use reduction perspective, each example has been associated with significant product-related environmental problems in the marketing of new products, choice of materials, and overall product impact.[78]

The discussion in this same text of nuclear power as "an industrial ecology that failed" provides an even more striking example of the ambiguities of the industrial ecology association of cleaner production with "greater control over waste streams and enhanced recycling of materials." This argument, put forth by Franz Berkhout, identified nuclear power as embodying core industrial ecology principles, since all its materials could be contained and utilized within a closed-loop system. The failure of nuclear power has not been its choice of materials or recycling capacity, nor its ability to achieve a closed-loop approach. Rather, nuclear power became a problematic technology due to its social, political, and psychological vulnerabilities, associated with such concerns as safety due to accidents or unintended uses related to nuclear proliferation, but not, according to this argument, due to the nature of the technology itself. Other environmental advocates argue, however, that, as in the case of nuclear power (or similarly, with bioengineered food products), the technology cannot be separated from its political, social, economic, and environmental contexts. Similarly, the industrial ecology argument has been criticized for failing to address the question of the hazards of particular production choices, whether for workplace health and safety or for communities and the environment.[79]

There are some industrial ecology advocates who do in fact argue that the social and political contexts for product and process choices, material flows, and technology selection need to be addressed as part of the analysis and the framework for implementation of an industrial ecology approach. However, the core of the industrial ecology argument is that industry *should* know best and that such choices should remain in the private sphere. Arguing that the current environmental policy system has created a kind of industry passivity in terms of regulatory compliance and even regulatory avoidance strategies, industrial ecology is said to offer industry the chance to "step forward to organize the debate," as Robert Socolow has argued. With industrial ecology, "industry becomes a policy-maker, not a policy taker." It is in fact through the industrial ecology pursuit of efficiency and technological innovation, its advocates further argue, that a new type of industry configuration is possible, led by new technology-oriented companies such as ATT, Mitsubishi Electric, Hewlett-Packard, Dow Chemical, and Monsanto. And these industrial ecology-oriented companies, Tachi Kiuchi, the managing director of Mitsubishi Electric has proclaimed, will constitute the next generation of "more profitable, high performance, and sustainable businesses," ushering in a new kind of global order.[80]

Extended Producer Responsibility   One of the earliest expressions of the industrial ecology focus on waste and materials concerned the problems of excess packaging, a key indicator of the inefficiency in how materials were used. While primarily identified as a nonhazardous solid waste issue, the problem of the design and inefficient use of packaging materials helped stimulate the development of the concept of "extended producer responsibility" to address a wide range of materials selection, use, and disposal or reuse issues. The term, first elaborated in the early 1990s by Thomas Lindhqvist, a Swedish professor of environmental economics, sought to place the burden of environmental responsibility for materials use upstream from the postconsumer stage to include both manufacturers and retailers. Thus, instead of placing the burden on municipalities and regional governments to manage waste streams at their end point, manufacturers of products, including packaging materials, would have to assume some of those responsibilities. In particular, as an economic strategy, Lindhqvist's concept sought to incorporate the costs of product disposal or recycling into the price of the product. That approach was given significant impetus with the passage of Germany's packaging ordinance in 1991 and the development of its Dual, or Green Dot system. The German program was designed to develop a method by which producers would have the responsibility to take back and recycle packaging waste.[81]

In the United States, the debates over how to effectively manage solid wastes and in particular packaging wastes had focused less on production choices about the materials used and products developed and more on the development of recycling and more efficient uses of such materials. The emphasis on packaging as a solid-waste management issue also meant shifting policy options, including a kind of crisis management approach. While the late 1980s and early 1990s had generated significant attention when the issue achieved crisis status, solid waste became less a focus of policymakers by the late 1990s, with the exception of the continued emphasis on recycling itself as a core environmental strategy. At the same time, the federal and particularly state policy mandates established in the early 1990s to reduce the solid-waste stream had evolved by the mid- and late 1990s into more voluntary approaches. Thus, when the extended producer responsibility concept began to be explored in this country, it largely focused on the "end-use" actions of the consumer, and the "shared responsibilities" between all parties, and not just the producer, to act in a more environmentally responsible manner.

The shared responsibility approach was elaborated by the President's Council on Sustainable Development (PCSD) which promoted, in its 1996 report, *Sustainable America*, the concept of "extended *product* responsibility." The extended product responsibility concept differed significantly from its predecessor by shifting responsibility from producers or manufacturers (responsibility as defined by production choices and outcomes) to one that focused on the shared responsibilities of designers, producers, suppliers, users, and disposers alike, with respect to all phases of a product's life. Instead of government action and specific requirements for producer action, the PCSD emphasized the need for consumer education and voluntary or incentive-based initiatives. Although this approach was seen as overcoming a fragmented system of regulation, the voluntarist emphasis did not provide a handle on how to reconcile these industry or consumer initiatives with existing regulations, nor establish a more holistic or throughput framework over a product's life span. The emphasis on product rather than producer only reinforced the elusiveness of implementation and identifiable outcomes.[82]

Design for Environment   One of the core strategies associated with the "next generation" concept has been "design for environment," a strategy derived in part from the "Design for X" industry systems. In these design modules, "X" refers to various desirable product characteristics such as assembly, testability, disposability, maintainability, or safety. In the early 1990s, the design for environment concept began to circulate among electronics and communications firms as a way to incorporate environmental factors such as efficiency into the existing product and process design framework. Under the leadership of AT&T's Allenby, the American Electronics Association established a task force on design for environment and produced a publication for its member companies that became a kind of founding primer for the development of industry DfE programs.[83]

Even more than industrial ecology, design for environment has been associated with the idea that environmental change should be seen as a function of industry adaptability while maintaining its core functions of profitability and growth. Design for environment, as one of its leading advocates defines it, "seeks to discover product innovations that will result in reduced pollution and waste at any or all stages of the life cycle, while satisfying other cost and performance objectives." These could include design for separability and disassembly, design for disposability, design for recyclability, product life extension or durability, and material substitution, among other design attributes. While no directed or mandated standards for design for environment have been established, various guidelines, ranging from a series of voluntary environmental standards established through the International Standards Organization (e.g., the ISO 14000 series) to informal company and consultant-related suggestions, have framed the design for environment approach. Most of these have been associated with the waste generation and material and energy use issues. Moreover, as a 1992 Office of Technology Assessment report pointed out, the U.S. policy focus on "regulating industrial waste streams," contrasted with the European interest in addressing the "environmental attributes of products." Even in the area of product design, such as material substitution, U.S.-based design initiatives have tended to address the recyclability of the material rather than its occupational or toxics use impact. Similar to the industrial ecology argument about Kodak's single use ("FunSaver Disposable") camera, design for environment advocates have focused on the engineering designs that can provide for the separation and ultimate recyclability of the camera's parts, but not the market-related considerations that emphasize the disposability of the product as a whole. Design for recyclability in this context becomes an add-on rather than front-end consideration.[84]

Design for environment has also become, for the U.S. EPA, a catch-all phrase for its "cooperative industry" projects that link EPA's technical assistance capabilities with the hope and expectation that industry can incorporate environmental choices in how it operates. In September 1992, the EPA established its own "Design for the Environment" program within the Office of Pollution Prevention and Toxics as a cornerstone for its "front-end" technical assistance approach, to be primarily oriented toward small businesses such as dry cleaning and printing. Like its industry counterpart, the EPA program established a range of "design" outcomes that were primarily "waste" oriented; that is, how to best influence a reduction of the end-point problems caused by an industrial process or product, without necessarily addressing the nature of the process or product itself.[85]

By the late 1990s, the proliferation of various "next generation" strategies provided an important new addition to the vocabulary of environmental change. Industrial ecology, to use Jesse Ausubel's metaphor of the "Copernican turn" or Paul Hawken's parallel concept of "natural capitalism," suggested a potentially radical restructuring of key production choices and most especially a focus on how materials were used. Extended producer responsibility proposed that the use of those materials was a production and not just a consumer decision, and identified the arena of "responsibility" as a public claim regarding those production choices. Design for environment offered a compelling argument that the way products were designed or processes identified could become subject to environmental and not just market-driven criteria.[86]

The development of these next generation strategies, however, could also be considered an extension of the "industry-knows-best" argument and the further erosion of the public role in production choices. Next generation strategies have remained voluntarist in nature and, for some, exclusively industry driven. Despite Paul Hawken's compelling plea to uproot a nonaccountable, profit-centered, and productionist-oriented market-driven system, most next generation advocates assume that aside from minimizing energy and materials usage and ecological impacts, these approaches have the goal of "maintaining the economic viability of systems for industry, trade, and commerce." Economic viability also means "globally competitive," a particularly significant objective for the high technology industries such as biotechnology, electronics, and communications that operate at the global scale both in terms of the dispersal of production and their global marketing reach.[87]

For some next generation industry-knows-best advocates, industry's need to be globally competitive and require a market friendly approach also reveals a concern about the public role in environmental change. Fears about public intervention run deep. Some have interpreted pollution prevention as a regulatory stalking horse or even a form of industrial policy. This could mean product bans or phasedowns, intervention around particular product, process, or technology choices, or establishing transitions for innovation or a change in production. The government, it is feared, could become an "industrial planner," exceeding even "the earlier excesses of end-of-stack regulation." In relation to this argument, the choice becomes clear: pollution prevention as "industry-knows-best" voluntarism or pollution prevention as a form of public planning, with a seat at the table for various community, environmental, and labor actors. The debate over the next generation of strategies for environmental change becomes then a debate over how change can be accomplished as well as the nature of the change itself.[88]

Breaking Boundaries

In a 1997 article for the *Harvard Business Review*, University of Michigan business school professor Stuart Hart identified three requirements for achieving sustainability: reducing population, lowering consumption levels, or changing the technology used to create wealth. While the first two issues had global and societal dimensions (and presented significant barriers for bringing about any substantial change) it was the third arena—the province of technology—that Hart defined as most appropriate for his *Harvard Business Review* audience. "Technology," Hart wrote, "is the business of business." Moreover, Hart argued, sustainable development itself potentially constituted "one of the biggest opportunities in the history of commerce." In a similar vein, Michael Porter and Claas van der Linde have argued that superior productivity and industrial competitiveness could be a function of environmental innovation and even regulation—as long as the promotion of innovation is associated with resource productivity; that is, maximizing the efficiencies in a production system.[89]

On one level, the strategies for environmental change in the last three decades can be considered a search for the "resource productivity" and technology base for what Marteen Hajer has called "ecological modernization." Such a system of modernization, based on waste reduction and new "green" technology developments, does not, Hajer argues, call for structural change in relation to production systems, but instead suggests "that there is a techno-institutional fix for the present problems." Hajer further sees this thrust toward ecological modernization as similar to 1970s' perspectives about efficiency, technological innovation, techno-scientific management, procedural integration, and coordinated management. Such perspectives also reflect "a fundamental belief in progress and the problem-solving capacity of modern techniques and skills of social engineering. . . . There is a renewed belief in the possibility of mastery and control," Hajer concludes, "drawing on modernist policy instruments such as expert systems and science." This "mastery and control" concept contrasts with the radical environmental perspectives of the 1970s that had focused on the need for clean production (for example, renewable energy resources) in conjunction with the need for "self-determination, decentralization of decision-making, and general human growth."[90]

At the 1998 Knoxville, Tennessee, meeting of community-oriented environmental justice and industry-focused environmental groups referred to earlier, the use of the concept of "clean production" was in part an effort to connect next generation approaches with the concepts of self-determination and the community related claims of environmental justice.[91] In this antiregulatory era, these and other environmental groups have continued to try to identify a new common ground for environmental action that can be at once community or place based, while confronting the issues of production choices and industry-related environmental outcomes. The absence of a public role represents one clear barrier to these efforts. However, another limiting factor in the discussion about clean production has been the inability to effectively address the issues of the workplace. Some of the next generation strategies had explored a "post-Fordist" or "post-Taylorist" emphasis that assumed greater worker engagement in creating more productive and efficient systems. But the labor advocates at the Knoxville gathering worried that the crucial pollution prevention arguments about the importance of the workplace were no longer as central as they should be. Others worried that the "clean production" concept did not successfully incorporate the community role in environmental change. The question remained: how could this new common ground be identified and how would change come about? Could the boundaries constraining environmentalism be broken, and production rather than technology become the basis for environmental action? And could a renewed environmental politics of place also seek to contend with the forces of globalization, the reduction of the government's role, and the long-standing decline of the capacity of civil society to bring about change?

The December 1999 demonstrations in Seattle protesting the meeting of the World Trade Organization were suggestive in this regard. The imaginative slogan at the Seattle events—"Teamsters and Turtles: Together at Last"—represented more than just the hope for a new kind of labor/environmental alliance. Several of the Seattle demonstrators focused on the industry-knows-best argument within the context of globalization and the disappearance of any public role regarding production decisions and their community, environmental, and workplace impacts. Some of the core issues at Seattle—food growing and production, workplace conditions, the hazards of certain technology choices, the unrestricted movement of capital—resonate in terms of the argument of this book about a new focus and expanded agenda for environmentalism. The Seattle demonstrators—Teamsters, environmentalists dressed as turtles, food activists, Third World NGO protestors, among others—had begun the process of crafting a new discourse and bringing the language of justice, workplace democracy, and livability into the streets. Once the pepper spray had dissipated and delegates and demonstrators had dispersed, the battles over discourse and opportunities for action would shift to the neighborhoods, regions, workplaces, industries, and production systems. Change was in the air again, and a host of social movements were in the process of being formed—or renewed.

II

Exploring Pathways

3

Dry Cleaning's Dilemma and Opportunity: Overcoming Chemical Dependencies and Creating a Community of Interests

Exploring Pathways

How can communities, workers, consumers, and the broader public or civil society contribute to social and environmental change? And how can these different civic actors help initiate such change when industry decision making and the role of government in environmental management appears so removed from providing for a public or civic role? The ascendance of the voluntarist and "third-way" perspective in environmental management assumes a role for civil society; indeed, one of the more influential books in the early and mid-1990s, *Civic Environmentalism*, emphasized that role. But how civil society actors can generate change is not readily discernible. Voluntarism, as a substitution for public intervention, may in fact mask how industry, sectoral, institutional, and cultural influences can erect barriers against such change. For environmentalists, establishing a movement for social and environmental change requires an understanding of those influences and barriers, and how the different players in civil society, by acting to identify a public or government role, can facilitate and ultimately become agents for change.

The next chapters seek to explore the limits and opportunities for change in three areas—cleaning clothes, cleaning buildings, and growing, producing, and accessing food. These areas are not case studies as such, but rather signposts for identifying a common vision and a set of strategies for change. The next three chapters explore whether and how that change can begin to occur.

The Small Business Dilemma

How does a small business like dry cleaning fit into the dominant environmental policy mode of bringing about "cleaner, cheaper, smarter" environmental change? For many small businesses, environmental change is often characterized as more expensive, not very smart (the proverbial 800-pound government gorilla imposing itself), and not even necessary (environmental problems are seen as exaggerated). Given these perceptions, there emerges an environmental policy conundrum for the "cleaner, cheaper, smarter" advocates: namely, how can environmental change be brought about voluntarily if the industry group involved is resistant to change? The answer is presumed to reside in the current hybrid policy system that exists today: part end-point mitigation and compliance oriented, part voluntary. With small businesses, however, that approach remains more problematic.

For more than two decades, policymakers have been forced to limit their environmental policy initiatives on the assumption that small businesses, such as metal finishers, printers, and dry cleaners, present significant obstacles in carrying out an agency mission of managing environmental impacts. Various environmental problems, such as the widespread use of chemical solvents for cleaning or inefficient use of materials or resources, have been identified as small business–related. Yet policymakers and regulators see small businesses as obstinate and often hostile to government regulation and even government assistance efforts. They are also seen as lacking in expertise and technical resources, financially unable to make the necessary technology-based improvements, or unable or unwilling to embrace the cultures of environmental efficiency and pollution prevention voluntarism. Since regulations are often politically contested—with small business capable of drawing on public support in its battles with government regulators—technical assistance and voluntary approaches are still assumed to be the most effective way that small businesses can be assisted in changing their most polluting and inefficient operations. Environmental agencies are constantly seeking ways to construct programs that are small business–friendly to overcome such hostility, while providing resources for compliance and better environmental practices. Programs such as small-business assistance centers (with their technical assistance manuals and small-business loan programs to help achieve regulatory compliance) often represent the beginning and end point in this effort. There is little focus on the actual situation—the institutional, cultural, historical, and sector influences and identities that establish the biases and potential opportunities for environmental change for particular businesses in particular settings.[1]

Small businesses are located at the margins of industry decision making, often subject to outside forces. The environmental problems they confront may be more reflective of their dependence upon manufacturers and suppliers in providing their products and shaping their processes or as subcontractors to larger businesses. Squeezed by their suppliers and fearful of regulation, small businesses represent a difficult set of issues for policymakers, even as their situation exposes the limits of the current hybrid policy system. Yet small businesses, particularly those that provide a customer or consumer service, may also be concerned with the public or community perceptions of their role and activities.

One small business frequently associated with such issues is dry cleaning. A neighborhood-based, customer-service-oriented business that generates more passion and more political conflict than any other small business today, dry cleaning has become symbol and substance of the small business dilemma for contemporary environmental policy. Its issues underline the need to identify the new constituencies and strategies necessary to bring about environmental and social change in the face of the obstacles this dilemma presents.

From Petroleum to Chemicals: The Search for a Solvent

Dry cleaning as a term was first derived from a mid-nineteenth-century discovery in France that camphene, a fuel for oil lamps, could clean garments such as gowns that had stains of an oily nature. This petroleum-based cleaning process, used commercially by dyers and scourers, was assumed to work best on fabrics like silks and wools by removing oil, dirt, and grease stains without altering the color or changing the garment. This use of a petroleum solvent contrasted with "wet" processes used by commercial laundries, which relied on water as their cleaning solvent, and which had also begun to appear in major cities in the United States in the mid- to late nineteenth century. The rise of the water-based commercial laundries and the petroleum-based dry cleaning operations reflected the development of specific garments for specific markets (for example, the introduction of the detached collar). It also identified a shift from family production (home laundering) to special-ized commercial services (for example, finishing garments through special steam presses). While the commercial laundries became depen-dent on the availability of new equipment and machinery, the rise of dry cleaning—a term introduced to refer to the use of these nonaqueous-based cleaning solvents—reflected the growing role of petroleum manu-facturers and, subsequently, of chemical companies in recognizing commercial garment cleaning as a market for their products.

As commercial dry cleaning operations took root in the early twentieth century, they were almost entirely centralized plants located in industrial zones or outside urban residential areas. These facilities relied largely on gasoline as their cleaning solvent, with fire hazards their major area of concern. While the sales volume of dry cleaning establishments remained volatile through the 1930s, important changes began to occur in this period that significantly restructured the nature of the cleaning process and the type of establishment associated with dry cleaning. The introduction of new garment fabrics such as rayon and other synthetics, and increasing population mobility combined with the larger numbers of women entering the workforce signaled that the size of the commer-cial garment care industry could be substantially increased. At the same time, a new class of chemical cleaning solvents were introduced as an alternative to petroleum, allowing more dry cleaners to become residen-tial based. The self-service operations and the small tailor shops that offered cleaning services as well as the larger centralized plants often located in nonresidential areas began to be replaced by the small urban neighborhood cleaner and medium-sized suburban commercial plants. A number of cleaners also located in dense metropolitan areas like New York City on the ground floor of residential buildings or in shopping centers (or, later in the 1970s and 1980s, in strip malls). These trends in turn established dry cleaning as a dispersed business in suburban as well as urban areas.[2]

The shift from the centralized plant to the neighborhood dry cleaner was primarily associated with the shift from petroleum to the use of chemical solvents for cleaning. Petroleum solvents required a central industrial facility to be located in nonresidential areas. Chemical-based cleaning, in contrast, provided opportunities for small businesses to develop their own on-site facilities and gain independence from the larger operators. Despite the higher cost of the chemical solvents and the smaller number of garments that could be cleaned by chemicals for each machine cycle, chemical-based facilities were still attractive because they required a smaller initial investment and less operating capital. This was particularly appealing to those who had little or no capital to begin with (e.g., veterans) and who might be interested in a start-up business where personal service and merchandising could become as important as operating the cleaning facility itself. "Retailer independence," a term encouraged by the chemical suppliers who recognized a new sales potential, became the slogan of the 17,500 "little businessmen [who had] started to do their own cleaning instead of farming it out," as a 1948 *Fortune* article put it. An entire new market and set of relationships had emerged.[3]

The first chemical cleaning solvent to be widely used was carbon tetrachloride, a chlorinated hydrocarbon that could easily evaporate without heat while effectively removing oil and grease stains. Carbon tetrachloride (or carbon tet as it was popularly known) had the advantage of eliminating petroleum's fire hazard problem and significantly reducing, due to its more effective cleaning properties, the amount of time required for cleaning. First used commercially in Europe during the late nineteenth century, carbon tet had already established itself within the U.S. garment care market by the 1930s and helped facilitate the shift to the faster-service, on-site cleaner. Widely used as a cleaning solvent and as an industrial degreaser, carbon tet had multiple other uses such as an ingredient in fire extinguishers, in the manufacture of refrigerants and insecticides, and as a nonflammable liquid cleaner for household consumers. But carbon tet also had a major downside already identified by the 1940s: it could be a significant health hazard for those who came in contact with the solvent.[4]

While there were some minor concerns about carbon tet's performance, including its tendency to corrode dry-cleaning equipment, carbon tet's health issues were its most problematic feature. There were substantial emissions within the plant, since carbon tet had a low boiling point and would evaporate rapidly. These emissions were magnified when machines were loaded and unloaded and when the filters and stills were cleaned or replaced. While certain precautions were advised to minimize human contact with the solvent, carbon tet's health impacts (ranging from cancer to loss of appetite, nausea, and various internal disorders) were becoming increasingly difficult to ignore as its use expanded.[5]

The more transparent the problems associated with carbon tet toxicity, the more widespread became the search for chemical alternatives by dry cleaners and the growing interest of chemical solvent manufacturers in establishing a stronger market presence for one of their products. One of these solvents, trichloroethylene (or TCE), was first introduced for dry cleaning in the 1930s. Touted as less toxic to humans and less corrosive for the machinery than carbon tet, the kinds of acute health effects associated with carbon tet were also not readily apparent. Indeed, a dry-cleaning industry manual distributed in the early 1960s suggested that those exposed to TCE at high concentrations might feel "light-headed," and that continued exposure to TCE could "cause a condition wherein the operator actually enjoys the fumes." "Apparently," the manual playfully commented, "they can become somewhat habit-forming as well."[6]

TCE's difficulties in establishing itself as the primary alternative to petroleum had more to do with its cleaning capacities, particularly its tendency to bleed dyes more than the other solvents on the market. This problem was exacerbated with even mild heat, thus precluding its use in machines where the solvent was reclaimed and clothes were dried. Another chemical solvent, introduced just a few years after TCE in 1934, appeared to answer many of the problems of TCE, carbon tet, and petroleum. This solvent, tetrachloroethylene (or perchloroethylene, also known as perc or PCE), did not appreciably corrode equipment, was more stable than TCE, could be used in any type of equipment without as great a danger of bleeding dyes, and had some excellent cleaning properties, particularly as a degreaser.

Perc, as it was popularly known in the industry, was also subject to voluntary occupational health standards (similar to those used for TCE). But since perc did not evaporate quickly, it was assumed that there were no significant ambient concentrations. For cleaners, perc's biggest drawback was its higher cost, more than $2 to $3 per gallon in the 1950s and 1960s when it began to displace TCE and carbon tet. But the higher price for perc also meant that cleaners were more inclined to try to recapture the solvent for further use, rather than simply venting it to the atmosphere and thereby increasing emissions. The ability to reclaim the perc was accomplished either through a transfer unit (which included a separate washer and dryer, where the dryer was used to reclaim the perc) or through a dry-to-dry machine (in which the washing, drying, and reclaiming of the solvent were all accomplished). For most cleaners, transfer units became the norm, since a greater number of loads could be cleaned within a given amount of time, thereby increasing dry cleaning's speed and overall volume. The transfer units also increased exposure levels (the perc vapors escaped into the ambient environment when the clothes were transferred between machines). But the health/exposure issue was not yet considered a significant problem by dry cleaners during the 1950s and early 1960s when perc began to be their solvent of choice. This absence of concern was due, in part, to regulatory attention elsewhere, the absence of any visible acute health effects, and the virtual absence of any monitoring of perc emissions.[7]

Through the 1960s, the dry-cleaning industry, now strongly associated with use of perc, was becoming a well-established segment of the broader commercial cleaning industry, which also included commercial laundries, rug cleaners, valet services, linen supply establishments, and diaper services.[8] Dry-cleaning operations were also increasingly associated with on-site cleaners and sole proprietorships, with increases in business due in part to the rise in discretionary income and the shift from blue-collar to white-collar employment in the overall job market. These trends were counterbalanced (though not entirely) by the introduction of certain synthetics and wash-and-wear fabrics, which could be home washed. Thus, dry cleaning, which had established itself as a growing though sometimes fluctuating industry due to changes in the garment industry, had become best known for its ability to clean with a chemical solvent, its defining characteristic.

Embracing Perc

Perchloroethylene, a clear, colorless liquid, has a variety of applications, including its use as a cleaning or degreasing agent, as an intermediate in the production of other chemicals (mostly fluorocarbons), as a deworming agent in animals, and as a grain fumigant. Since its widespread introduction during the 1950s, perc has depended heavily on sales to dry cleaners, representing at its peak as much as 90 percent of the market, according to one 1964 estimate. Between 1970 and the early 1990s, perc sales to dry cleaners generally fluctuated in the 50–70 percent range of the solvent's overall sales, with more significant declines occurring in the mid- to late 1990s when regulations stimulated greater efficiencies in use.[9]

During this forty-year period, the chemical industry/dry cleaning link, based on the growth of perc use as dry cleaning's principal cleaning agent, has expressed itself through an increasing web of relationships. During the early 1960s, for example, the flattening of perc sales, both in this country and in Europe, became an issue not only for dry cleaners but for their chemical suppliers. As a result, some chemical suppliers devised strategies to strengthen their downstream markets. In England, the country's largest chemical supplier, Imperial Chemical Industries Ltd. (ICI), put up the initial funds and identified other partners (e.g., the country's largest manufacturer of dry-cleaning equipment) for an advertising campaign promoting dry cleaning. They did this on the basis that "what was good for the [dry cleaning] industry would be good for ICI."[10]

When dry cleaning's fortunes improved, the chemical suppliers were also quick to take advantage. For example, a rebound in sales in the 1960s for perc was attributed directly to the widespread introduction of coin-operated dry-cleaning establishments that relied on perc as the cleaning solvent. When first promoted in the early 1960s, coin-ops were considered a significant boon by perchloroethylene suppliers such as Dow Chemical. The suppliers rightly anticipated that at least some petroleum-based dry cleaners, who still controlled a smaller though significant share of the market, would switch to perc in order to capture part of the coin-operated business.[11]

By 1970, nearly half of all coin-operated facilities had added a dry-cleaning component, while dry cleaners in turn were also adding their own coin-operated machines. For the chemical solvent manufacturers, those changes not only meant new markets for their product, but also helped in the overall promotion of perc-based dry cleaning itself. Most significantly, by increasing the trend toward the neighborhood cleaner, which the coin-ops had also facilitated, it had the effect of reducing the market share of the large chains with centralized plants that relied in whole or in part on the use of petroleum solvents. At the same time, the obvious possible health concern associated with coin-ops—the addition of a significant new source of perc exposures for both operators, workers, customers, and the environment—was casually dismissed by both the chemical industry and the dry cleaners. "The experience of small drycleaning shops situated in business districts and using perchloroethylene," argued one chemical industry publication, "demonstrated the safety of this solvent."[12]

By the early 1970s, perc manufacturers had come to recognize that dry-cleaning operations represented a baseline from which to measure future output and profits. Other factors remained significant, such as fluctuations in the fluorocarbon market or regulatory pressures on other solvents. But despite those factors, the bottom line in terms of perc sales continued to be its ability to achieve growth or at least stability in the dry-cleaning market. Based on that core consideration, perc producers, led by Dow Chemical and including such other chemical giants as PPG Industries, DuPont, Diamond Shamrock, and Stauffer Chemical, hoped for a continual increase in production based on the consumption patterns of its downstream users. If dry cleaners were becoming wedded to perc, the chemical manufacturers had embraced dry cleaning as the greatest source of growth in their sales of this high-volume solvent.[13]

Changing Labels, Shifting Patterns

For dry cleaners, the 1970s and early 1980s came to represent a period of change and increasing problems. The industry found itself subject to significant uncertainty due to the nature of the garment-care market itself (what clothes would or would not be taken to the dry cleaners). The changes in fiber mix, including the increase in "easy-to-wash" fabric materials, paralleled the significant shift in production strategies among garment manufacturers. This included a highly dispersed, transnational structure of production, increasing subcontracting of functions, and changes in technology and marketing. A diverse array of garment types and materials became available to consumers that either required special cleaning, including dry cleaning, or, with certain fabrics, no professional cleaning at all. One consequence of this change meant that dry cleaners, as well as consumers who sought to do home laundering for their clothes, often found themselves, in the words of consumer advocate Betty Furness, "without the slightest idea of how to take care [of the clothes]" being purchased and worn.[14]

During the 1960s, the concern over the changing nature of garment manufacture and its implications for cleaning focused on the information available through labels. In 1966, President Johnson's consumer affairs adviser, Esther Peterson, sought to develop a voluntary approach among garment manufacturers and retailers to ensure adequate care labeling information. Peterson's group, the Industry Advisory Committee to the Government on Textile Information, consisted of representatives of fiber, fabric, and garment manufacturers, retailers, dry cleaners, and commercial launderers. Though publically lauded by Peterson as an important example of consumer-related industry/government cooperation, the effort was also seen by this committee as a way to head off care labeling legislation through voluntary action. In 1968, the group issued a report that included a glossary of garment-care terms that could be used for labeling purposes. It also offered a plan for continuing voluntary action on the part of the garment industry to develop permanent garment-care labels for the full array of clothes and fabrics. Although a major trade organization, the American Apparel Manufacturer's Association, endorsed this voluntary approach, there was only limited participation by manufacturers.[15]

The lack of labeling information was compounded by the increasing globalization and segmentation of production taking place throughout the garment industry. A single item of clothing brought to market could have multiple transnational origins. The fabric material could be produced in India, the dyeing process could take place in Singapore, garment construction could occur in Thailand, the retailing and marketing operations could originate in the United States, and the sale of the garment could take place in multiple outlets in this country and elsewhere. The problems with cleaning and caring for such garments were compounded by the variety and differences in the materials, processes, and production units involved. Who would do the labeling was as confusing as how the labeling would occur. This globalization process also meant that the concept of "identical" garments was itself becoming increasingly problematic. For example, two retailers, offering the same type of garment, presumably constructed in the same manner, with the same size and color, might be offering garments produced in different places with different characteristics that could influence how successfully the clothes would be cleaned. Even as late as 1996, twenty-five years after formal care labeling information had been established, our own research found that as many as 10 percent of the "identical" garments ordered from the same company were produced in different countries. In 1968 and 1969, when the care labeling issue was being debated, the problem of globalization was more directly identified as the problem of "imported textiles," when in fact the distinction between imported and domestic garments was already becoming problematic. The making and wearing of garments, like the growing and production and consumption of food, as discussed later in this book, was losing its local or place-based identity in face of the powerful trends toward a global system of production and restructuring of needs.[16]

In 1969, under growing pressure from consumer organizations, the Federal Trade Commission initiated steps toward a formal care label rule when it published a proposed trade regulation for care labeling of textile products. Manufacturing interests continued to oppose any formal rule making, arguing that labeling added costs, and that manufacturers did not have the knowledge nor want the responsibility for knowing the care requirements of the clothing they produced. The most important issue for the dry cleaners was the need for additional information on fabric type and cleaning instructions that could conceivably reduce the number of manufacturer-related problem garments. This included, for example, information about dye processes or the finishing of the garment, or the bonding of different fabric types, identified as responsible for as much as 50 percent of the problems experienced by dry cleaners. When the "care labeling" rule was adopted in December 1971, dry cleaners and their chemical suppliers hoped the change would facilitate further stabilization and maturation of the industry, and provide information that could further support the cleaning process itself.[17]

Along with the change in the care label rule, the dry-cleaning industry during the 1960s and 1970s continued to experience changes in its structure and functions. Productivity gains in dry cleaning, for example, had increased more than threefold during the 1960s, even as overall market share and output remained relatively flat. This meant in part a shakeout and/or slight decline in the number of businesses (although not necessarily increased concentration, since individual coin-operated facilities were also on the increase during that period) and a decline in employment as well. By the 1970s, the focus on productivity meant more a focus on machines and equipment rather than the skill level and capabilities of the presser, spotter, or cleaner. Growth, however, remained modest. Trends like the comeback of natural fibers such as wool, silk, and cotton, the increases in the number of women in the workforce, and the development of new marketing strategies such as locating dry cleaners in retail stores or strip malls, still failed to produce any significant growth spurt. While dry cleaning had matured (in terms of technologies and processes as well as defined markets), it remained subject to the influences of garment trends and supplier and upstream relationships often beyond its control.[18]

Enter the Koreans

One new group attracted to dry cleaning during the mid-1980s, despite the industry's uneven performance, were first-generation Korean immigrants. During the late 1960s and early 1970s and again in the late 1970s and early 1980s, there had been a major influx of Korean immigrants into large metropolitan areas such as Los Angeles, New York, and Chicago.[19] Some of these immigrants had modest capital resources to invest and were interested in identifying business opportunities where family labor could be utilized (in part to stretch capital resources) and where there were few language or locational barriers for entry. Three industries in particular appeared most attractive—small grocery stores, seafood stores, and dry-cleaning operations. While the Korean immigrants attached significant importance to social and cultural values (family, religion, community ties), the extensive dispersal by location of dry-cleaning facilities did not appear to pose a barrier, since the importance of place remained more a cultural than geographic factor. The continuing shakeout within the industry, which included the closing of as many as 6,000 marginal plants in the 1980s, also provided an additional opportunity for new operators to enter the business by purchasing plants on the verge of closing.[20]

The Korean entry into the dry-cleaning world was rapid and dramatic, from a nearly nonexistent segment of the industry in 1980, to as much as 50 percent in major metropolitan areas such as Los Angeles by the mid-1990s. For the Koreans, the attractions to the business were clear. Capital requirements, while greater than an initial investment in a grocery store, were still not substantial. Operating costs could also be reduced by reliance on family labor. This had two advantages: the economic benefit (less "contract" labor, the ability to stretch operating capital), and the "cultural" benefit (family-centered business, maintaining intact families, the ability of immigrant families to survive and adapt). Dry cleaning was also initially considered a "clean" (*kkakkuthan*) and stable type of business. There appeared to be fewer risks, and profit margins were considered more attractive than comparable businesses. The management of the business seemed less labor intensive (e.g., providing a day off on Sunday for families), which contrasted with other easy-entry businesses such as the small grocery stores that often required longer hours and had greater risks.[21]

The Korean entry into dry cleaning also occurred while other crucial changes in the industry were taking place. The issues of fluctuating sales and the volatile nature of the garment-care market due to shifts in apparel and textile production were creating conditions of increased segmentation in the dry-cleaning industry. This segmentation was reflected in different pricing strategies, emphasis on quality versus speed in the cleaning service, and the subtle but increasing pressures associated with environmental factors. Though the Koreans were present within each of the different segments of the industry, several non-Korean trade association and industry officials tended to identify Koreans as more significantly associated with the "lower quality"/discount end of the industry, often operating at its margins. This characterization in turn was associated with the assumption that language barriers were an important problem area for the Korean cleaners and would reinforce any difficulties they might have in providing "quality service" at the counter. "The language difficulties of the new immigrants," one prominent dry cleaner argued at an industry-related conference, "leads to a cookie-cutter approach to dry cleaning and undercuts any perception of drycleaners as 'professionals.' "[22]

Environmental factors also played a role in the evolving demographics of the industry. The discount or lower-end cleaners were seen as more reluctant to purchase the new closed loop machines. These machines, more expensive to purchase and maintain than their predecessors, included a number of features that increased efficiencies and reduced environmental impacts. Unlike the earlier generation of transfer machines, where clothes would be cleaned and then transferred to a dryer (and where perc emissions would be increased through the transfer process), the new dry-to-dry machines were able to perform both cleaning and drying functions. This also allowed for increased amounts of the used or spent perchloroethylene to be recaptured and reused. The machines were also equipped with pollution control devices such as refrigerated condensers and carbon adsorbers, to further mitigate dry cleaning's significant environmental impacts. Thus, the machines, which had been introduced in the late 1980s and 1990s in part to meet anticipated environmental regulations, were seen as the industry's most effective response to more stringent regulation. Their environmental advantage was due to greater efficiencies in perc use (that is, less solvent used per garment cleaned) as well as reductions in air emissions that had become the primary focus of environmental regulation.[23]

However, despite their recognized efficiencies and environmental advantages, significant numbers of the new machines were not being voluntarily purchased by either the discount cleaners or even by many of the high-end cleaners due to their expense. Those presumed to be most resistant to such purchases were often portrayed as cleaners who dumped perc-laden wastewater illegally into the toilet or down the drains, and who were most inclined to hold on to the older, less expensive transfer machines and were therefore most reluctant to purchase the new, more expensive equipment. These assumptions, however, were never directly evaluated, reflecting anecdotal information and gossip rather than the gathering of data on performance and activity.

Assumptions about differing approaches also contributed to the further segmenting of the dry-cleaning industry, exacerbated by the increasing pressures regarding environmental issues. A series of divides within dry cleaning had emerged: high-end versus low-end, Korean as opposed to non-Korean. Korean cleaners created their own trade associations and publications, establishing further language and cultural fragmentation within the industry by the early 1990s. Interestingly, it was the chemical suppliers, more than the government regulators or the non-Korean segment of the industry, who moved quickly to adapt to the changing dynamics of the industry. Chemical industry materials were more routinely translated into Korean and "Korean" staff positions were also established in such organizations as the Halogenated Solvents Industry Alliance. This contrasted with the non-Korean segment of the industry as well as with environmental regulators who seemed ill-equipped to address the rapidly changing industry demographics and various divides. By the 1990s, it was the question of the environment as well as these cultural and institutional fissures that generated the strongest passions and represented the most significant influence determining the nature and future of the industry itself.[24]

Regulatory Battles

At the heart of the environmental question facing dry cleaners have been the issues of risk—for those who work in dry-cleaning establishments, to the environment and surrounding communities, and even for those who wear the clothes that are cleaned. The dry-cleaning industry's increasingly elaborate ties to their upstream chemical producers, moreover, have had major implications for dry cleaners associated with the industry's potential liabilities and its future position.

During the 1960s, perc had escaped any significant regulatory attention in comparison to its chief solvent competitors, TCE and carbon
tetrachloride. Nevertheless, studies of worker exposure during the 1960s
indicated that high levels of exposure to perc could create significant
health problems, such as impaired liver functions.[25] A series of episodes
in the mid-1960s, including the death of a Stockton, California, dry-
cleaning employee due to exposures to perc fumes and several incidents
where cleaners were found unconscious behind the equipment in their
shops, drew public attention to the hazards associated with perc use. In
response, industry groups and regulators sought to establish warnings to
avoid what were considered unnecessary exposures (e.g., improperly ven-
tilated areas). However, even these modest efforts were opposed by the
owners of the coin-machine dry-cleaning facilities, who had become a
significant force in the industry.[26]

During the 1970s, perc received increasing attention from both
researchers and regulators. A key mid-1970s study by the National
Cancer Institute (NCI), which identified perc as a liver carcinogen in mice,
touched off a flurry of chemical industry activity. This included indus-
try-sponsored research that was used to counter regulatory initiatives as
well as studies by such research bodies as the NCI and the National Institute of Occupational Safety and Health. For example, during the late
1970s the Consumer Products Safety Commission (CPSC), based on the
NCI and NIOSH studies, sought to classify perc as a carcinogen, as part
of a broader "cancer" policy then being developed within the Carter
administration at the CPSC, EPA, and OSHA. Perc manufacturers successfully blocked the CPSC policy through legal action, contending that the
cancer policy would ultimately lead to a ban or restriction in use, seeking
to embrace the dry-cleaning industry as a dependent downstream source
that would be most threatened by such action. CPSC chairman Susan
King, who had led the effort to develop the new cancer policy for her
agency, angrily attacked the chemical industry counteroffensive. "Certain
segments of industry are violently opposed to the government even beginning to implement programs to protect the public from cancer-causing
agents [such as perc] in the workplace, schools, the home, and general
environment," King told one chemical trade publication. The chemical

industry groups, King argued, had "fought every [government] department and every agency on this front."[27]

During the 1980s, the regulatory battles further intensified, particularly those within the EPA, in response to the growing body of scientific
studies on perc carcinogenicity and the debates related to the interpretation of those studies. In the early 1980s, the Reagan administration
EPA sought to slow down the regulatory pressures, deferring, for
example, any action on toxic air contaminants required by provisions of
the Clean Air Act. But the pressures still continued to build. In 1985, the
International Agency for Cancer Research (IARC) identified perc as a
probable human carcinogen, which in turn triggered a new round of
debates over risk and toxicity. This in turn placed new pressures on the
EPA to establish standards to regulate perc as a possible water, air, and
land contaminant. With a Democratic Congress upping the pressure
through new legislation, a handful of new regulations impacting dry
cleaners was developed. The prohibitions of land disposal of hazardous
wastes in municipal landfills elaborated in the 1984 Hazardous and Solid
Waste Amendments to the RCRA, for example, now forced dry cleaners
to dispose all their used solvent. It also required that the muck and filters
from the recovery of the solvent be sent to a (more expensive) hazardous
waste facility for recycling or incineration.[28]

But it was the threat of air regulations at the federal and state levels
that was creating the most significant and potentially the most onerous
pressures on dry cleaners to develop more expensive control and recovery systems to reduce emissions. These problems were compounded by
the fact that the transfer machines had largely become the industry norm,
since the level of exposure was increased as clothes were transferred from
one machine to the next. The amount of perc emissions from the
operations associated with dry-cleaning transfer machines alone was a
substantial contribution to the overall load of perc in the ambient environment, particularly for such a small business. One industry figure estimated that as much as 400 gallons of perc were needed per 10,000
pounds of garments to replace the amount that was exhausted to the
atmosphere. And since dry cleaning used such a high percentage of perc,
the issue of controlling or eliminating exposures became an issue of how

to regulate dry cleaning. Thus the risk debate—how toxic was perc—was joined with the debate over endpoint uses—how to regulate the activities of a small business rather than focusing on the source of the hazard.[29]

Through the late 1980s, the interpretation of the IARC listing and parallel debates occurring within EPA's Science Advisory Board—that is, whether perc should be considered a probable or possible human carcinogen or something in between—came to be reflected in the debates over the 1990 Clean Air Act Amendments. It was becoming clear not whether but how dry cleaning—a dependent, downstream user of perc—would be regulated. For the dry-cleaning trade organizations, these became defining battles for the future of the industry, given perc's ascendance as the dominant cleaning agent. "You will be darn lucky to find [any future] dry cleaning," William Fisher, then assistant general manager of the International Fabricare Institute, complained to a chemical industry publication, referring to the possibility that the EPA might classify perc as a probable human carcinogen.[30]

These issues came to a head with the passage and subsequent implementation of the 1990 Clean Air Act Amendments. The use of the Clean Air Act to regulate perc emissions from dry cleaning had been explored by the EPA as early as 1980 regarding provisions related to volatile organic compounds (VOCs). The EPA, however, hesitated about initiating any action until a suit brought by a community group in Oregon forced the agency to begin the process of establishing new source performance standards for perc use in dry cleaning. Negotiations to settle the suit led to a 1990 Consent Decree agreement stipulating that the EPA would need to propose such rules within a year of the agreement and within two years of passage of the Clean Air Act Amendments then being debated in Congress.[31]

The 1990 amendments represented the most significant legislation to pass during the Bush presidency, eclipsing by far the limited regulatory focus of the Pollution Prevention Act. Its provisions for regulatory review of a large body of hazardous air pollutants (189 different substances, including perc) emerged at the time as the single most important regulatory battle the dry-cleaning industry and its chemical suppliers had ever faced. Perc in fact became the first of the hazardous air pollutants to be

reviewed, a decision ultimately welcomed by both the solvent manufacturers and dry-cleaning trade organizations who wanted sufficient regulatory certainty.[32]

The key issue was how the Clean Air Act's NESHAP (National
Emissions Standards for Hazardous Air Pollutants) for perc would be
established. The NESHAP standard required a review of health risk and
feasible technologies (both control technologies and, potentially, any substitute or pollution prevention technology) available to meet a proposed
standard. This process immediately became highly charged for the EPA,
as well as for industry and environmental groups who sought to monitor
it. Prior to the 1990 legislation, only a handful of NESHAP standards
had been promulgated after 1970 (when the NESHAP process was first
established through the 1970 Clean Air Act). This was largely due to the
ambiguities associated with the risk analysis and standard setting procedures as well as strong industry lobbying around each specific chemical under review. But the buildup of environmental concern about
toxicity issues during the 1980s placed the onus on the EPA to demonstrate a capacity to act—and to act quickly.[33]

Unlike the 1990 Pollution Prevention Act, which provided no explicit
regulatory framework for pollution prevention other than the promotion
of voluntary action and the use of information tools, the 1990 Clean
Air Act Amendments offered a possible pollution prevention outcome.
Provisions in the act could identify available pollution prevention
technologies as a potential method of compliance to meet the standard
to be established. However, the interpretation of what constituted a "best
available control technology" as well as the determination of what constituted a level of "acceptable risk," which established the threshold for
how much exposure would be allowed, was central to the outcome of
the standard setting process.[34]

For the chemical manufacturers, the question of interpretation was
compounded by the growing campaigns of antitoxic groups for eliminating the use of chlorinated solvents like perc. In this case, a pollution
prevention approach, as advocated by the antitoxics groups, meant the
elimination of the use of perc in dry cleaning, an approach that had been
challenged by the chemical industry and dry-cleaning groups as threat-
ening the very foundations of dry cleaning itself. Instead, groups such as

the Halogenated Solvents Industry Alliance, which represented the perc manufacturers, focused on the use of control or mitigation-related add-on equipment, such as carbon adsorbers or refrigerated condensers, rather than a prevention or substitute technology. Those control measures, in fact, became the centerpiece of the new NESHAP for perc, as opposed to substituting for perc itself, the direct pollution prevention route. Such control measures were designed to increase the efficiency of perc use and reduce spills, leaks, and various types of fugitive emissions in dry cleaning. For the chemical industry, while such measures would conceivably reduce perc sales to existing cleaners due to greater efficiencies, the costs of such measures would be borne by the dry cleaners and not the chemical manufacturers.

Once the NESHAP had identified an end-point control technology approach, pursuing a pollution prevention outcome appeared to be foreclosed. For one, the cost of these new control technologies (the now required "dry-to-dry" or nontransfer machines with various add-on equipment such as refrigerated condensers) could raise the price of a dry-cleaning machine to $50,000 or more. This further undercut any pressure for developing substitute technologies that would eliminate the use of perc as the cleaning solvent, once the huge capital investment in the required control technology was made. In commenting about the NESHAP rule process, Paul Cammer, the president of the Halogenated Solvents Industry Alliance, put it bluntly: as far as potential replacements for perc in dry cleaning, Cammer proclaimed, "there is nothing!" That remained the crux of the argument around the NESHAP and the crucial issue in the years to come, as the NESHAP came to be implemented. Peace would reign again, with dry cleaning's long-term dependence on chemicals remaining intact.[35]

The Search for an Alternative

As the Clinton administration prepared to take office in the winter of 1992, perc use in dry cleaning had already emerged as one of its more visible environmental issues. After several years of contention and poor regulator-regulated relationships, the Clean Air Act's NESHAP process had appeared to resolve the issue of perc regulation. To put the issue behind the agency, EPA staff had already decided that the best approach was through a roundtable discussion to remove or at least diminish the conflicts (particularly for small businesses) associated with pollution control implementation. Absent any perc-based alternatives, the focus would likely remain on control technologies, whether carbon filtration, solvent recovery, or any other approach that would keep perc "out of the environment and in the dry-cleaning machines," as a Dow Chemical publication characterized the goal.[36]

The problems of dry cleaning had achieved significant visibility, due in part to the campaign launched by Greenpeace, an environmental organization that thrived on direct action and high-visibility campaigns. Greenpeace's call for the phaseout and ultimate banning of chlorine-based products had created new pressures for an administration that was seeking to carefully craft an environmental identity. The Clinton/Gore approach sought at once to capture the support of the environmental groups, while appealing to industry groups through its advocacy of the nonregulatory third-way solutions such as Design for the Environment. Among the established Design for the Environment stakeholder groups, dry cleaning had seemed particularly appealing as a way to turn aside or deflect controversy and conflict, the Greenpeace chlorine campaign notwithstanding.[37]

One of the problems with third-way stakeholder approaches, according to its critics, was the absence of any specific designated outcome, regulatory or otherwise. Indeed, when the initial stakeholder roundtable meeting took place in May 1992, the absence of any such outcome seemed likely. The meeting was convened by Ohad Jehassi, a young EPA staff member who had been given the assignment because no one else at the agency wanted to be involved. "I was 22 and clueless," as Jehassi recalled the moment. The roundtable discussions had originally been conceived as a way to identify a common ground between the dry-cleaning industry and their chemical suppliers and the various regulators and technical experts. Together these stakeholders would try to establish consensus about how best to implement the NESHAP by determining what constituted acceptable control technologies.[38]

At the roundtable, two presentations arranged by Jehassi unsettled the chemical and dry-cleaning industry participants. The first involved the findings by the California Regional Water Quality Control Board of perc contamination of more than 260 wells in California's Central Valley, primarily due to dry-cleaner discharges into the sewer system. The second involved a New York State Department of Health investigation of potential exposure of residences located above dry-cleaning establishments. The groundwater study raised the troubling issue of potential liability of dry cleaners under Superfund and/or related state cleanup measures. The New York study pointed to the possibility of regulation of dry cleaners located below residences (of which there were significant numbers in places like New York City and San Francisco), as well as possible consumer concerns about perc contamination. Aside from these presentations, the roundtable failed to explore possible alternatives to perc use, despite the EPA's description of its Design for the Environment process as including the evaluation of "comparative risk information on potential substitutes." It was the lack of any discussion or presentation about alternatives that caused the environmental groups to play a relatively minimal role in the roundtable, similar to their absence in the NESHAP process.[39]

Alternatives, however, were available. Shortly before the May roundtable, Jehassi received a call from a Greenpeace member about a promising new alternative approach that had been developed in England, but Jehassi didn't pay much attention to it. "I would get lots of calls from vendors about this or that process or some new pollution prevention technology, but before I would do anything more, I'd ask to see the numbers." Such numbers should reveal, Jehassi reasoned, how a technology held up in terms of performance, economics, or in assessing any possible environmental impacts, intended or unintended. The person promoting the new technology, a third-generation cleaner named Richard Simon, had decided to market his new approach in the United States under the name "Eco-clean." Such an evaluation of this process, then, could also provide a real-world dimension to the understanding of whether such an alternative should be considered.[40]

The discovery of Richard Simon's process added a new dimension to the regulatory debates about dry cleaning. The origins of the process were not new. Richard Simon's grandfather had first established University Tailors in 1926 as a valet service in South London. It appealed to upper-end customers who had once had an in-house valet staff, but now sent their clothes out for cleaning. Founded prior to the widespread introduction of perc and carbon tetrachloride, the business established a reputation for its "traditional family methods" based on a combination of hand and machine washing, spotting, drying, and tumbling, with water as the primary solvent. Simon, who had entered the family business at sixteen, had begun by the 1980s to market the family operation as an environmentally oriented business. This environmental profile included the use of returnable wooden hangers, returnable bags, and organic and biodegradable soaps. But most significantly, there was no chemical solvent used in the process. When a London Greenpeace staff member, Beverly Thorpe, first approached him to allow Greenpeace to promote his family's operation as an alternative to perc-based dry cleaning, Simon decided the U.S. market offered the best opportunities to establish and ultimately franchise his "Eco-Clean" business.[41]

Responding to Jehassi's desire for numbers and evaluation, Greenpeace contracted with the Chicago-based Center for Neighborhood Technology (CNT) to undertake a preliminary evaluation of Simon's Eco-Clean process. For the CNT, a grass-roots-oriented think tank and policy advocacy group with a strong interest in alternative, pollution prevention–oriented technologies for small business, their engagement in dry-cleaning issues provided an opportunity to evaluate—and advocate for—new kinds of environmental and economic links to help sustain communities. The CNT's lead staff member on these issues was Jo Patton, who had a long history in addressing environmental and industry issues. The CNT in turn contracted with an engineer, William Eyring, to go to London and undertake an assessment of the effectiveness of the process, its ability to be replicated, and the degree of customer satisfaction.[42]

Eyring's report, released in July 1992, immediately caught Jehassi's attention. In evaluating the comparative performance of dry cleaning and the Eco-clean process in removing stains and cleaning clothes, Eyring found that the alternative process was "at least as effective as conventional dry cleaning for most fabrics and stains." The CNT report also argued that any increase in costs would be "more than offset by increased safety to the workers and the public by the improved wear of fabrics."[43]

During the winter of 1992, the EPA began its own evaluation of this new "multi-process wet cleaning" system. EPA volunteers wore two sets of identical garments over a two-day period. The garments were then taken to a test lab associated with the dry clean industry-based New York Dry Cleaning School. Performance evaluations and an economic modeling were undertaken to see if this new technology could be considered comparable to perc-based dry cleaning. The results of the EPA's limited study, Jehassi told the press, seemed "promising."[44]

However, the "discovery" of Richard Simon by Greenpeace, associated as it was with Greenpeace's antichlorine campaign, situated the promotion of wet cleaning for the dry-cleaner groups as part of a radical environmentalist plot. The dry-cleaning groups, moreover, had long sought to distinguish their industry as nonwater based, to establish the use of chemical solvents (and petroleum before it) as a professional method of cleaning, as opposed to home laundering and commercial laundering of garments not specifically labeled "dry clean only." Once the use of chemical solvents had been embraced, dry cleaners also tended to identify themselves as *chemical users*, a relationship strongly reinforced by the myriad of relationships that had developed between supplier and user.[45]

The association of the new multi-use wet-cleaning process with Richard Simon's unorthodox approach and Greenpeace's advocacy also situated the new process as a "low tech," hand labor, and even a counterculture style of garment cleaning. Such associations further exacerbated the mistrust of an industry sector that had long sought to establish respectability through its professional use of chemical solvents and advanced machinery. However, by the early 1990s, several European manufacturers of new "higher-tech" professional wet-cleaning machines had established U.S. distribution outlets and eagerly sought to enter what increasingly appeared to be a wide-open market for new technologies. This "machine wet-cleaning" approach, as distinct from the multiprocess wet-cleaning process, was primarily based on the notion that computer controls for the washing and drying machines could substitute for the hand labor in controlling for key factors that influenced how certain, more delicate garments could be cleaned. These computer controls could account for the amount of heat, agitation, and tumbling in the machine, and length of time for machine and air drying in relation to each particular garment fabric, whether wool, silk, rayon, or some combination of fabric.[46]

The advent of "machine wet cleaning" created the opportunity to develop a Design for the Environment focus on alternatives as more central to EPA's dry-cleaning sector approach. Working closely with CNT's Jo Patton, the DfE program set aside funding for the CNT to undertake an evaluation of a commercial wet-cleaning facility. This would be designed to significantly extend the EPA analysis and place it in a real-world context where a cleaner had to confront the realities of an actual garment-cleaning business. After exploring the idea of using the funds to establish its own operation, CNT identified an investor interested in the new technology who in turn hired a long-time dry cleaner, Ann Hargrove, to open "The Greener Cleaner," a new "100 percent" wet-cleaning facility in southwest Chicago. With Richard Simon's Eco-Clean operation (now called Eco-Mat) taken over by a new set of investors with ambitious plans for franchising their process, and The Greener Cleaner opening its doors in 1994, it appeared that a pollution prevention alternative might be available at last. The question was, would it be viable? And if viable, would it become acceptable to cleaners, regulators, and customers as well?[47]

Evaluating Alternatives

In 1994, while the Chicago CNT wet-cleaning evaluation was being designed, Deborah Davis, a San Diego businesswoman interested in developing an environmental business, contacted our Pollution Prevention Center (PPERC). Davis was interested in our undertaking an evaluation of a new commercial alternative wet-clean facility she wanted to launch in the Los Angeles area.[48] PPERC would review this facility, which Davis called "Cleaner by Nature," in three areas. We would analyze *cleaning performance* (how successfully the clothes would be cleaned and how satisfied the customers would be),[49] *financial capability* (what costs were involved and whether the business could operate profitably),[50] and *environmental criteria* (what environmental impacts could be identified).[51] The evaluation was designed as a case study of a real-world commercial cleaner operating a new technology with pollution prevention claims. At the same time, a broader comparative evaluation of dry cleaning and wet cleaning in each of the areas involved would also be undertaken.

The PPERC evaluation began the day Cleaner by Nature opened its doors in February 1996.[52] From a pollution prevention perspective, the results, following on the heels of the CNT evaluation, were striking. The cleaning performance lab evaluation indicated broad comparability, while the customer surveys indicated more satisfaction with wet cleaning in certain areas such as the feel for the garment. Performance difficulties had been encountered at Cleaner by Nature in its first several months of operation, but quick improvements occurred with more experience and training on the job. Overall, the differences between the two cleaning processes were not significant in terms of their ability to successfully clean the garments that came over the counter. It was also clear that wet cleaning, as a new technology, required more training, was more labor intensive, and could conceivably require more knowledge about garments and greater skill in how best to clean them.[53]

The financial evaluation reinforced the findings from the performance evaluation. The capital costs associated with wet cleaning (including the purchase of the washer and dryer and special pressing equipment) were actually lower than the costs of the new dry-cleaning machines (including various add-on or pollution control technologies required by state and federal regulations). Although operational costs were basically equivalent, there were interesting differences in terms of which costs were greater for each cleaning method. For wet cleaning, labor costs were greater, given the greater attention and skill required for particular components of the cleaning process, including the pressing of the garment. For dry cleaning, there were increasing costs associated with regulation (permitting fees, liability insurance, etc.) as well as the costs of the chemicals and the need for the pollution control equipment. Not quantified among the dry-cleaning costs was the "peace-of-mind" factor associated with the stresses of liability and regulation. Thus, while overall costs were equivalent, the trade-off between the two processes involved the need for more skilled or semi-skilled labor in wet cleaning and greater chemical and regulatory costs (as well as the peace-of-mind factor) in dry cleaning. The implications of the trade-off in this case were even more pronounced: more skill as opposed to more risk.[54]

In the environmental area, there were no significant differences in relation to water and energy use, but there were critical differences associated with the use of chemicals that identified the substantial environmental or pollution prevention benefits associated with wet cleaning.[55] Aside from the concerns around soil and water contamination (the core of the liability concerns for dry cleaners), there were major issues associated with air quality, occupational hazards, and hazardous waste generation related to dry cleaning (and no equivalent concerns in wet cleaning). Though dry cleaners were almost invariably small operations, the substantial use of perc by cleaners (even when new machines with control equipment reduced the amount required), still created enormous problems associated with its use. Dry cleaners accounted for 60 percent of all perc emissions in the region, and thus represented, despite the size of the business, a major regulatory concern for the local air district. Hazardous waste generation was also significant and issues over worker exposure remained an area of contention. The shift to wet cleaning therefore represented a major environmental breakthrough, at once eliminating a major source of environmental and occupational problems and the protracted regulatory battles and "peace-of-mind" and liability factors that had become the number one concern for dry cleaners in the 1990s.[56]

The evaluation presented clear outcomes; wet cleaning was judged comparable (and viable) in performance and financial terms and was clearly superior as a pollution prevention alternative in environmental terms. Moreover, despite the fact that overall differences were small in the financial and performance areas, they resonated in terms of the arguments developed in this book. While increases in the need for skilled labor and training were identified as economic costs in wet cleaning, they could also be considered a "benefit" in terms of the social goals associated with job creation and the reskilling of work. The additional economic costs associated with dry cleaning, such as greater chemical and regulatory expenses, could also be considered significant social costs, in relation to the occupational and environmental consequences to workers and communities.

These evaluations reinforced, and, in some areas, extended the evaluations of the Center for Neighborhood Technology, EPA, and Environment Canada (the Canadian equivalent of the EPA). Based on these studies, it appeared that a choice was available, a choice that potentially had significant environmental, health, and social benefits for cleaners, customers, workers, communities, and regulators alike. The question was whether the regulators—and cleaners—would be willing to make that choice, given the constraints imposed by regulatory paradigms, industry dependencies, and the idea that the best outcomes were exclusively technological in nature.[57]

By the spring of 1998, at the occasion of an EPA Design for the Environment conference, the issue of alternatives had entered center stage. With commercial wet cleaning still representing only a small share of the garment-cleaning market, the focus of the meeting tended to dismiss wet cleaning as a full or "100 percent" alternative, and explore instead the range of "advanced" (albeit still only potential) technologies. One particularly attractive alternative for environmental technology advocates both inside and outside the EPA was liquid carbon dioxide, a solvent substitute that had been previously developed for a number of other cleaning applications. CO_2 was attractive because it appeared to offer no significant environmental problems. Unlike wet cleaning, it had also attracted a handful of large corporate interests (e.g., Hughes Environmental Systems, which, along with its parent company, was subsequently sold to the Raytheon Corporation). CO_2 advocates also touted it as a full alternative to perc in terms of its capacity to clean the full range of clothes. Yet CO_2 also had some formidable obstacles. The cost of the machine, for one, was extremely expensive for the small neighborhood cleaner, as much as three times the cost of a new dry-cleaning machine and more than five times the cost of a wet-cleaning machine and pressing equipment. CO_2 advocates also were concerned that wet cleaning, which had already been evaluated as a viable commercial alternative, could capture the "alternative to perc" market before CO_2 could be effectively introduced and made attractive to cleaners.[58]

Unlike the agency's ambivalence toward wet cleaning, EPA officials looked approvingly at CO_2 and were willing to work on potential policy instruments, such as a tax-credit program, that could hasten the introduction of the technology. While wet cleaning continued to be situated in relation to its presumed "low tech" and "environmentalist" associations, CO_2 offered the classic "technological fix," a "high-tech" substitute technology with major corporate and R&D backing. The problem, however, remained timing. CO_2, by the end of the 1990s, still seemed several years away from commercialization. Its costs remained prohibitive and therefore required a major restructuring—and centralization—of the dry-cleaning industry if it were to be accepted as a 100 percent alternative, and it still remained unproven in terms of whether it was indeed viable in relation to its performance.[59]

In June 1998, after a number of delays, the EPA released its "Cleaner Technologies Substitutes Assessment (CTSA) for Professional Fabricare Processes," the major outcome of its Design for the Environment process. Though the CTSA was criticized by dry-cleaner organizations and environmentalists alike, its conclusion—that "several technology alternatives to PCE dry cleaning are available for commercial fabricare"—highlighted the fact that wet cleaning especially had achieved "viable alternative" status. The issue of the transition from perc use had become the order of the day. But how was such a transition to occur, particularly since the structure of regulation and the role of the agencies remained focused less on a pollution prevention transition than how to handle the morass of problems associated with the existing structure of end-of-pipe regulation?[60]

Liabilities and Uncertainties

Even as wet cleaning made its first appearances in the United States and the issue of alternatives emerged as a major topic of discussion, the squeeze on perc only continued to intensify. This created a series of pressures that weighed heavily on the more than 85 percent of dry cleaners who continued to use perc. Among perc's problems, the issue of liability was particularly significant, with the Superfund process emerging as a major area of contention, in part because of the large number of sites where contamination was linked to dry-cleaning activities and court rulings had extended liability to owners of contaminated sites. This meant, for example, that real estate companies owning a shopping center or mini-mall where a dry cleaner was located or banks that provided the loan for the purchase of such a site could be held liable if contamination was discovered at the site. As a consequence, the focus on perc use as a health and environmental problem, previously limited to various government agencies and a handful of consumer, workplace, and environmental groups, now directly concerned realtors and lenders, who had become unwitting though significant players influencing the future direction of dry cleaning.[61]

In response to these liability concerns, dry-cleaning interests packaged legislation, introduced by Texas Congressman Joe Barton, to modify or eliminate Superfund's liability provisions that were increasingly affecting cleaners. But despite an enormous preoccupation with this legislation in the dry-cleaning press and among some trade organizations, the Barton Bill stalled through three Congressional sessions. The exhortation and disappointments around the Barton Bill only underlined the growing recognition that dry-cleaner vulnerability, the "perc stigma," as one environmental engineering consultant put it, had become increasingly linked to potential liability action.[62]

While the Barton Bill stalled, dry-cleaner groups sought to develop some relief at the state level, specifically through legislation that established funds to help pay for cleanup and thereby possibly limit dry-cleaner vulnerability. But these "liability exposure reduction" programs carried with them a cost to the cleaners; new expenses in the form of fees assessed for dry cleaners, taxes on perc sales, and, in some states, an overall gross receipts tax for a perc-based cleaner. In several states, dry-cleaner organizations became divided over the magnitude of the fees. Several dry-cleaning organizations, including some of the Korean groups, feared that the charges, which could amount to more than $10,000 annually for a small cleaner, were far too high. In some states, dry-cleaner fees and perc taxes were not sufficient to fund the state cleanup funds, and increases were subsequently instituted, causing some dry cleaners to wonder whether they needed to repeal legislation that they had initially helped establish. The realtor and lender groups, on the other hand, argued that the funds through the state programs were too small to significantly reduce the liability of the deep-pockets players. The costs of cleanup, the realtors contended, would quickly outstrip the availability of cleanup funds secured through the dry-cleaner fees and taxes.[63]

But it was the fears associated with losing one's business due to liability, now especially exacerbated by lender and realtor actions, which had ultimately created an extended uncertainty for dry cleaners. Dry-cleaner publications continued to report horror stories involving cleaners forced to go out of business because of liability actions. Similarly, increasing numbers of realtors were indicating that leases to dry-cleaning tenants would not be renewed or that they would restrict any future dry-cleaner tenant from their properties, decisions that were often put in place when a property was put up for sale. By the end of the 1990s, the CEO of the International Fabricare Institute, the major dry-cleaning trade group, was warning that "arguably, virtually every plant in the industry has some contamination liability."[64]

Aside from the liability factor, regulatory pressures associated with perc use also continued to mount. Dry cleaners were increasingly aware that it was becoming difficult to imagine a *status quo ante* regarding questions of regulation and the environment. "Every dry cleaner in the country now has, like it or not, a junior partner in the form of government agencies that scrutinize their operations in ever greater detail," one cleaner publication lamented in an editorial commentary. And even though end-of-pipe regulations facing dry cleaners had been designed to create more certainty, the problem of *compliance* still remained paramount.[65]

For the environmental agencies, compliance was an unhappy reminder that even with greater regulatory certainty in its end-of-pipe focus, there was still uncertainty as to whether those standards were being met. Monitoring for compliance raised the spectre of confrontation with a small business; thus agencies were also focused on how compliance could be more easily achieved. Toward that end, the EPA endorsed a voluntary compliance program from the state of Massachusetts, the Environmental Results Program (ERP), that would also serve as one of EPA's Project XL programs. The Massachusetts program focused on small businesses such as printers and photo-processors as well as dry cleaners. The key to the program was identifying the basis for compliance of existing standards through a voluntary inspection system conducted by (in relation to dry cleaning) certified dry cleaners. In effect, compliance became a type of industry-based (as opposed to government agency-based) certification program to monitor and evaluate compliance.[66]

When the Massachusetts XL program was launched for the dry-cleaning sector, 100 of the state's 1,000 cleaners participated in the first year. But of those facilities, only 6 were in actual compliance. The Massachusetts results paralleled those in New York and California where extraordinarily high percentages of cleaners—98.5 percent in the New York case and 95 percent for the southern California air district—were found not in compliance with specific regulations such as the NESHAP. Thus, the concept of mitigation through controls—expensive, time consuming, and ultimately serving as a barrier or at least a constraint in the shift toward alternative approaches—was problematic even on its own terms. This extraordinary lack of compliance, establishing a disconnect between regulation and practice, represented a little-discussed but essential problem for both the voluntarist and end-of-pipe mitigation approaches. It also provided another "peace-of-mind" problem for the dry cleaners.[67]

With the low rate of compliance, expensive end-of-pipe regulations, mounting liability problems, and risk levels and standards that seemed designed to be broken, it had become clear that a shift in the framework for regulation needed to occur. One opportunity in that direction was the consideration by the Federal Trade Commission of a change in its garment care label rule. Responding to the evidence that an alternative such as wet cleaning to perc use could provide a significant environmental benefit, the FTC drafted a rule in 1998 that provided for garment manufacturers to provide a professional wet-clean label. Despite the FTC's initial embrace of what could be considered a pollution prevention approach in a key area influencing customer demand, a series of obstacles also emerged over how to define wet cleaning and how to establish a test protocol.[68]

Though it suffered delays and the outcome of the rule change remained uncertain at the end of 1999, the proposed FTC wet-clean label rule demonstrated that regulatory shifts were possible and that a reorientation in rule making could be identified along a pollution prevention path. But to be effective, any change in rules, any shift toward a transition to a pollution prevention outcome, also required a change in how the garment care industry itself responded. Such a change would certainly be influenced by regulatory or liability actions and by demonstrating that a new pollution prevention technology or process was viable, as both the PPERC and CNT studies had identified. But any shift from dry cleaning to wet cleaning also needed to address the complex social and institutional dynamics that had emerged in this industry, including the chemical dependencies and the very definition of how the cleaning was to be accomplished. Along with these regulatory and technology issues, the question remained: who could best facilitate a transition to a new type of garment care?

Identifying a Community of Interests

For cleaners to pursue such a transition, the question of chemical dependency needed to be addressed, given the extent to which dry-cleaner/chemical industry ties had for so long dominated the structure and leadership of the formal trade organizations and key industry lobbying groups and publications. The role of perc distributor R.R. Street is a case in point. During the critical period from the late 1980s through the 1990s, Street's corporate vice president Manfred Wentz was one of the most visible figures within the dry-cleaning universe of trade and lobbying groups. This was primarily due to his role as a kind of unofficial "technical gatekeeper" in assessing various technologies and commenting on potential industry directions. Wentz had a long history associated with the dry-cleaning industry, dating back to the early 1970s when he was director of research at the National Institute of Drycleaning (which subsequently changed its name to the International Fabricare Institute). In 1998 he became the executive director of the International Drycleaners Congress. And when Wentz accepted an appointment to establish a research department for the Textile and Textile Maintenance industries at North Carolina State University, the Street company became a key financial supporter of the new program. Another well-known figure in the dry-cleaning industry, *National Clothesline* columnist Stan Golumb, spoke of the revolving door historically between dry-cleaning trade organizations and R.R. Street, characterizing the Street official as a friend of the dry cleaner, a person "who knows where you [the dry cleaner] are coming from." Golumb himself had held a position with Street before becoming a columnist for the dry-cleaning trade publication.[69]

The chemical industry's importance has been both substantive and procedural in shaping the response of the dry-cleaning industry to regulatory pressures as well as the continuing embrace of perc in relation to those pressures. During the 1980s, for example, when the risk assessment debates were at their most intense, groups such as the HSIA took the lead in attempting to influence the outcome of both the scientific and regulatory review processes. Chemical industry spokesmen, such as HSIA's Steve Risotto, participated as major players at dry-clean industry conventions and other gatherings, and chemical industry association press releases were reprinted verbatim as news items in dry-cleaner trade publications. Chemical industry spokespeople even took the lead in identifying how dry cleaners should respond to consumer complaints about chemical use and formulate an effective public relations response equating dry cleaning with chemical use. Indeed, one dry-cleaning gathering that identified "insiders" and "outsiders" on various panels placed chemical industry representatives as "insiders."[70]

However, stresses over regulation and liability fears also began to take their toll during the late 1990s. In 1997 those fears were realized when the former owners of Pilgrim Cleaners, one of the largest independently owned chains of dry cleaners in North America, brought suit against a number of chemical suppliers and distributors. Pilgrim, with sixteen of its cleaning facilities in Texas subject to liability actions under Superfund, decided to sue to shift responsibility and ultimately recover its cleanup costs. The Pilgrim suit immediately became the subject of controversy and contention among cleaners, given the suit's implications (reversing the long-standing alliance between the solvent producers and suppliers and their dry-cleaner clients) and timing (in a period when perc use for garment cleaning was being challenged on several fronts).[71]

Similar to experiences of other dry cleaners, the former Pilgrim owners first became vulnerable in September 1994 due to a sale of their forty-year-old business. The sale had triggered a site review and the subsequent discovery of perc contamination at numerous Pilgrim sites. The former Pilgrim owners claimed that their chemical suppliers had failed to adequately inform their customers of the hazards and proper procedures associated with perc use. Worried about the implications of the suit on dry-cleaner/chemical supplier relationships, several of the defendants, with the exception of perc distributor R.R. Street, decided to settle the case out of court. Street, however, fought the suit and prevailed on an 11–1 vote of the jury, which held that the owners and operators of dry-cleaning businesses and not the suppliers were liable. But the judge in the case subsequently found Street liable on one count related to provisions of the Texas Solid Waste Disposal Act, with Street ordered to pay $1.5 million to help defray cleanup costs. A Street press release warned that the suit was playing "right into the hands of those wishing to take unfair advantage of our industry's exposed vulnerabilities."[72]

While dry cleaners were becoming increasingly concerned about their legal and regulatory vulnerabilities, the dependence on perc—the idea that garment cleaning was ipso facto chemical solvent cleaning—continued to be a powerful obstacle in the shift to a pollution prevention alternative. This was particularly true with wet cleaning, which also generated dry-cleaner concerns that water-based methods, with their home laundering association, would be considered nonprofessional. Even the clear success of Cleaner by Nature (Deborah Davis's facility had expanded to several drop-shop sites and the business had nearly tripled in only three years) tended to be dismissed, since Davis had previously not been a cleaner prior to opening her facility in 1996. In fact, some of the earliest attempts to establish wet-cleaning businesses had been associated with other, community-related goals, such as Davis's desire to establish an environmental business.[73]

The association with community goals also extended to the nonprofit as well as the for-profit/environmental business arena. In 1995 a Brooklyn, New York, community development corporation (CDC), the Fifth Avenue Committee, decided to expand its activities beyond the traditional CDC focus on housing, to move more in the direction of job creation and the development of community enterprises. The Fifth Avenue group, based in the Park Slope section of Brooklyn, a mixed-income neighborhood, was interested in setting up a business that could provide jobs for community members but would also become self-sufficient and sustainable over time. Similar to other initiatives among nonprofit organizations, the Fifth Avenue group wanted the business to be developed and operated by community-based organizations to "generate a significant *social return* in addition to *financial return*."[74]

In exploring the type of community enterprise it might establish, one of the board members brought to the group's attention the Eco-Mat franchising business originally based on Richard Simon's "Eco-Clean" concept. A connection to Eco-Mat was particularly appealing to the community development group. The group preferred a franchising business, since an established operation, it was assumed, could provide training and expertise to allow a new business such as wet cleaning to get off the ground. Also, the environmental benefits combined with the financial trade-off (increased labor costs but no chemical and regulatory costs) were clearly attractive. The machine-based wet-cleaning business thus provided environmental and community benefits (including the additional jobs that were needed), without creating any significant additional financial burden.[75]

As negotiations proceeded during 1996 and 1997 with the Eco-Mat group, the Fifth Avenue Committee concentrated on identifying a suitable location, raising the capital through traditional means, and putting together a team to run the operation. This included several "unskilled" community members recently on welfare who were to receive a living wage (above the minimum wage) and training in a new line of work that could conceivably represent a career path. The training and expertise provided by Eco-Mat were considered crucial, since wet cleaning represented a new technology and required operator skills and knowledge of spotting, how to evaluate which machine controls should be used for washing, the amount of time required for drying, and the skill and method of pressing.[76]

When the Fifth Avenue Committee opened its doors for its new Eco-Mat franchise in September 1997, the enterprise got off to an auspicious start. The facility was situated across the street from the Park Slope Food Coop and was immediately able to draw on an environmentally conscious constituency within a mixed income community. The Fifth Avenue group, moreover, was a well-recognized and respected community organization that also had deep roots in the community.

In the first several weeks after its opening, the volume of customers was far beyond initial expectations, due to the credibility and reach of the organization, the location, and the interest in a nonchemical cleaning process. However, this early success also proved to be problematic. The training from Eco-Mat never materialized, as that franchising operation began to experience problems of overreaching and disorganization, compounded by its desire to establish a "monopoly" or near-monopoly status for a new technology. Similar to the experience of other wet-clean operations, including Cleaner by Nature, the start-up period proved to be its most difficult in terms of both performance (for example, problems with shrinkage and dye runs) and the management of the operation itself. Moreover, the problems of the Eco-Mat franchising operation (it would ultimately declare bankruptcy in late 1998 after prematurely going public and seeking to raise capital for its expansion plans just two years earlier) were becoming more of an albatross than an advantage to the young operation. However, despite the difficulties of seeking to establish a business based on a new process and a new framework for a community enterprise, Park Slope Cleaners was able to survive its first year and significantly improve the performance and financial aspects of the enterprise. Most important, for the Fifth Avenue group, the higher costs associated with the additional—and more skilled—jobs that the operation required, was a significant advantage in terms of the social goals of the enterprise. The community enterprise approach was most directly associated with job creation as well as the environmental benefits that together established the operation's public interest or "community of interests" value.[77]

While the Brooklyn initiative (and similar explorations in other communities such as Philadelphia, Oakland, and Boston), could be associated with a community of interests in relation to providing a "service" to the community (job creation, environmental benefits), it failed to answer the question of how to develop a community of interests among *cleaners*. That question was of particular importance to dry cleaners who were deciding whether to shift to this new pollution prevention technology. Among dry cleaners, some of the strongest bonds were among Korean cleaners, as discussed earlier. Ethnic ties, the importance of family, and shared knowledge and lessons were each powerful motivators.

The first significant attempt within the Korean community to address the question of wet cleaning was stimulated by a community group rather than a cleaning organization. The Korean Youth and Community Center (KYCC) in Los Angeles, which had ties to Korean community groups around the country, first developed an interest in wet cleaning and dry cleaning in the wake of the Los Angeles riots in 1992. Korean storeowners, particularly liquor store owners in South Central Los Angeles, had been a specific target during the civil disorders. Some stores had been burned to the ground, and when the disorders spread north past the 10 freeway, Korean individuals and groups stocked weapons and otherwise prepared for the events to spill over into the Koreatown section bordering the largely immigrant and ethnically diverse Pico Union area. Tensions between African-Americans and Koreans appeared to be particularly intense, and had been or would soon become the subject of articles, organizational dialogues, and even the subject of one of Spike Lee's films (*Do the Right Thing*).[78]

The KYCC (formerly the Korean Youth Center) had formed in 1975 to address issues affecting Korean youth, including jobs, education, and problems of violence in the community. During the 1980s, the group broadened its agenda to provide a more comprehensive set of services to its youth constituency. Though more "mainstream" than militant Korean activist groups like the Korean Immigrant Workers Association, which focused on issues of economic justice and workplace abuse of Korean immigrants, the KYCC nevertheless had emerged as a strong community voice that clearly articulated a social justice agenda for the community. The group was particularly effective among the "1.5" generation of young Koreans who had migrated to the United States at an early age or had been born shortly after their parents had migrated.[79]

As with a number of community groups, the 1992 civil disorders represented a pivotal moment for the KYCC. The organization decided to significantly restructure, adding "community" to its name, and expanding its arena of activity to include affordable housing and community economic development. The KYCC, like a number of other community groups, was also particularly interested in how environmental justice and community economic development issues could be joined. A major opportunity along these lines presented itself in the wake of the riots. A contentious debate had erupted over whether liquor stores in South Central Los Angeles that had been burned, many of them owned by Koreans, would be allowed to reopen. Community organizations that had taken the lead around reducing the number of liquor stores in the community, such as the Community Coalition Against Substance Abuse, wanted to identify ways in which the liquor store owners would not reopen their stores but instead establish more community-oriented businesses. The KYCC, which had an economic development and business retention program, immediately became drawn into the discussions about the Korean-owned liquor stores. The notion of a shift from a liquor store to a dry-cleaning operation immediately presented itself, given the high percentage of Koreans in dry cleaning and the interest in a business where other Koreans had already established a presence. But the KYCC had also become aware of the problems around perc use, and they worried that such a transition would lead to criticisms that they were advocating the switch from one hazardous (to the community) business to another.[80]

The KYCC first learned about wet cleaning at the point when Deborah Davis was beginning to pursue her idea of an environmental business. Working closely with our center, the KYCC carefully monitored the results of our operation, and had parts of it translated into Korean and distributed to the Korean language press. After successfully applying for one of the limited number of U.S. EPA's "environmental justice-pollution prevention" grants, the KYCC sought an existing Korean dry cleaner who would be willing to switch to a wet-clean process and then serve, similar to Cleaner by Nature, as a demonstration site for other Korean cleaners.[81]

In 1998, the KYCC was approached by Joe Whang, a Korean cleaner who operated a small dry-cleaning facility on-site in a declining shopping mall in Cypress, California, a mixed-income community on the Los Angeles/Orange County border. The location for Whang's facility had also become a concern, as the large supermarket next door had shut down. Whang's own children had been urging him to find a less hazardous business, thus all the issues of regulation and peace of mind were paramount in his consideration as well.[82]

However, Whang, like other small cleaners, was skeptical that wet cleaning could "work" and worried that his customer base would shrivel when customers discovered that clothes were being cleaned in water. With the KYCC offering to partner with him on the basis of a more limited version of the evaluation/demonstration that had been undertaken at Cleaner by Nature, Whang cautiously agreed to switch to wet cleaning at his same site, fearing he had no other options. With only limited training, given his concerns about timing and budget, Whang installed the wet-cleaning equipment over one weekend, but never closed the facility during the transition for fear of losing customers. The name of his facility, Cypress Plaza Cleaners, also didn't change. In fact, the fear of customer loss caused him to only display a small sign inside the store explaining that he had switched to a less hazardous system. Whang did not assume he'd have an "environmentally conscious" group of customers, either in terms of his existing customers or any new customers that might be attracted to the business. This contrasted with Cleaner by Nature's strategy, which actively sought to distinguish itself as a new, environmentally oriented business.

To Whang's surprise the transition went far smoother than anticipated. He maintained his customer base while increasing his business by 10 percent over the course of the first fifteen months after his switch to wet cleaning, though his location in a depressed mall continued to hurt efforts at attracting new customers. While there had been a few performance problems during the first months of the switch, the learning curve that came with experience both reduced the problems and gave Whang far more confidence about his transition to a 100 percent wet cleaner. At a workshop that our center hosted, Whang spoke of the shift as a relearning process, both in terms of how to clean, the role of pressing, and a more intimate knowledge about garments. Aside from the trade-off in costs (more labor costs, no regulatory or chemical costs), Whang also identified a shift or trade-off in his working hours (more skilling, less paperwork, more peace of mind). With his increase in customers and revenues, Whang's confidence as a quality wet cleaner handling a full range of garments over the counter had increased considerably as well.[83]

As a demonstration site, the KYCC was also able to arrange tours for a number of Korean cleaners, several of whom expressed an interest in exploring a switch to wet cleaning but wanted help (whether technical assistance, new wet-clean care labels, or loans and/or other financial help). To obtain such support, what in effect amounted to a transition planning framework, was critical to a pollution prevention approach. But the absence of any systematic government or industry programs to facilitate a transition represented a significant pollution prevention barrier. Nevertheless, the idea of such a transition—establishing a new type of community of interests based on pollution prevention, reskilling, and the survival, if not the flourishing of a small business sector—seemed feasible, even as the planning for such a transition had yet to occur.[84]

Rethinking Industry

When the CNT and PPERC identified wet cleaning as a viable pollution prevention technology in 1996 and 1997, the threshold issues for pollution prevention—does the technology work, is it profitable, are the range of environmental impacts taken into account?—appeared to have been answered. However, conflicting signals from the agencies and a skewered policy and regulatory framework worked against, from a pollution prevention perspective, what has often been presented as a technical question—is the technology viable? The Korean factor, as well as the interest of the nonprofit groups, also identified cultural and institutional issues—and potential environmental and social justice approaches—associated with how an industry transition might play itself out. Supplier-related and upstream issues—chemical dependencies and garment manufacturing choices about garment fabrics, garment construction, and garment labeling—continue to remain paramount in influencing such a transition. To say that the market will decide—the mantra of voluntarism—is in fact misleading. What choices customers may make—wet cleaning, dry cleaning, or any other system—cannot be divorced from the range of factors—from regulatory actions to upstream and supplier influences—that have been described in this chapter.

The argument about the "community of interests" provides a social overlay to the issues of the market. Upstream-downstream and supplier/distributor issues are also crucial in understanding both the barriers and restructuring issues involved. A new set of production-based relationships need to be established, as well as a redefinition of production objectives and production values (e.g., work reskilling, environmental outcomes, job creation, better-constructed garments). Upstream issues also have to enter in. Pollution prevention in this context is a production rather than just a technology question. One needs to identify a "garment chain," suggesting an extension of the industrial ecology argument to address the processes of garment manufacture and cleaning as well as the materials used and the wastes generated. And while the garment care/dry-cleaning question can also be considered a "small business" issue, the outcomes of this transition are emblematic of broader issues.[85]

Can one say then that there is an intersection of pollution prevention and environmental justice regarding the issues of garment cleaning? Does a strategy that identifies a community of interests—potentially cleaners, community groups, neighborhood residents, and customers— represent an effective starting point for a transition? And do such issues associated with ideas of the transition also represent a pathway for social change?

The enormous pressures that have come to bear on dry cleaners have only been exacerbated by the absence of a pollution prevention policy framework. Dry cleaners need peace of mind—and the ability to function as a business serving neighborhood residents. An alternative such as wet cleaning offers not simply a technology choice but can also reinforce those community ties essential to the business itself. It is also clear that what most distinguishes wet cleaning from dry cleaning—aside from the significant environmental differences and pollution prevention benefits— has to do with the work process itself. "You need the knowledge of the garment," Joe Whang commented at a wet-cleaning technical issues workshop. "You have to do more categorizing, more distinguishing, more understanding of how the cleaning process will work for each particular garment."[86] To successfully wet clean involves not just increasing the skill level around such work tasks as pressing or sorting garments, but also elevating knowledge of the overall work process itself in what has been considered a low-skill industry sector. In terms of this and other factors related to the public and community role with this industry sector, the transition to pollution prevention may well have the potential to represent a transition to a new set of workplace and community relationships. But can this idea be extended, if the work is more degrading and the community of interests more difficult to assemble?

Core to that question is the importance of social movements and community action as well as worker and workplace engagement in restructuring industry sectors and workplaces and identifying the actors most crucial to environmental, community, workplace, and industry outcomes. The role of the KYCC and its desire to establish a community of interests for a transition to wet cleaning for Korean cleaners comes to mind here. For garment care, ending chemical dependencies through pollution prevention solutions only represents one part of the answer. How those dependencies are ended—and how a transition is accomplished—provides important clues about whether a new pathway for change can be identified.

4

Janitors and Justice: Industry Restructuring, Chemical Exposures, and Redefining Work

Patagonia's Dilemma

The cleaning crew at Patagonia was in a bind. These janitors, employed by the firm that had contracted with Patagonia to clean the Ventura County, California, headquarters building of this environmentally sensitive apparel company, had specific instructions to complete a set of defined cleaning tasks within a strict time frame. Arriving at 4:30 in the afternoon when most of the Patagonia staff left work, the cleaning crew had until 8:00 P.M. (when Patagonia shut down for the night, partly as an energy savings as well as a "quality-of-life" initiative) to complete their work. While Patagonia had not established a specific program involving the selection of less hazardous or environmentally preferable janitorial cleaning products, the contractor was aware that the company prided itself on having instituted its own in-house environmental program to complement its overall environmental goals. These goals included the use of recycled plastic and subsequently of organic cotton in producing its line of outdoor garments, and identifying the least energy-intensive form of transportation of garments to their outlets. The in-house environmental programs division, led by a former Greenpeace staff member, had also established a series of initiatives for the head-quarters building aimed at reducing energy use as well as minimizing solid-waste generation, among other objectives. Cost considerations for the company continued to be a major factor in deciding which programs to pursue. Moreover, there continued to be a separation of functions between the environmental programs, including the in-house initiatives, and the company's facilities management operation, with its concerns about cost, efficiency, and quality of life for Patagonia employees.[1]

Despite its strong environmental profile, Patagonia's contractor had been primarily selected for cost and performance reasons. There had been a brief effort to have a previous contractor incorporate selection of less hazardous cleaning products, but the products used were considered ineffective. While the new contractor had no specific instructions, there was an informal understanding that Patagonia would prefer that environmental criteria enter into the choice of at least some cleaning products. This interest on the part of a client was not a new situation for the cleaning contractor. Several other clients had in fact established a "restricted use" list of ingredients in cleaning products considered to be potential health risks, and therefore not to be used in cleaning their buildings. As a consequence, the contractor had a set of "regular" products used on most of the buildings it cleaned (including chemicals that showed up on the "restricted use" lists), and a second set of compounds (the less hazardous products) due to client preference.[2]

When the contractor began work for Patagonia, it was assumed, though not made explicit, that the less hazardous products would be used. For the particular cleaning crew, several of whom worked double shifts for the contractor, these were less familiar products, unlike the products used on their other jobs. The "green" products were seen as more time consuming in their application (more labor required); they had to be used in ways that required a different kind of application (a different training process); and they often produced a different kind of result (different definitions of what constituted a "cleaning" outcome). None of those factors came into play at Patagonia.

One day, a Patagonia employee complained about a chemical odor in the bathroom. The problem was traced to the bathroom cleaner, a strong chemical that was one of the contractor's regular or more hazardous chemicals. This cleaning agent had simply been brought to Patagonia by a second-shift employee from his earlier job. He had done so to complete the required cleaning tasks by the 8:00 P.M. deadline. In the search for an environmental approach, the janitors in this case were seen as contributing to the problem, not the solution. But was this a problem of the janitors themselves constituting a barrier for environmental change? Or was this a problem of how the janitorial work was defined, and the relationship of the conditions of work with work outcomes?

Wages and Chemicals

How do considerations of work, environment, and social change come into play in relation to janitorial work and the commercial cleaning sector? Consider the case of cleaning products for commercial buildings, most of which are used for janitorial cleaning work. A number of studies have indicated that various ingredients found in cleaning products used for janitorial service or building maintenance work in commercial establishments such as office buildings, hotels, and restaurants, may represent significant occupational and environmental hazards. But the strategies to identify and prevent such hazards are nearly nonexistent. Current regulatory policies and the most common ways of evaluating risks to workers (for example, examining relationships associated with exposure to toxic ingredients in cleaners) have been difficult to design and develop. These difficulties are associated with the nature of the businesses involved, the small number of workers exposed in any particular workplace, and the types of cleaning products used, where multiple ingredients and low concentrations often preclude the current types of regulatory interventions or evaluations of the risks involved. At the same time, there has emerged an increasing interest in developing new kinds of environmentally preferable cleaners that could reduce exposures to potential hazards and provide additional environmental benefits for all of the parties or stakeholders associated with janitorial cleaning services. But, as with the case of the dry cleaners, how would such a shift in use take place? Can the mobilization of janitors to improve the conditions of their work and establish a "living wage"—the cry for justice—become the basis of an environmental mobilization as well? Similarly, can the focus on environmental risks strengthen the movement for different working conditions, including better wages, union representation, and greater respect for the work and the worker?

These issues are all related to the nature of the work itself. Janitorial work in commercial buildings is often perceived as low-skilled, low-wage "dirty" work—scrubbing toilets, stripping floors, sanitizing baseboards. Whether the products used in such work are "toxic" or hazardous, and thus represent possible occupational and/or environmental risks, is largely ignored by the industry and even the workers themselves. As in the Patagonia case, janitorial or custodial workers may be reluctant to use products that are perceived as less potent. These biases in turn are embedded in how the industry has evolved and the pressures associated with the squeeze on labor costs, which has been a fundamental industry issue. Similarly, the choices that distributors, building owners, and contractors make regarding why certain products are used, the type of training (if any) that is offered in how to use such products, and how work tasks are structured and labor costs are controlled, all become related matters.

Janitorial cleaning services often refer to interior building cleaning services, the type of work typically associated with janitorial or custodial cleaning work. The nature of the work associated with these services also represents multiple tasks. For example, a study of ergonomic hazards in custodial cleaning work in the Los Angeles area conducted by the UCLA Labor Occupational Safety and Health (LOSH) program identified as many as six work classifications for janitorial services: maid; utility work; bathroom cleaning; vacuuming; collecting/disposing of trash; and a waxer. Janitors in fact often perform work in more than one classification. This work may occur in office buildings, banks, retail and department stores, supermarkets, schools and colleges, independent professional offices (doctors, lawyers), factories and industrial plants, warehouses, airport terminals, hospitals, nursing homes, and other health care institutions, among numerous other commercial or retail facilities. Hotels and restaurants represent a separate, though parallel set of businesses, cleaning tasks, and work categories.[3]

In recent years, janitor shifts for building cleaning services have typically been at night, with two ten-minute breaks and a thirty-minute lunch break, or one fifty-minute break during an eight-hour shift. Janitors usually work alone or in pairs to perform all tasks on a floor, or in small crews, where each member of the crew is responsible for a different job, depending on the amount of square footage to be cleaned. Janitors working for nonunionized service contracting companies can (and often do) have work shifts shorter than eight hours, with no breaks, a trend magnified by the use of part-time and temporary labor. The area cleaned by each janitor, per work shift, varies by building and task, but has been increasing in recent years due to budget cutbacks and reductions in the employment of a full-time janitorial workforce.[4]

Janitorial work has been categorized into two types of employment relationships: in-house employees and contract hires. In-house janitorial employees are full-time support staff of the companies that own and/or occupy commercial buildings. They almost invariably have better wages and benefits (including medical, vacation, and sick leave), than contract hires. In-house jobs, however, are obscured within the larger economic goals and business objectives of the companies for which the work is performed; hence, information on this segment of the industry frequently has far less detail than equivalent studies or reports on contract cleaners. What has become increasingly prevalent today when discussing janitorial services is a near-exclusive focus on the contract cleaners; that is, companies whose main purpose is to provide cleaning and building maintenance services on a contract basis.[5]

The ethnic composition of the workforce (both the contract and in-house employees) also reveals rapidly changing patterns that can significantly influence industry activities, such as wage levels and work definitions. For example, the current ethnic composition of the janitorial workforce in Los Angeles County is more than 80 percent Latino in the private commercial sector, which represents a considerable change during the past two decades. However, ethnicity patterns in public sector employment differ considerably from the private or contracted workforce. A far higher percentage of African-Americans are employed as janitors in the public sector. Nationwide, the majority of janitors are women, and this pattern can also be found in the Los Angeles County figures as well.[6]

While the nature of the workforce has changed, so has the cleaning service operation itself due to the shift toward contracting. While cleaning buildings can be done in-house, even the largest building owners or institutions that establish in-house operations are still dependent on a range of contract services, whether purchasing cleaning products or providing for ancillary services or other products. Though still an important segment of the market, the trend of both small and large building operators has shifted toward contracted services, a function of a squeeze on labor costs and the changing dynamics within the commercial cleaning industry.[7]

Nationwide, the contract cleaning industry is large, diverse, and fragmented. There are more than 50,000 commercial building service companies employing nearly one million people. Building service companies range from one- or two-person part-time businesses to multinational corporations. Franchises account for only a small amount of revenues, about 4 percent. However, there has also been a significant consolidation in this segment of the industry that has occurred in the past ten to twenty years. In particular, a few diversified companies, such as One Source and American Building Maintenance (ABM), have assumed a dominant position among contractors servicing large building owners.[8]

One Source, for example, which is headquartered in Belize, took over International Service Systems (ISS), a major cleaning contractor that had more than 100,000 employees in twenty-nine countries. ISS began as a Danish security firm almost a century ago and entered the cleaning market in the 1930s. By the 1990s, ISS had decided to emphasize even further its cleaning operations, which included both cleaning services and a major cleaning product manufacturer (Spartan Chemicals), which provided the line of products used in the United States for ISS, now One Source, contracts. The name change to One Source, moreover, reflected the company's assessment that more and more building managers would turn to outsourcing, and that the new company could then become "the One Source for facilities services solutions," as Ray Gross, the president of One Source, put it.[9]

ABM, which has revenues greater than $1.4 billion and more than 50,000 employees in the United States and Canada, also emerged as a broad-based building maintenance operation, with janitorial, mechanical, electrical, pest control, and business security operations, among other building services. Within its Janitorial Services group, ABM has had an integrated relationship with Easterday Janitorial Supplies, its product manufacturer subsidiary. More than 90 percent of ABM's cleaning supplies are purchased through Easterday, while ABM accounts for

32 percent of Easterday's sales. Similarly, while independent cleaning product manufacturers also differ significantly in size and diversity of operations, the independents with the largest market share (that is, companies without a direct relationship to a cleaning service operation), tend to be medium- to large-sized companies. Thus, while the sector can be characterized as primarily small business, there are important integrated operations that are large as well as product supplier influences (similar to dry cleaning) that directly impact the environmental (and workplace-related) outcomes associated with the cleaning activity.[10]

The rise of the contract cleaning industry coincided in part with the national economic spurt in building construction of new office and retail buildings during the mid- to late 1980s, with contract firms experiencing an average yearly increase of 14 percent in revenues from 1984 to 1989. At the same time, declining union strength and an influx of primarily Central and South American immigrants who emerged as the primary low-wage workforce in this segment of the industry paralleled the continuing squeeze on wages as the dominant factor influencing the comparative position of firms and services. As a consequence, cleaning workers joined their counterparts in industries like fast-food operations as among the lowest-paid workforce in the country. This represented a dramatic shift in the wage structure for workers who had previously constituted one of the higher-paid areas of entry-level unskilled or semiskilled work, particularly for those workers and firms represented by unions. In 1996, by way of example of this downward pressure on wages, as many as one-third of all contract cleaning employees would have needed a pay hike for their employers to comply with a $5.15 minimum wage. Even in the midst of the boom of the late 1990s and the tight labor market, the average starting wage of a janitor employed by a contract cleaner was still only $6.42 in 1999, not much higher than the minimum wage. This increasing trend toward low-wage employment was further impacted by a parallel trend toward part-time/no-benefit jobs, particularly again among the contract cleaners. In fact, the debate over whether municipalities or states should legislate a "living wage" for workers, employed by and/or contracted through public agencies, was driven in part by the rapid wage restructuring that had taken place in janitorial cleaning services during the 1980s and 1990s. Those downward pressures on wages were particularly acute among the highly competitive contract cleaners.[11]

Wage declines and the increased use of an immigrant workforce have also shifted the way the nature of the work and the skills involved have come to be characterized. Janitorial or custodial work had previously been considered semiskilled, due to skills gained by experience about the cleaning process, familiarity with cleaning products and their use in relation to specific cleaning tasks, and knowledge about what constituted appropriate cleanliness (that is, how clean is clean). But those work definitions began to erode during the work and wage restructuring that took place in the 1980s and 1990s. Such wage and work pressures, as well as the decline in union representation, contributed to a type of deskilling of the job and the work involved. As one industry official characterized this shift, "There is no definition of skill used in the maintenance industry. Someone with no experience can easily replace someone with eight years' experience." This trend was also reinforced with the shift to contracting and its significant use of part-time labor, which also lessened or eliminated any connection or knowledge of the place of work or areas to be cleaned and maintained.[12]

While this wage and work restructuring was taking place, profit margins remained relatively small for many contractors, often ranging between 4 and 6 percent. Given the percentage of labor costs in relation to overall costs, the reduction in wages, and the increase in amount of square footage cleaned on an hourly and daily basis, wage restructuring associated with "increased productivity" (that is, more area cleaned at a lower wage rate) became the difference in the degree of profit and competitive advantage for janitorial employers. This trend was also reinforced by the increasing tendency toward integration and consolidation among the largest firms handling the largest contracts.[13]

This wage and work restructuring and speedup in the cleaning process occurred even during periods of expansion during the 1980s and 1990s. For example, during the late 1970s and 1980s, commercial building vacancy rates in Los Angeles dropped considerably, with rentable downtown square footage doubling between 1976 and 1988 due to new construction. Net operating income for contractors increased 60 percent in real terms from 1981 to 1987, rising from $7.67 per square foot to $12.33. Meanwhile, janitors' hourly wages failed to keep pace with inflation, dropping to a median of $4.50 per hour in current dollars for the Los Angeles–Long Beach area. Due to the loss of union contracts, many of these janitors also lost medical, dental, life insurance, and other benefits.[14]

Yet the recession of the early 1990s, which created a significant decline in the business climate (including commercial building construction), only heightened these trends further. In Los Angeles, building owners faced their highest vacancy rates in twenty years and had rents drop as much as 40 percent. Increasingly, companies and building managers in the private sector sought to eliminate in-house janitorial staff and began contracting for these services to save money; for example, by the mid-1990s, an estimated 90 percent of the commercial private sector in the Los Angeles area was contracting for cleaning services. These trends continued even after the recovery in building construction and office development kicked in during the mid- and late 1990s.[15]

At the same time, during the 1980s and 1990s, the contract cleaning industry became even more competitive, while also experiencing some consolidation and integration among the largest contractors. Through much of this period, the top fifty companies captured only slightly more than 25 percent of annual industry receipts, with average revenue for each establishment increasing gradually from $277,000 in 1987 to $329,000 in 1995. Entry into the industry by small contractors continued to be relatively easy, with a continued influx of cheap migrant labor, low start-up costs ($6,145 national average in 1990), and low long-term debt (12.5 percent of total liabilities in 1990). These start-ups have increasingly found a niche in cleaning new small professional offices that have grown as a result of the entrepreneurial booms that occurred during and subsequent to the recession of the late 1980s and early 1990s and the economic upturn of the mid- and late 1990s. As a result, many of the smaller contract cleaning companies have continued to experience low profit margins, while, at the same time, maintaining a large share of the overall market for contract cleaners. Moreover, the competitive nature of this segment of the industry, influenced by the rapid entry of small start-up companies, especially the small "mom and pop" establishments that have gross sales less than $100,000, has also reflected the decrease in the payroll per establishment. Thus, as this segment of the industry has become more competitive, contract cleaning companies, both large and small, have attempted to reduce their costs of doing business by fierce underbidding for contracts. For these companies, the way to remain competitive and reduce costs has been the restructuring of janitors' wages and benefits. This has included the shift to part-time and temporary labor, as well as through work-task restructuring (i.e., by securing a greater amount of square footage cleaned per unit of labor cost and in relation to overall production costs).[16]

As this competitive trend has developed, more and more of the largest building maintenance companies have reorganized themselves into full building service companies to remain competitive. This in turn has reduced the need for building managers to contract with several different companies for all the necessary services, which in turn lowers the costs of building owners by their ability to contract with just one company. For example, American Building Maintenance Industries (ABM) has increasingly promoted itself as a one-stop contractor for large commercial buildings. The company, whose Web page motto is "One Contact—One Contract—One Invoice," highlights its ability to run parking lots and garages, fix elevators, and provide pesticide services, as well as offer janitorial services, which still account for two-thirds of its revenue stream and three-quarters of its profits. This has allowed ABM, and several other large contractors, to more effectively weather downturns in the industry better than other companies. A parallel trend has also occurred among industrial supply firms that have sought to expand their product lines to include janitorial supplies. This more competitive market then also allows customers—the building owners—to demand more services for the same amount of money.[17]

Aside from wages, cleaning chemicals have become the largest single expenditure for contract cleaning companies. As a consequence, contractors and even building owners become wary of any change in cleaning chemicals perceived to be more expensive. This presents a significant barrier (both in perception and various cost criteria) in selecting "green products" (i.e., those products that are marketed as environmentally preferable). For the distributors, moreover, profit margins on cleaning chemicals are potentially significantly greater than most of their other items.[18]

Thus, the single most dominant issue that has emerged in the past two decades in the building cleaning services area has been wage restructuring and the increasing importance of reducing labor costs for competitive purposes in a highly fragmented market. Reducing labor costs has become a function of a changing labor market (i.e., the shifting nature of the workforce, particularly among contract cleaning services) with its emphasis on greater productivity, as measured by the amount of square footage cleaned per unit of labor cost. The increased role of small start-up firms, many of which rely on a part-time, unskilled, or untrained immigrant workforce, has also been significant. These trends have influenced the ways in which the nature of the work task (what should be cleaned and how it should be cleaned) is defined. It has also influenced what kinds of products (chemicals that are designed to accomplish varied tasks that also help reduce the amount of labor required) get used.

The issue of chemical use can then be seen in the context of the structure of work, including the set of relationships between product manufacturers and distributors, the cleaning services, and the clients or building owners. Such an analysis of industry structure in turn facilitates the discussion of pollution prevention opportunities and barriers as well as environmental and social justice–related issues in the janitorial cleaning industry. More specifically, it becomes important to determine how occupational and environmental hazards can be reduced, given the influences and trends that exist within the industry, and the players who can contribute to or undermine a pollution prevention outcome. Those questions relate to the discussion of the hazards themselves.

Identifying Hazards

As job skills and wages have declined and various segments of the industry have restructured, the type of cleaning chemicals utilized has become an increasingly significant factor in how the industry operates. Janitorial work itself has become more a reflection of the potency of the chemicals used than the knowledge of how best to achieve a standard of cleanliness applicable to each particular place.

How to achieve such an identifiable standard of cleanliness is not simply a health-based consideration or aesthetic judgment. The concept of how clean is clean has also been historically and socially constructed. In the first decades of the twentieth century, women and children were often employed to clean commercial establishments and public buildings, with the job requiring "long and hard work in unsanitary buildings," as one immigrant worker put it. The work of these janitors contrasted with the janitorial labor defined in relation to custodial work or building maintenance. Janitors, or "custodial workers," the term preferred by long-term janitorial workers, have had historically, despite the "dirty" nature of the work, an association with the places to be cleaned or maintained. Custodians were considered "place keepers," workers who maintained buildings and grounds, an association particularly relevant in places like schools. The grounds for the first urban school garden established in 1891 at a school site in Roxbury, Massachusetts, for example, was prepared by the school's janitor as part of that "building and grounds" custodial function. The janitor as custodian also had a gender connotation, since male workers were also presumably considered more skilled in knowing the correct chemical or cleaning compound and method to use on different surfaces. "Maintenance of large buildings is no longer a job for the inexperienced and unreliable," the 1947–1948 Annual Report of the Los Angeles County Janitors Department noted. "It requires thought, training, and constant application to duty."[19]

By the 1950s and 1960s, coincident in part with major office building construction booms, the job of janitor—and the structure of janitorial work—began to significantly change. This not only involved a change in the association with place-based "custodial" functions, but with new kinds of hazards on the job, associated as much with cleaning compounds as cleaning tasks. Although hazardous cleaning products such as lyes, caustics, and acids have long been used for various cleaning tasks, there has been an increasing use of new and more hazardous cleaning chemicals during the past two to three decades—coinciding with the dominant trends of wage and work restructuring. Many of these products now contain a higher proportion of various toxic ingredients, such as phenolic compounds and petroleum-based solvents. While the hazardous nature of such products has not emerged as a major environmental concern, the impact on worker health has become a critical component of the outcome of industry and job restructuring.

While the issue of the occupational hazards for janitors has remained an important subtext with respect to industry restructuring, environmental issues associated with chemical cleaning products grew significantly during the 1970s and 1980s. Concerns about chemical exposures in cleaning compounds were most identified with two other arenas distinct from the question of occupational exposures. These have involved household and consumer exposures and indoor air quality concerns. The focus on household chemicals, such as cleaning products and lawn pesticides, has touched on key elements of environmental advocacy since Earth Day 1970—the search for a green consumerism and the concept of individual responsibility in contributing to, and resolving, environmental problems. A cottage industry of publications, charts, magazine tips, and other popular sources of information has identified the hazardous nature of many common household chemicals and the search for alternative products, which have included such traditional cleaning agents as vinegar and baking soda. When the organization Green Seal developed its environmental labeling and certification program for cleaning chemicals—the first of its standards to be developed in the early 1990s—it was established for household cleaners rather than for cleaning agents used for commercial buildings.[20] The types of exposures and environmental concerns associated with the use of household products were not an insignificant issue, such as the problem of stormwater contamination due to outdoor household pesticide use. But the products themselves, particularly the household cleaners, did not pose as significant a problem of exposure as those associated with the use of commercial cleaning chemicals or industrial cleaning solvents. But as a *residential* as opposed to *occupational* concern, household cleaners as distinct from commercial cleaners became a major area of activity for a wide range of environmental groups precisely because of the appeal of consumer action and the focus on individual responsibility. Invariably, publications that spoke of "fifty things to do to save the planet" identified a shift to less toxic household cleaners as high on the list for opportunities for environmental change. As those appeals intensified, a number of consumer product companies sought to shift their own marketing strategies to incorporate "green" messages about their products, even when such changes in product ingredients were minimal or had not taken place. As a consequence, the focus of consumer environmental action shifted even further to include the growing concern about inaccurate or phony "green marketing" for consumer products rather than a focus on how to change the design and manufacture of those products.[21]

The issues associated with indoor air quality, including those in commercial buildings, also emerged as an environmental concern in the context of individual exposures. Buildings that had poor ventilation, that had carpeting that created exposures from off-gassing, or were cleaned with products containing hazardous cleaning chemicals, significantly increased the exposures for the people who worked in those buildings as well as for those who cleaned them. Concerns about "sick building syndrome" due to the nature of the indoor air quality of the building itself as well as how it was cleaned, and chemical sensitivity problems for building occupants, created a new focus of environmental action in relation to building occupants. Building owners, similar to the consumer product companies, became more focused on negative feedback from occupants due to exposure problems. This, in turn, led to increased attention by the building owners about their potential regulatory and liability vulnerabilities as well.[22]

Though the focus on household cleaning products and indoor air quality does not provide a direct occupational focus, it has nevertheless created a focus on the use of chemicals for cleaning purposes, including for commercial buildings. Ingredients such as phosphoric acid, sodium nitrilotriacetate, ethylene oxide, methylene chloride, various phenols and glycol ethers such as 2-butoxyethanol, and hydrochloric acid are the kinds of hazardous substances that can now be found in cleaning chemicals. Hazardous chemical lists, identified by agencies like the Agency for Toxic Substances and Disease Registery (ATSDR) or through legislative action (the Toxics Release Inventory) or due to public action (California's Proposition 65) have also become more widespread. But the most serious occupational and environmental hazards for such cleaning product ingredients or solvents as glycol ethers, toluene, or trichloroethylene have been primarily associated with their use in industrial applications. While several of these same hazardous chemicals have been used in the cleaning of commercial buildings, office buildings have not required the same type of cleaning applications such as for degreasing. Nevertheless, the hazardous nature of chemicals found in commercial building cleaning products are as much the rule as the exception, particularly for such products as heavy-duty floor strippers, bathroom cleaners and disinfectants, and various insecticides and pesticides.[23]

A case study of four office buildings in southern California by our Pollution Prevention Center, which evaluated the most common chemical ingredients in the cleaning products used at each of the buildings, underlined this issue. The most common ingredients at these sites were various glycol ethers, compounds that have also become targets for community and workplace action in relation to their use in industrial applications. In the PPERC study, adverse health effects were also identified for several of the other chemicals used, including those associated with eight different chemicals listed as carcinogens and a number of hazardous compounds with significant adverse impacts such as allergens or dermatitis-causing agents. For the glycol ether products, health impacts such as reproductive toxicity (associated with ethylene glycol methyl ether and ethylene glycol ethyl ether) were associated with the listed ingredients. The most widely used compound, EGBE (or ethylene glycol butyl ether), was contained in 22 percent of the products used, and has been identified as acutely toxic and suspected as carcinogenic. Moreover, these buildings, which included both a commercial high rise as well as a government office building, did not include any operations or uses that would have presented any different cleaning requirements than those commonly found in most office buildings.[24]

The PPERC office building study also pointed to the dearth of information about the nature of the potential hazards and the means to address those hazards. The environmental and occupational hazards associated with the commercial building cleaning chemicals are, to begin with, poorly regulated and not well understood, particularly in the degree of exposure as well as the question of multiple exposures from various chemical ingredients, sometimes from the same product. Janitorial cleaning companies place a low priority on environmental health and safety expenditures, including for compliance where regulations do exist, since, as one industry consultant put it, "there is no obvious return on the investment." Part of the difficulty in addressing the issue of hazards for janitorial cleaning products is the nature of the information available for such products. Information tools, such as the material safety data sheets, or MSDS, required by the Hazard Communication Standard, potentially represent a source of evaluation and guidance in the development of an environmental approach across the different segments of the janitorial cleaning industry. But how the MSDS are used as well as the nature of the information available underlines the difficulty in constructing an environmental and occupational health approach.[25]

Mandated requirements for MSDS were first established through the federal Hazard Communication Standard that was promulgated on November 25, 1983.[26] The standard was associated with the provisions in the 1970 Occupational Safety and Health Act, which called for "appropriate forms of warning as are necessary to insure that employees are apprised of all hazards to which they are exposed."[27] However, OSHA was slow in developing and implementing specific regulations governing hazardous materials, despite evidence gathered in the late 1970s from the National Occupational Hazard Survey. This survey indicated that as many as one in four workers was exposed to chemicals identified as hazardous by the National Institute for Occupational Safety and Health (NIOSH). After the release of the survey, the Secretary of Labor, Eula Bingham, argued forcefully that a new labeling standard was needed to insure adequate information to workers about exposures. The push for a standard, especially one that could be designed to make information available to workers and employers alike about the chemical hazards and risks associated with specific products used in workplace settings, was also influenced by the national debate then taking place concerning worker and community right-to-know issues. A number of localities and states had instituted, during the late 1970s and early 1980s, right-to-know ordinances designed to increase community and worker knowledge of the risks and hazards they were subject to. In fact, by the mid-1980s, twenty-nine states had established right-to-know ordinances; of these laws, twenty-five addressed worker right-to-know provisions.[28]

The 1983 Hazard Communication Standard, despite early opposition from the Reagan administration, was developed in part to address this growing right-to-know movement promoting the concept of providing risk and hazard information for both workers and communities. As originally designed, the Standard covered only certain industries and included provisions for labels and other forms of warning, information and training associated with such warnings, and material data safety sheets for each hazardous chemical produced or imported by the manufacturer. It also had specific provisions to insure worker access to such information and employer programs to address the problems associated with the use of hazardous materials. The Hazard Communication Standard was subsequently expanded under court order to include all businesses, including those associated with the manufacture or use of janitorial cleaning products. Whether to include janitorial cleaning products, in fact, had been an issue of some contention that was ultimately resolved in favor of the more inclusive reporting requirements.[29]

However, most information tools, including but not limited to MSDS, have not become specifically associated with producing an environmental or pollution prevention outcome. For example, the Toxics Release Inventory or TRI, established through Section 313 of the Emergency Planning and Community Right-to-Know Act of 1986, can be considered the most popular environmental information tool currently available. The TRI, however, has not been directly associated with pollution prevention-related process or product changes at TRI reporting facilities, even as the TRI has at times functioned as an incentive to bring about such changes, given its high public profile. Similarly, requirements for MSDS (the closest occupational equivalent to the TRI) were developed in part with the same intent to establish a broad information tool for workers. MSDS information about chemical hazards and safety requirements, it was hoped, could establish a source of data that could in turn influence training requirements and procedures and the types of products used as a result of worker and public knowledge. It was also hoped that the TRI information could become a factor for the downstream chemical user (in the case of cleaning products, the supplier, service company, building owner, or cleaning contractor) who might seek different kinds of products on the basis of such information.[30]

However, unlike the TRI, the MSDS never achieved the interest and visibility that the TRI acquired. Without that visibility and use among those affected, MSDS have played only a limited role as information tool and source of influence in the decision making about cleaning products. Problems with the MSDS have been multiple. They include the lack of clear and understandable information, compounded by the failure to translate for workers who would require such translation (particularly relevant for the janitorial workforce as it changed during the 1980s and 1990s). Nor is the availability of MSDS information associated with any type of intervention, whether in terms of worker training or limited product use. And while the TRI became, despite its limitations, an effective tool for community environmental advocacy, particularly in developing greater public visibility about the hazards identified, the MSDS had almost no role in toxics use and occupational health advocacy. At most, MSDS became the occasional tool of unions and other worker advocates who decided to focus on the health and safety implications of specific cleaning product hazards at a particular location.[31]

Unlike the release of TRI information, which has generated significant interest and activity, MSDS became a largely underused and often unknown source of information and opportunity for seeking to change the way in which products have been used. But despite the limited role of such information tools, and the parallel absence of any significant form of regulatory intervention, the issue of chemical hazards and the search for alternatives has still resided at the edges of the industry.

Identifying Alternatives

At the June 1996 annual meeting in San Francisco of the Cleaning Management Institute and the trade publication *Cleaning and Maintenance Management*, a workshop was held on the topic "Selecting Environmentally-Friendly Cleaning Products." Attendance at the workshop was small, and nearly all of the audience represented public agencies, such as the building maintenance supervisors from the Los Angeles

Unified School District. As members of the audience pointed out during the discussion, their concerns about possible chemical hazards associated with janitorial cleaning work was partly in response to broader public concerns about chemical hazards, particularly hazardous household products, rather than industry-specific pressures from regulators, workers, or even clients.

In contrast to this session, there was a packed audience for a subsequent workshop on the topic "Managing in a Downsized Environment." This session focused primarily on how to insure minimal absenteeism. However, absences from work were not considered by the workshop participants to be in any way a function of the workplace environment, such as acute health-related impacts from chemical use or other work-related issues. Instead, both the speaker and the workshop participants focused their attention on how to prevent workers from abusing company sick-time policies. Concerns about absenteeism, expressed by the building managers, contractors, in-house employers, and others in attendance, could also be seen as reflecting the dominant industry focus on the squeeze on labor costs, an approach that has also sought to maximize worker productivity and chemical use efficiencies.[32]

Despite this separate focus on labor costs and worker productivity (as distinct from occupational and environmental concerns), environmental issues have nevertheless begun to emerge as an important, even if inconclusive, area of attention among janitorial cleaning product manufacturers, distributors and suppliers, cleaning contractors, building owners, and in-house employers alike. It has also emerged as an issue for unions, such as the Service Employees International Union and the Hotel and Restaurant Employees Union. This is reflected in the growing number of articles regarding chemical hazards in industry trade publications concerning such issues as indoor air quality, in the wide range of efforts to develop occupational health and environmental criteria in the selection of cleaning products, and in union newsletters. There is, moreover, a growing interest in identifying ways to establish a greater environmental profile among industry participants, partly by demonstrating a willingness to explore environmental approaches, including pollution prevention. Yet there continue to be barriers for adopting environmentally preferable cleaning products, including the need for greater information about how to evaluate product toxicity, and the prevailing bias against using products that might be less hazardous (based primarily on concerns that cleaning and labor costs will go up). In addition, there is a widespread assumption among different industry players that the workers themselves—in response to pressures to work more quickly and efficiently—will be unresponsive to a shift to less hazardous products. Such "green products," many of the contractors, suppliers and distributors, and building owners have decided, will simply not work as well. Thus, the productivity factor (that is, the need to cut the costs of cleaning per square foot) will continue to influence the work routines dictated by the employer search for lower wages and speed up of the cleaning process.[33]

Despite this limited interest or willingness to establish an environmental framework for product selection among the suppliers and contractors, environmental concerns among the building owners, such as in the Patagonia case, have begun to be perceived as a growing factor in the selection of janitorial contractors and cleaning products. Trade publications such as *Cleaning and Maintenance Management* have devoted increasing attention to the green products field. They have also covered environmental regulatory issues and the problems associated with chemical exposure, such as indoor air pollution, worker absenteeism due to occupational exposures, and building occupant feedback on such issues as odor (chemical smell) and acute health risks (such as skin irritation). Pressure for change has been particularly associated with indoor air pollution issues, given that products are likely to be used—and exposures can occur—in enclosed spaces, often with little or no ventilation. Many of the pressures, as one chemical industry publication noted, have been largely driven by the concerns of the clients—the building owners—who are "increasingly demanding cleaner, greener products, free from the taint of certain chemical regulatory lists."[34]

As these pressures for change have emerged, there have been a variety of initiatives to develop environmentally oriented criteria for product selection and use. One of the first involved a joint effort by the EPA and the General Services Administration to establish "guidance for including environmental attributes in purchasing decisions." The GSA initiative had emerged out of a series of discussions between the GSA and EPA about how to identify a method for the purchase by federal agencies of environmentally preferable cleaning products for the buildings owned or leased by the federal government. Although those discussions predated Clinton's election in 1992, they intensified after the election due to the administration's interest in using the federal government itself as a staging ground for environmental purchasing initiatives.[35]

Toward that end, on October 20, 1993, President Clinton issued Executive Order 12873, "Federal Acquisition, Recycling and Waste Prevention." The order, which was designed to make federal agencies accountable to the waste management and prevention measures in the Resource Conservation and Recovery Act, also sought to demonstrate that the federal government could be an "enlightened, environmentally conscious and concerned consumer." Defining "environmentally preferable" as "products or services that have a lesser or reduced effect on human health and the environment," the president mandated that the EPA guide various government entities to help them decide to purchase such environmentally preferable products. This "environmental preference" purchasing order became part of an overall EPA-based Environmentally Preferable Products (EPP) Program designed as a nonregulatory "guidance" in the effort to facilitate the purchase of environmentally preferable products and services.[36]

During this same period, both the EPA and GSA began to respond to Vice President Al Gore's "reinventing government" initiative. Introduced during Clinton's first term, the initiative had included a directive to the GSA to develop the appropriate laboratories as part of the reinvention process. However, the change in Congress after the 1994 elections slowed down the process of "reinvention," and created some caution about establishing criteria for an environmentally based purchasing program. Nevertheless, the GSA and EPA, which had joined forces to develop baseline criteria for the purchasing executive order, sought to answer the question of whether environmentally preferable products could perform adequately. A pilot project was developed to evaluate nineteen cleaning products considered environmentally preferable to be used to clean a courthouse in Philadelphia. A subsequent written and verbal survey of the workers regarding the performance of these products provided enough preliminary information for the EPA staff to at least begin to identify such preferences through a Product Selection Guidance.[37]

By 1996, the GSA-EPA effort had led to the development of a ranking system for seven specific "environmental attributes" for various cleaning products like floor finishes and carpet care products. Missing among the attributes, however, were such key issues as chronic toxicity as well as other significant health impacts. These omissions were due in part to the limited information available about the identity and concentration of product ingredients. This problem was compounded by the fact that cleaning product manufacturers, who were opposed to the initiative, would not provide such information on the basis of the "confidential business information" provisions in the Toxic Substances Control Act.[38]

While the federal government agencies faced difficulties in establishing environmental guidelines for purchases of cleaning products, local and statewide governments also developed their own initiatives during the 1990s for selecting vendors and cleaning products. The state and local governments were influenced in part by the 1992 Agenda 21 program at the Earth Summit in Rio de Janeiro and subsequent efforts to establish "sustainable city" and state pollution prevention programs in the United States. At the same time, environmental groups and coalitions, such as the Washington Toxics Coalition in Seattle, the Community Environmental Council in Santa Barbara, California, and the Environmental Health Coalition in San Diego also sought to identify pollution prevention programs and purchasing strategies for local and state government entities. Several of these initiatives went beyond the GSA-EPA programs to identify criteria regarding product toxicity and potential health hazards. While occupational health issues didn't factor directly into these efforts, they nevertheless were viewed by industry associations, such as the Chemical Specialties Manufacturers Association and the Soaps and Detergents Association, as potentially undercutting the market for their chemical cleaning products.[39]

One of the first such initiatives was developed by the state of Massachusetts, which created environmentally preferable purchasing criteria through the state's operational services division. These included a series of product ingredient specifications that disqualified any product or ingredient that was listed as a carcinogen, ozone-depleting substance, phosphate, or volatile organic compound. One key list of such hazardous ingredients had been created through the state's Toxics Use Reduction Act (TURA), one of the first state pollution prevention acts that had been supported by community, public interest, and mainstream environmental groups. The Massachusetts cleaning products program had also been aided by the technical evaluations provided by the Toxics Use Reduction Institute (TURI), the research program at the University of Massachusetts at Lowell that had been established through the TURA legislation. TURI's work in testing and evaluating alternative cleaning systems, and the state's program to encourage the use of alternatives, was also attacked by the state's chemical industry trade association (the Massachusetts Chemical Technology Alliance). The industry groups instead promoted the concept of using existing chemicals "in a way that's safe and that minimizes risk."[40]

Other state and local initiatives, such as those in Minnesota, Texas, California, and Florida, were developed in partnership with community groups and environmental advocates. In Minnesota, a 1996 pilot project conducted by a nonprofit coalition of community organizations was designed to evaluate how to reduce waste and toxics use at the Saint Paul City Annex, including the substitution of nontoxic cleaning products. The success of the pilot program led to the development of a scoring system established by the state that assigned a value to specific environmental characteristics of a product, with an overall rating based on the total number of points identified. City programs also sought to establish criteria for product selection, with several established in conjunction with community and environmental group input or as direct outcomes of efforts to establish a pollution prevention program.[41]

Opposition to the programs from the chemical industry trade organizations became increasingly pronounced and ultimately resulted in the modification or elimination of several initiatives, including those in Texas and in Oakland, California. One such program involved efforts to establish a standard for cleaning commercial and institutional buildings through the American Society for Testing and Materials (ASTM), perhaps the most significant of the voluntary standard–setting bodies associated with industry activities. ASTM had been organized at the turn of the twentieth century to establish a framework for developing standards for materials based on a "consensus" process, particularly between producers and users of such materials. During the 1970s and 1980s, the ASTM, which came to be criticized by environmental and community groups as industry dominated and for seeking to divert and ultimately substitute its pronouncements for government-issued standards for particular products, sought to add consumer, occupational, and environmental components to its voluntary-consensus standard-setting process. A new ASTM "environmental assessment" committee was formed (and in turn established subcommittees to address such issues as storage tanks, real estate commercial transactions, and the review of environmental regulations), partly in response to the increased pressure of regulation and from community action. In that context, an ASTM effort, through its "pollution prevention and environmental stewardship" subcommittee, developed a process to address the construction and design or "greening" of commercial and institutional buildings, including the operations and maintenance of such buildings.[42]

This ASTM process was headed by Steve Ashkin, at the time a vice president of the Rochester Midland Corporation, a chemical manufacturing firm that had established an office of environmental technologies headed by Ashkin and that had also developed a green products line for cleaning chemicals. During the mid-1990s, Ashkin had become a leading advocate for a stronger environmental profile among cleaning product manufacturers and suppliers. Ashkin had also actively sought to open communications with community and environmental advocates as well as state and local officials who had been at the forefront of the development of environmentally based purchasing criteria for cleaning products and services. Ashkin hoped to develop a successful consensus process for the ASTM committee, which would include the chemical industry associations; employee unions; federal, state, and local government representatives involved in the issue; academic researchers (including our center); and community and environmental advocates. At the same time, Ashkin had become the most visible industry figure advancing the cause of environmentally preferable cleaning products, a process that Ashkin saw as "rooted in the ethic of pollution prevention."[43]

The ASTM subcommittee task force established for the consensus process did involve many of the players that Ashkin had hoped would participate, and some of the early discussions involved lively exchanges between committee participants of the value and nature of such a standard. At the beginning of the process, Ashkin had included in a first draft several of the existing environmental standards and attributes, including the EPA-GSA program and an innovative project being developed by the city of Santa Monica in California. From that first draft, debates among the participants began to occur. This included whether the term "environmentally preferable cleaners" should be highlighted or even used, as opposed to a more process-oriented notion of "stewardship," which would focus less on the criteria to evaluate products and more on such issues as worker training. "In some ways," Ashkin recalled, "we were facing what I would call the 'Rodney Dangerfield complex'; namely, 'anybody can be a janitor,' or 'you can be pretty dumb to do that type of work.' It was essential for us to overcome that bias." Similarly, there were distinctions concerning the issue of worker involvement in the process of planning and implementing a cleaning program. As the discussions about crafting a standard continued, several of the community, environmental, and union participants either abandoned the effort or remained largely absent from the deliberations. This was partly due to frustration with the increasingly dominant role of the chemical industry organizations as well as the problem of inadequate resources to continue participation.[44]

The key to the development of the standard was how to address the question of criteria for purchasing environmentally preferable cleaning products. At one point, language was introduced that criticized the use of the existing environmental attributes and "environmental preferability" criteria as "flawed," stating that it was "impossible to establish overall environmental preferability." Moreover, any references to "environmentally preferable purchasing" and even reference to EPA's own "guiding principles" document were deleted to insure that the ASTM standard would not lend credibility to a federal EPP procurement program that had been based on pilot projects. Although, as the chemical industry participants had urged, there was emphasis on worker training, the concept of worker participation in decision making as a necessary component of "stewardship" was also eliminated. The final result was an ASTM standard initially designed to delineate a set of approaches for addressing environmental impacts from cleaning products and processes that had instead become a guide for a broadly framed "stewardship plan." This stewardship standard in turn had eliminated any of the language specifically identifying why and how to switch products and change processes. Ashkin, however, saw important advances in the development of the standard, including its use by some union locals in contract language and in relation to the growing interest of the building owners and office occupants regarding environmental and health-related impacts. But the ASTM process also indicated that the ability to bring about environmental change in the use of cleaning products continued to be a moving target.[45]

Worker Participation and Environmental Change: The Santa Monica Experience

In 1992, two of my students undertook a pollution prevention audit of hazardous materials, including janitorial cleaning products, used by City of Santa Monica departments and facilities. The study identified a wide range of hazardous materials and products used by the city that represented possible occupational as well as environmental hazards. The study's findings, in turn, helped bolster the case for a municipal "toxics use reduction" program in relation to city activities. To implement that goal, the city decided to commit to a program for identifying environmental criteria in the selection of janitorial cleaning products. In the course of instituting such a process, the city also began to explore the link between environmental issues associated with the selection of the cleaning products with how those products were to be used.[46]

In crafting this program, the city's environmental staff relied on Santa Monica's Sustainable Cities Program, which had been adopted by the city council in September 1994. The program required the city government to incorporate environmental objectives for its own governmental operations or the services it provided the community. These ranged from water conservation goals to reduced hazardous materials use, as well as broader directives to develop alternative strategies that could sustain and benefit the community and local economy. Armed with our center's report and the guidelines provided by the Sustainable City Program, particularly those related to hazardous materials use, the city's Environmental Programs Division staff were given the green light to establish a pilot program.[47]

Two core objectives were identified: establish environmental criteria for purchasing cleaning products and supplies, and collaborate with the janitorial workforce to help in the selection and use of any alternative products introduced as a result of this shift in purchasing. The city had a custodial staff of about sixty employees, including older, longer-term employees as well as younger staff, several of whom had worked for only a short period of time. The workers belonged to an in-house employees association and wages were above the industry norm. The cleaning products that had previously been used by the city included product ingredients associated with both acute and chronic health concerns.[48]

As a first step, the city contracted with S.A.F.E. Consulting, an environmentally oriented group of former cleaning workers from Moran, Wyoming. The SAFE group, which immediately established a rapport with the city employees, was hired to help assess existing cleaning product use and identify opportunities for testing alternative products. The group also facilitated a series of intense discussions, focus groups, and eventually workshops with the janitorial staff. All of the city's existing products, identified as potentially hazardous by the city's environmental programs staff and the PPERC audit, were reviewed. These sessions were at times explosive, as the janitors recounted a number of episodes when they had experienced various acute health effects, such as breathing problems and skin irritation. Toxicity issues were also raised. Although the workers knew about the Material Safety Data Sheets provided for each product in use, they had little ability to translate or use such information in the context of how their jobs were defined.[49]

Though these sessions were helpful in identifying problems with various products, suspicions among the workers still remained. Why was this process being initiated? Would the janitors really play a role in deciding what products were to be used? Just what did it mean to be a "stakeholder"? Ultimately, a decision was made to allow the workers to identify the products of concern and evaluate the alternative products on the basis of how well they performed. The workers would also participate in the discussions regarding what standards of cleanliness were to be upheld and whether the work routines needed to be changed.

During this six months' test period, custodial staff were able to identify which of the alternative products performed successfully and which of the products could not accomplish the cleaning tasks required. The workers also identified which of their tasks might need to be reviewed in relation to overall performance standards. New training procedures were also discussed and implemented. Health concerns were included in the feedback sessions, including, for example, odor issues associated with some of the citrus-based terpene products that had been used as substitutes.[50]

In September 1994, the city established its new bidding process. Bidders included some of the largest product manufacturers such as 3M, as well as a number of small firms that exclusively identified themselves as "green product" companies. A number of firms, including some of the "green product" companies, were immediately disqualified for not providing information (e.g., on VOC content) or failing to meet the pass/fail criteria. One company even submitted MSDS information with product names that did not match the names of the products submitted. For the city, the process was elaborate and time consuming but ultimately considered worthwhile in obtaining far more information than any of the other existing environmental-oriented purchasing programs and providing a first step in seeing if this type of program could work.[51]

The key remained the implementation process itself, including how the janitors themselves would respond. During the test period, the consultants to the program had provided product recommendations and information about the quantities of the new product needed for a particular application. Once products were introduced, they also provided assistance with storage configurations and product dilution and gave on-site demonstrations of product use. In addition, workers were given notebooks containing specific instructions for product use and product substitution. The janitors later indicated in a focus group session with our center that the training provided by the consultants and staff made them feel as if they were starting a new job.[52]

Through their training, the custodians had also received information on health and environmental effects of each of the products involved— both the former products, and, after the new products were selected, for the new products as well. As it turned out, information that could be provided clearly, intelligibly, and on a continuous basis provided an important motivation for the janitors to commit to the new products. At the same time, the possible damaging health effects of the old products also emerged as a topic that most of the workers felt they could now openly discuss. In the process, several of the janitors indicated that they had crossed what could be considered a pollution prevention threshold. The more familiar they became with the new products, the more the janitors began to feel a sense of control and ownership of the cleaning process. "Those toxic products would burn your ass," one worker commented at the focus group session, while others pointed to their fewer headaches and less nausea. Their own comfort level in being able to acclimate to more hazardous products—the type of conditioning that caused dry cleaners to lose any sense of smell the more they worked with perc— had changed. One of the janitors commented at one point how he had entered a building outside the city where cleaning products were being used and the odor from the products had caused him to "feel sick to my stomach."[53]

In preparation for a second round of bids, the city significantly streamlined its process, although maintaining core categories and establishing what amounted to a more generic set of "pass/fail" criteria, including acute toxicity, chronic toxicity, biodegradability, and air quality. However, the issue of janitor participation and feedback, the city staff realized, had also become central to the environmental purchasing process itself. In one interesting episode that occurred after the second round of bids went out, representatives from the 3M company complained that they wouldn't bid in the new round. Among their issues was the lack of preference for an automated delivery system (e.g., premix, ready-to-use) that 3M had developed and touted as "environmentally preferable" since, the company claimed, it lowered exposures. The company contended that the mixing process itself increased exposures because janitors frequently used improper ratios; they would dilute less because they didn't understand the process or because they wanted to use stronger solutions. "We had a fundamental difference that had to do with the role of the janitors in this bid process," Santa Monica staff member Debbie Raphael recalled. 3M had presumed what Raphael characterized as a "low IQ factor" and Steve Ashkin had called the "Rodney Dangerfield complex." Even though janitors would be using more concentrated and therefore more hazardous product ingredients, there would be, 3M claimed, lower exposures overall since the janitors were not required to do any diluting or evaluation of appropriate dilution levels. "We operated on a different set of assumptions," Raphael recalled, "since we preferred that the janitors establish a greater awareness of the products they were using."[54]

Five years after the program had been introduced, the Santa Monica initiative in purchasing and implementation had become a model for local and state agencies alike. The two key figures directing the city's efforts, Brian Johnson and Deborah Raphael, would soon become major participants in the national debates regarding environmentally preferable cleaning products. Johnson served for a time as vice chair of the ASTM subcommittee addressing this issue, and both Johnson and Raphael became a primary source of information for other developing initiatives at the local, state, and federal government level regarding purchasing criteria. The Santa Monica model also pointed to the significance of worker training but in a context of worker participation. It also became clear, after the second round, that without continued feedback and retraining (e.g., annual or biannual review and evaluation sessions), the tendency to revert to traditional practices remained a possibility. While the performance factor—how the products cleaned—would be central to the evaluation of alternatives, the issues about "practices"—how the products were used—remained a crucial, albeit less tangible, area to establish both criteria and changes in the nature of the work itself.[55]

In judging the overall performance of the alternative products, the janitors had indicated, based on their experience, that a period of time was needed before a comfort level could be achieved with product use. "We were comfortable with the old products, that's what we were used to," one of the workers stated at the focus group session. Problems with purchasing through the city's warehouse also reinforced some of the earlier patterns. At the same time, it was also clear that the productivity factor had to be redefined: new products required more application time, more training and information sharing, more feedback and interaction. But after the second generation of alternative products was introduced, product performance continued to improve, and worker satisfaction with the substitution process improved as well. The productivity of the workers had decreased if judged by the amount of labor required for *participation* but had increased if defined by the outcomes associated with product performance, job satisfaction, and the desire to succeed in performing the tasks commonly agreed upon.

Toward the end of our focus group session, one of the custodians asked in a slightly sardonic tone, "How many times have I been told I'm a pioneer?" It was clear in the session, however, that he and other staffers were proud of the "pioneer" label. Along those lines, the Santa Monica Environmental Programs Division presented the custodial staff with plaques containing the cover of a *Maintenance Supplies Magazine* feature article on the City of Santa Monica Janitorial Services. It was the first time these workers had been formally recognized by their employer for a successful model, even if it still remained outside the norm of industry practices and employer-employee relationships.[56]

The Dignity of Janitors: Union Struggles and Worker Co-ops

On one level, the Santa Monica experience could be characterized as an experiment in promoting worker dignity. The association of the concept of "dignity" and "janitors," however, was more directly elaborated in the emerging campaigns for "justice" that also became a central rallying cry for a new type of labor movement. One of the most promising and dynamic initiatives within the labor movement during the past two decades in fact has been the "Justice for Janitors" organizing campaign of the Service Employees International Union (SEIU). Launched in the mid-1980s, Justice for Janitors was designed to stem the loss of union contracts and membership that had become endemic within the labor movement as a whole. For the SEIU and its constituency of janitors, these issues were also compounded by the restructuring of the industry itself. This included the shift to contractors, the fluctuations in the office building market and the push for labor-related cost-cutting strategies, the growth of the commercial building industry in the South and in the suburbs, and the sudden surplus of unskilled labor in certain regions associated with the immigration of Central Americans and others. "While their brothers and sisters in manufacturing agonized over concessions versus plant closings," one SEIU official commented on the need to identify new strategies, "building service locals faced the nightmare of seeing union jobs converted into nonunion jobs right before their eyes."[57]

Understanding that a building-by-building organizing strategy was limited, if not impossible in an era of contracting, the SEIU sought to construct a strategy that was focused on speaking for the entire community of janitors (including nonunionized workers). At the same time, it hoped to mobilize politically on behalf of this newly energized social movement. While still focused on the issue of obtaining contracts and increasing membership, Justice for Janitors, unlike some of its industrial union counterparts, decided to identify a new type of appeal about worker dignity. This was particularly true in major urban centers like Los Angeles with its young, largely female, and immigrant (predominantly Latino) workforce. The union organizing strategies were infused with some of the newer campaign tactics that focused on this need for community support and on the ability to utilize "corporate campaign" tactics. J4J, as Justice for Janitors was also known, was effective at using creative and at times outlandish tactics. Such tactics included disrupting traffic in front of targeted buildings. It involved guerilla theater-type demonstrations such as parading a giant toothbrush in front of the building of a Philadelphia company that had required its janitors to use—and purchase their own—toothbrushes to clean toilet bowls. Similar tactics were also used to focus on broader social and economic themes, such as begging for food in a building owner's neighborhood to dramatize the lack of a living wage. Through these tactics, Justice for Janitors sought to introduce the issues of social justice and worker dignity related to the conditions of work, the lack of living wages, and the treatment of workers as part of its overall argument about the need for union representation.[58]

The development of this approach in the mid- and late 1980s was essentially a renewed effort to identify opportunities for increased membership and, where possible, industrywide "master contracts." The J4J campaigns primarily concentrated on large urban areas like New York, Chicago, Los Angeles, and Pittsburgh, which had the largest employers and where building owners were assumed to be most vulnerable to the union's tactics. The organizing drives were also most successful in constructing a movement among the immigrant workers (many of whom had had experience with labor organizing and social movements in their home countries), in addition to enlisting activists and immigrant communities to the campaign's cause. Parallel "living wage" campaigns were also organized in places like Los Angeles, in part as an extension of the union's strategy to ensure that union and non-union building owners or contractors alike would need to pay the equivalent or near equivalent of the higher union wages. This in turn would preclude a two-tier structure of wages and contracts that could undermine the Justice for Janitors initiative as well as the parallel campaigns that had been developed by other low-wage sector unions, such as the Hotel and Restaurant Employees Union.[59]

The success of the Justice for Janitors campaign was reflected in the increase in union representation in several cities during the 1990s. This culminated in the dramatic and successful three-week janitors' strike in Los Angeles in April 2000, whose outcome one publication characterized as "how the janitors changed Los Angeles." The success of the strike—and the campaign—was in turn based on the union's ability to mobilize its members as part of a broader community and link the organizing focus to the issue of the workers' dignity.[60]

Although the issues of wages and representation had remained dominant in the Justice for Janitors campaigns, there had been some modest interest within the union in addressing work conditions, including cleaning product hazards as well as the speed-up process that was also linked to the types of cleaning products used. But the union's core concerns in an era of industry and workplace restructuring continued to be wages

and representation. To say a janitor had dignity—and the Justice for Janitors campaign proudly displayed the term "janitor" as essential to its
campaign message—meant being paid a living wage and having a union
to back up that demand.

The question of dignity was also central to an initiative that was
launched in the late 1990s to establish a cooperative of janitors based in
Pico-Union, one of the immigrant neighborhoods of Los Angeles. This
cooperative venture, called Pueblo Nuevo Enterprises and founded by
Philip Lance, a priest from a local Episcopal church, was organized to
establish a social justice–oriented community program. Lance had been
influenced by the arguments laid out by Jed Emerson and others that
"social entrepreneurship" provided an effective tool in community and
worker "capacity building." At the same time, the Pueblo Nuevo group
wanted to provide stable jobs for those who needed the employment and
where the conditions of work could be influenced and ultimately controlled by the workers themselves.[61]

The Pueblo Nuevo cooperative divided between the worker owners
and the "management" group that included Lance and a couple of other
assistants who helped with the finances, scheduling, and outreach or
marketing of the cooperative. The co-op's board of directors included
several of the janitors themselves. To join, worker-cooperative members
paid a small fee, which was then returned after they departed. The cooperative was organized in some ways similar to a for-profit small contracting firm, with a major focus on increasing revenues and developing
large and stable contracts for cleaning buildings. Wages were higher than
most of the small contracting firms' wages and also included benefits
(which many contracting firms did not), but they were lower than for
some of the large contractors with union representation. If the cooperative returned a profit at the end of the year (as expected in 1999 after a
couple of years of modest losses made up by grant funds), then actual
"dividends" would be returned to each of the members.[62]

Ultimately, what most distinguished Pueblo Nuevo was its self-
conscious "worker-owner" identity. The janitors got to vote on the
procedures and policies associated with the work. There were also feedback and review processes established to evaluate whether a worker was
not performing. In the one case where a performance issue emerged, the

management team along with the members worked out a new "proba-tion" period, which allowed the worker to keep his job and decide whether to work "harder" or more effectively.[63]

In terms of cleaning products, there was some interest in less haz-ardous products and the group's Web site also emphasized its interest in a "clean environment." But the cooperative had not been able to estab-lish an effective environmental assessment process. A review of the MSDS of the cleaning products provided by its supplier in fact indicated similar kinds of hazardous product ingredients as those identified in our center's evaluation. As with the case of Justice for Janitors, the worker cooperative model helped establish a social justice framework, but without a clearly articulated environmental and pollution prevention framework.

A Justice/Prevention Link?

During the 1990s, the city of Santa Monica's Toxics Use Reduction Program for janitorial cleaning products evolved into a successful model in relation to product substitution (using what could be considered pol-lution prevention criteria). But the program's success resided also in how the program was adopted and implemented (using what can be consid-ered environmental justice criteria). At the same time, the Justice for Jan-itors campaign and the Pueblo Nuevo worker cooperative model provided direction for renewing a social justice agenda. Both approaches enabled workers in a low-wage sector, subject to increasingly untenable outcomes associated with the restructuring of the work, to develop more security and dignity related to the job. Each of the models in turn estab-lished different starting points: environmental objectives; union organiz-ing objectives; community and job development objectives. Each of these approaches are significantly embedded in their own distinctive language and discourse. Environmental officials and activists focus on the lan-guage of risk and procurement standards even as new methods are explored to bring about environmental change, including the role of workers in bringing about such change. Union organizers talk the lan-guage of contracts and campaigns, of wage agreements and membership, even as they focus on the conditions of work, conditions that can include the occupational and environmental hazards that workers constantly encounter. And "social entrepreneurs" and community development activists focus on the survival of their small businesses, and talk of job creation strategies and worker and community empowerment goals, even as they seek to identify the implications of such new workplace and community relationships.

Environmental change and social change; the social and the ecological; community, workplace and environment—each represents aspects of discourse and points to new forms of action that should be connected but often are not. But are there arenas where these concepts can become more interchangeable, and where a broader, more integrated discourse and guide to action can emerge?

5

Global, Local, and Food Insecure:
The Restructuring of the Food System

Seeds of Change

In the spring of 1992, a group of my students prepared to launch a year-long, environmental justice–related research project. The students were most interested in identifying how environmental issues connected to social and economic concerns in low-income neighborhoods and how an environmental justice approach effectively addressed community needs. The discussions at first seemed rather abstract. But then the events of April 30 though May 2 in Los Angeles—the riots, or civil disorders—erupted. These events not only reoriented the students' discussion but made far more compelling the importance and immediacy of assessing—and addressing—community needs.

The background to the riots during the previous two decades had been the restructuring and eventual decline of the region's manufacturing industries and source of higher-paying jobs. Those industrial changes had in turn extended Los Angeles's "widening divide," as one 1989 study characterized the social and economic gap between communities and racial and ethnic groups. A similar study of "negative land uses" in East Los Angeles that same year provided a local perspective regarding the burgeoning field of environmental risk discrimination. If the widening divides and risk burdens were becoming endemic, then the riots became both the occasion and the cautionary explanation of the powerful social, economic, and environmental insecurities that were unraveling the social fabric and further exacerbating the divisions between communities and neighborhoods within the region.[1]

Deeply affected by the riots and wanting to be more purposeful about their research, the students sought to select a set of core community issues, through a case study approach, that could identify both problem areas and opportunities for community action. For the case study, they selected a neighborhood in South Central Los Angeles that reflected the region's evolving demographics (more immigrants, increasingly Latino) and widening-divide characteristics (lack of jobs, inadequate transportation, etc.). The students then undertook a needs assessment (an evaluation of core problems and community needs) in conjunction with some community and church groups. They asked residents to identify their most urgent issues, and assumed, despite their own "environmental" orientation, that it would more likely be in the area of "community economic development" in the context of the civil disorders.

The needs assessment produced surprising results for the students. Above and beyond the need for jobs or even housing or transportation, food issues were identified by many of the residents as their most immediate and widespread concern. This included problems of food access, food quality, and food price. At the same time, the residents and community groups surveyed revealed a strong interest in developing alternative approaches, such as a neighborhood farmers' market, a community or school garden, or a new community-oriented supermarket for the area. These were approaches that potentially encompassed a set of alternative strategies for how a neighborhood or a community could meet its food-related needs. What the students were discovering about this significant interest in food issues also paralleled what other needs assessment surveys of low-income communities were indicating in Los Angeles and elsewhere.[2]

The year-long research undertaking, which culminated in the publication of the report, "Seeds of Change: Strategies for Food Security for the Inner City," provided three key insights. The food issues identified in the students' South Central Los Angeles case study area highlighted the study's overarching concept of *community food security*. Those issues were then placed in the context of a *food systems analysis* that could identify the structures and outcomes related to how food is grown, processed and manufactured, distributed, marketed, and sold. These two ideas or approaches—community food security and food systems analysis—in turn had powerful environmental implications. In relation to the interests of the students (and the arguments in this book), they also helped identify ways in which questions of community economic development and the environment, or the social and ecological, could be joined.[3]

While community food security referred to the problems and possibilities for action at the regional, local, or neighborhood scale, food systems analysis identified the environmental, economic, and macro- as well as micro-related issues and policies that shaped the way those problems were experienced. Both community food security as an action strategy and food systems analysis as a conceptual framework pointed to a more dynamic environmental approach, integrating issues around land use, sustainability, production, and community life. Indeed each of these areas—community food security and food systems analysis, and their implications for a broadened environmental agenda—could also be seen as potentially contributing to the development of a new kind of environmental discourse.

The concept of food security had first emerged in the 1970s and early 1980s in the international development field. In the Third World setting, the ability to achieve food security was often used interchangeably with the need for hunger intervention in communities and nations experiencing high poverty rates. But *community* food security emerged as a different type of concept, both in the Third World context, and, increasingly during the late 1980s and early 1990s, in the U.S. context as well. Community food security, for one, was primarily community rather than individually focused, the way most hunger intervention programs were oriented. The initial definition of community food security elaborated in *Seeds of Change* and other documents during the early 1990s—"all persons obtaining, at all times, a culturally acceptable, nutritionally adequate diet through local, non-emergency sources"—was careful to distinguish the goal (culturally acceptable, nutritionally adequate, local sources) as well as the form (nonemergency) of intervention. As it subsequently evolved, community food security came to be seen as a strategy for community empowerment (with its focus on increased access, cultural specificity in food choice, and food self-reliance) and prevention (with its focus on dietary and nutritional considerations and sustainable food production). Community food security analysis also sought to evaluate the nature of the resources available, both community and personal (the "basket of strategies" for sustainable livelihood that development analyst Robert Chambers identified in the Third World context), as well as how such resources were to be made available. Community food security indicators could include income levels, transportation factors, availability of storage and cooking facilities, food prices, nutritional and dietary issues, and the cultural appropriateness of food choices. Community food security issues might also refer to food safety, environmental hazards, patterns of ownership, production and processing methods, food sources, and the nature of the food product itself.[4]

Such community food security indicators, in turn, can be primarily associated with food system–influenced outcomes. Achieving community food security can also be predicated on the possibility of building what Harriet Friedmann has termed *alternative food regimes*, or what has more popularly become known as community food systems. Food systems analysts have contrasted regional or local food growing and marketing arrangements or "food regimes" with what has become a dominant, long-distance, industrialized, highly concentrated, and globally reorganized system of food growing, processing, manufacturing, marketing, and selling. Food systems analysis has also explored the shift from the local to the global, where corporate restructuring has increasingly come to influence and alter the very definition of what is meant by food. And both community food security and a food systems analysis potentially provide, in environmental terms, a broader community-based focus for environmental justice, and a production-oriented or seed-to-table framework for pollution prevention.[5]

By the time the *Seeds of Change* group published its results in 1993, interest among U.S. groups in a community food security approach oriented toward building regional or community food systems had grown considerably. This was magnified by the realization that the indications of community food *insecurity* had increased almost exponentially during the 1980s and 1990s in the United States as well as in Third World regions and communities. The *Seeds of Change* study, for example, indicated that the case study area, the majority of whose residents were recent immigrants from Central America and Mexico, was significantly food insecure. Twenty-seven percent of the residents surveyed through a random sample said they experienced hunger an average of five days every month—in effect, continually dropping in and out of hunger.[6] A food price survey reinforced earlier findings that these residents did indeed pay more for food than their middle-class counterparts, even as they paid a much higher percentage of their available income on food.[7] The study also identified a lack of fresh and high-quality food. It indicated that diet and nutrition and physical health issues were prominent (obesity among school-age children had become an "epidemic" according to several studies, including one study of children attending several Los Angeles low-income schools). Children were sometimes eating not enough calories and more often calories without sufficient nutrients or with excessive fat, salt, or sugar content.[8] It pointed to food access problems (which had become even more severe with the steady abandonment of low-income communities by full-service food markets). This latter problem was especially exacerbated in those neighborhoods (such as the *Seeds of Change* study area) where average car ownership was substantially lower than regionwide averages. It also pointed to the range of negative environmental impacts—from pesticide use and pollution and waste problems, to urban sprawl and loss of farmland, and urban land use implications—associated with a changing food system.[9]

As part of the evaluation, the students traced the rise nationally of a new type of movement addressing community food security issues and the corporate restructuring of the food system. This food movement had first exploded on the scene in the late 1960s and early 1970s as part of the search for alternative lifestyles and a more radical approach to the problems of agriculture and the rural economy. The food movement subsequently shifted its attention to questions of hunger and poverty and the continuing decline of the small family farm in the Reagan era in the 1980s. By the 1990s, the food movement began to focus more directly on issues at the regional and community level. Several of these regional groups talked about sustainable food systems and community approaches to meeting food needs, and had identified a new kind of community food security politics as central to their agenda for action.

Such community food security groups are not "environmental" in the way that the term "environmental" has been commonly understood. Nor were they "environmental justice" or "pollution prevention" groups in the way environmental justice and pollution prevention issues have generally come to be defined. But in their analysis of an evolving food system, in identifying how food issues are experienced at both the regional and neighborhood scale, and in their embrace of these new movements, the students had begun to pursue their original goal: namely, identifying an environmental link to community needs. Similar to both environmental justice and pollution prevention, community food security, they argued, could be considered a new type of social and environmental change movement. Such a movement could also potentially extend boundaries. It could address questions of production as well as place, and it could seek to describe and challenge the dominant food system, the better to know how to change it.

Changing Food Systems

To describe a food system involves identifying the stages of a production system, from planting and growing, to the development of food products, and the marketing, selling, and consumption of those products. The changes in the U.S. food system (a system that has increasingly assumed a global character) have occurred throughout the twentieth century and have been especially pronounced in the last several decades. These changes have been experienced at each of those stages of production, and they have had powerful outcomes in terms of availability and access to food, and the very nature of the food product that is produced and consumed. This chapter explores those changes in production, while describing key outcomes that have influenced the way we think about food. While those changes have been systemic in significant ways, they are not, however, inexorable. The food system remains a fluid system of production, an area still potentially subject to alternative strategies of production and consumption as explored in the next chapter. And, within the universe of social and environmental action, food issues, which have come to represent a significant opportunity to construct a new type of environmental and social agenda, have also become the place where the local meets the global.

Farmers: Growing Food

The ways in which food is grown are an obvious starting point in understanding and situating food system changes. Already by the post–World War I period, growing food, at least in certain parts of the country, was becoming less of an avocation, with its Jeffersonian connection to the land. The central focus on productivity as a crucial outcome in food growing became the agricultural counterpart of the Taylorist concepts for industry. Indeed, one key productivity advocate, agricultural economist Edwin Nourse, argued in 1929 that while farming had not become a full-fledged factory operation, it was important to determine "just how far the scientific management point of view will go in agriculture."[10]

Already by the 1930s, how far farming might go was becoming, in places like California, more apparent from the changes in food growing. Led by California, the emphasis on productivity was associated with the related trends of long-distance delivery due to technical innovations such as refrigerated rail transport, more effective storage, and single-crop specialization, as well as an increasing lack of diversity in those crops. The California mode of farming was also characteristic of "industrial production." This took the form of greater mechanization, more expensive inputs, more reliance on capital, and a system dependent not just on farm families rooted in communities, but on workers who had to migrate to follow the work. By the time Carey McWilliams described this system in *Factories in the Field* in the late 1930s, the California industrial mode, with its "diminishing income for farm workers, mounting rural poverty, [and] displacement caused by mechanization," could be clearly differentiated from the family-based farm operation with its much-celebrated connection to the land.[11]

These changes were magnified in the period during and following World War II, with the rapidly expanding use of fossil fuel–based fertilizers, agricultural industrial chemicals, antibiotics for livestock production, and irrigation water (which, among other inputs, replaced land, diversity, and labor as the principal components of agricultural production). Combined with changing manufacturing, marketing, and distribution pressures, the very notion of agricultural inputs had become almost entirely an "off farm" rather than "on the land" source. Pesticide use alone increased fortyfold between 1950 and 1980, a trend that also included significant increases in application of chemicals per acre of cultivated land as well as in the overall quantities of pesticide use. The changes in production due to mechanization after World War II— changes that had led to crop specialization or mono-cropping—had also helped lay the groundwork for the rapid and quite extraordinary jump in the use of pesticides, since mono-cropping only exacerbated the problems of pests. In addition to pesticides, nitrogen-based fertilizer use also increased exponentially, as the major chemical companies such as Dow, Shell, and DuPont eagerly exploited the new agricultural markets for their huge bomb-related inventories of nitrogen that had accumulated during World War II. Even as early as the 1950s, this California model of the industrial reorganization of agriculture had emerged as the defining pathway for U.S. agriculture as a whole.[12]

The loss of farmland that began to accelerate during the 1950s and 1960s also became characteristic of the reorganization of farming as an economic activity, as the number of farms continued to decrease while the average size of farms increased. As cities sprawled, farming continued to be pushed further outward along the city's edge, and efforts to save urban fringe farmland became more associated with open space and environmental amenities than sustaining a small farm or regional farm economy. By the 1980s, farms accounted for one-fourth of the revenue generated just for the actual food grown on the farm, let alone for the revenues generated by processed or value-added food products. Moreover, one 1986 study indicated that more than 90 percent of all U.S. household purchases of food and beverages were for manufactured or processed food items as opposed to nonprocessed food items, a shift that was also exacerbated by a rapidly growing percentage of meals consumed away from home. Farming was becoming more and more a part-time occupation, particularly for small farms whose income was nominal, if nonexistent. One University of California economist even argued that farming in the United States could ultimately disappear within the context of a global food economy and the continuing pressures to convert farmland in the United States to urban development as its highest or most profitable use.[13]

The growing of food was also changing in terms of where as well as how the food was grown. An increasingly globalized food system was taking root, with an international division of labor in food growing as a first step in a highly differentiated and segmented system of processing and production, along with a globally linked, integrated marketing of products. This global shift had been facilitated during the mid-1950s by the passage of Public Law 480. This law was designed to use food as a political tool in the Cold War, while at the same time establishing new international markets for U.S. food products and U.S. grown commodities such as wheat. It also enabled U.S. food companies "to gain entrance into a market at the smallest expense possible," as one company executive put it. In addition, PL 480 complemented the U.S. government's "green revolution" policies, which were designed to restructure food-growing patterns outside the United States that made them more dependent on off-farm businesses such as manufacturers of agricultural equipment and chemicals. Each of these strategies sought to shift non-U.S. food consumption patterns, a constant theme of a food system that thrived on its ability to manipulate taste and product. The PL 480 programs especially promoted the idea that the residents of other countries, like those in the Far East, should now "eat the [U.S. produced] wheat who didn't eat wheat before," as one USDA official commented. This global restructuring of food economies also created dependencies in Third World countries that had considerable social and environmental consequences such as land use, water contamination, or population migration. Such transformation of food growing and food choice, described in such popular 1970s books as *Merchants of Grain* and *Diet for a Small Planet,* could be seen as a direct result of the international food aid and technology export programs initiated in the 1950s. Food growing and producing, as Frederick Buttel put it, had come to be "increasingly coordinated across the globe through world markets, domestic farm policies, and foreign trade policies."[14]

Despite these changes, the shift in growing toward a more industrial mode has been uneven, and, even in California, incomplete. This form of uneven development is linked in part to issues associated with the seasonality of food growing, at least for certain crops. Despite the continuing, and much commented upon, decline of the family farm, the seasonal aspects of food growing have, in some limited ways, constrained a wholesale shift to large-unit capitalist production in the growing of food. This has included difficulties in recruiting and maintaining a seasonal labor force, as well as securing the type of capital investments in technology and equipment that could be used for only a limited (seasonal) period of time. As a consequence, farming has also evolved into what can be considered a dual type of enterprise, differentiated between the large farms that have exhibited some or many of the characteristics of "industrial production" and what Buttel has called "small, subfamily farms." Some areas of agricultural activity, such as animal production (cattle, poultry, hogs), have become more fully industrialized and represent the most concentrated, capital-intensive forms of production associated with the growing of food. It is these areas of farming that have been perhaps most extensively molded by the "off-farm" interests, and where the survival of the family farmer has become most problematic. Yet the problem of the survival of the family farmer, a continuous theme in the policy arena and in books, movies, and songs, has often been examined outside the framework of an evolving food system and how that food system influences its front (farming) and back (consuming) end points. At the center of the food system reside the middle players, including those "off-farm" interests, for whom the very concept of growing simply represents an input into the producing or "remaking" of food into a product, and where everything from the seed to the finished food product has been reconstructed and transformed.

The Middle Players: Brokers, Processors, and Manufacturers

The shift from farming based on fixed land resources to food grown as raw materials for manufactured products established what economists came to call the agribusiness market economy.[15] This emerging food economy has been shaped and influenced by the most powerful segment in the food system, the assortment of middle players situated between grower and consumer. It has been the middle players who have extended their influence throughout the food system. This occurred as early as the turn of the twentieth century when for the first time the majority of food items purchased no longer came directly from the farm, and the way food was grown began to be influenced by businesses that provided inputs such as credit and farm equipment. With the shift toward a chemical-based, more industrialized agriculture, these middle players have dictated the terms in which those shifts have come to be adopted, and who have brokered, processed, manufactured, and reconfigured food into the kinds of products we know today.[16]

Since World War II, when the agribusiness market economy fully extended its reach both regionally and globally, this process of industrialization became particularly pronounced for a number of specific commodities, including those in the area of animal production. In terms of poultry production, for example, chickens up until the 1950s were raised in family yards in every region of the United States. Thousands of hatcheries, feed suppliers, and processors of eggs and broilers supplemented the large numbers of farmers raising poultry, many of whom were involved in a wide variety of other farming activities as well. But changes in the poultry industry beginning in the 1950s and 1960s fundamentally restructured the industry. Changes were spurred by the development of huge packing lots and the use of antibiotics to increase the number of chickens squeezed into the lots, while reducing the amount of feed required per chicken. These changes also became emblematic of the movement toward the integration of the production, processing, and distribution functions within single firms.[17]

During the 1970s and 1980s, the vast majority of chicken farmers felt it necessary to sign contracts with such integrated firms as ConAgra, Archer Daniels Midland (ADM), and Tyson Foods due to this industry restructuring. The "integrators" would own the hatching facility, the feed mill, and the processing plant, and contract with the poultry growers for the chicken to be grown as raw material. Growers effectively became employees of the integrators. Huge numbers of chickens were hatched, housed, fed, medicated, and slaughtered in industrial-like conditions. The four largest firms, increasing their share of the market from 23 percent in 1980 to upward of 55 percent of overall poultry production by the 1990s, extended their influence through various joint ventures or related arrangements with retail and fast-food outlets. "Value-added" and branded poultry products became key segments of the poultry business. These items were sold to the supermarkets, who were then able to conduct elaborate advertising campaigns (based on these branded products) that became common during the 1980s and 1990s in order to increase sales.[18]

Fast-food outlets were even more influential in reorienting the poultry business. Chicken McNuggets, introduced by McDonald's to offset the growing popularity of the fried-chicken outlets such as Kentucky Fried Chicken, became the most significant of the value-added products influencing industry restructuring. "Their impact was so big," the president of ConAgra commented about the chicken nuggets phenomenon, "that it changed the industry." In 1980, just prior to the ascendance of the McNugget, one of every ten chickens had become a processed product. Just a few years later that number jumped to one in three, with the production of chicken nuggets alone accounting for as much as $6 billion in poultry-related sales. By the 1990s, the revolution was complete. The poultry assembly line, managed through contract labor, now generated a wide variety of processed products. A "chick to table" system of control that integrated poultry farming into the larger food production system had become the defining characteristic of what had previously been a relatively small-scale, low-tech, family-farm activity, with a regionally dispersed set of middle players.[19]

Further restructuring of certain commodity sectors such as pork production also occurred even where the middle players had maintained a long-standing role in the industry. Cured or processed pork products, such as bacon and ham and sausages, had long represented the major aspect of hog production activity, with as little as one-third of a hog carcass ending up as fresh cuts. But similar to poultry, hog farming was widely dispersed, often as just one aspect of farming activity. Prior to the 1950s, some of the huge meatpackers, such as Armour and Swift, controlled a significant portion of hog slaughtering operations. Nevertheless, the areas of production that required the largest capital outlays, such as the manufacture of by-products, represented a smaller portion per pound of the live weight of the animal than in beef or even sheep. Thus, hog growing and production tended to remain relatively dispersed both regionally and in the ownership and control of different aspects of the hog production chain.[20]

The decline of small-scale hog-farming enterprises already began to take place in the period prior to World War II when as many as 4 million hog farms could still be found in several regions of the country. But the most significant changes were associated with the industrialization of production, or what University of Wisconsin food systems analyst Bruce Marion called the "second phase of structural reorganization," which occurred in the 1970s and 1980s. These changes were related to the large-scale confinement of hogs and the substitution of capital for labor in the processing sector. Large capital investors were also encouraged to enter the industry by a 1981 change in the tax laws when hog raising buildings were made eligible for investment tax credits at a five-year rather than fifteen-year schedule of accelerated depreciation. As a consequence, the size, scale, and nature of hog production changed. This occurred most dramatically in places like South Dakota and other midwestern states where individual farmers had earlier raised their piglets from birth until they were ready to be butchered as adults. The squeeze on local farmers from processors like Cargill intensified the trend toward standardized contract farming, similar to the shifts taking place at the same time in the poultry industry. This also led to specialization of function as well as concentration in the size of the hog farms. Hog integrators owned the breeding, gestation, and farrowing facilities, contracted out other aspects of production, such as the nursery and growing stages, and were able to control each stage of production through their specifications. As a result, small-scale hog farming became an endangered activity. During the 1990s alone, hog farmers with fewer than 1,000 pigs who had accounted for a third of the nation's pork production at the beginning of the decade represented just 5 percent of all hog farmers by the end of the decade. These large, contract-based hog operations were also generating enormous environmental impacts, with hog wastes on some sites equivalent to the wastes of a large city.[21]

The shifts toward an industrial model that took place in poultry and hog production were even more pronounced in beef production, even as its reputation as a corporate-dominated business had been set more than a century ago. Meatpacking, in fact, provides one of the earliest examples of the prominence of the middle players in relation to the food system. In the nineteenth century, changes in transportation (including the development of refrigerated railroad cars) and the developing relationship of urban centers like Chicago to its resource-producing hinterlands changed the nature of meat production. Centralized in the new meatpacking stockyards like Chicago's "Packingtown," meat production became highly concentrated. It emerged in little more than three decades as a classic tight oligopoly (with five companies controlling more than 80 percent of the slaughtering and 90 percent of the branch houses and transportation systems), the subject of major antitrust action, and an eventual Federal Trade Commission Consent Decree in 1920. The FTC action in fact was aimed at breaking up a system that the FTC argued could make the five companies "such a power in the marketing of foods" that they could ultimately "take over any food area they chose to enter."[22]

But despite the dominance of the meatpackers, meatpacking activities still remained dependent on the number and size of the packing facilities located near the centralized yards. After the Consent Decree, a period of deconcentration in the industry occurred between 1920 and the early 1960s. Though some of the big meatpackers continued to control a significant share of the market, their operations remained largely separated from feed and livestock regions, and significantly less than half of the cattle was sold directly from the feedlot to the big meatpackers. Reduced barriers to entry into the business also occurred due to improvements in refrigeration and transport technology, as well as the development of a system of federal grading of meat, which lessened the impact of the brand name in beef marketing and distribution. A number of new specialized plants, established with smaller capital outlays and utilizing more efficient in-plant technology, extended this process of deconcentration.[23]

The beef industry began to experience dramatic changes in the 1960s when cattle feeding shifted out of the eastern cornbelt and a new generation of meatpackers, led by Iowa Beef Processors (IBP), began to locate their beef slaughtering plants in cattle feeding areas like Colorado. The development and transport of boxed beef (as opposed to whole carcasses) in refrigerated trucks in the 1960s, along with the location of the processing plants near giant feedlots, significantly restructured beef slaughtering and processing activities. These changes led to the collapse of hundreds of smaller companies. Unlike earlier periods, when the carcass was shipped to the retail outlet, it was now cut up into primal and subprimal cuts in the packing plant and delivered in vacuum-sealed bags in cardboard boxes to wholesalers and supermarkets. The boxed-beef system created transportation and labor cost savings. Meanwhile, the retail markets no longer relied on butchers but instead created a category of employees known as "meat managers."[24]

The shift toward boxed beef helped reestablish the trend toward concentration among the processors. From 1979 to 1984 alone, the market share of boxed beef jumped from 44 percent to 77 percent, with 90 percent of the cattle sold directly from the feedlot to the packer where the boxed-beef operations were located. By the late 1980s, meat processing, the fourth largest manufacturing industry in the United States, established itself as the largest single segment of the food industry in terms of total assets, value added to product, total employees, and total business receipts. The grazing of cattle also accounted for one-third of all the land in North America, while one-half of the U.S. cropland was devoted to the production of livestock feed. By the 1990s, more than 85 percent of beef production had once again come to be controlled by just three companies: IBP, Cargill, and ConAgra. Each of these companies, in turn, perfected a system of production that included the use of growth hormones, union-busting strategies, and the push for a "world marketing system . . . where everything will be bought and sold on a world basis," as one ConAgra executive put it. These changes, noted Bruce Marion, were unprecedented in terms of the rate and structure of the concentration that had occurred in just a few decades.[25]

Similar to poultry and hog production, beef production also created an enormous range of environmental and occupational hazards. These were associated with the scale of production, the assembly line system established in processing, the range of hormones, chemicals, and fertilizer used and released into the environment, and the dumping of excess manure leading to fish kills, toxic algae growth, and contaminated streams and wells. At the same time, the rise of the three-company oligopoly extended union-busting activity and transformed what had been a rather well-paid, largely unionized workforce into a low-wage occupation utilizing a vulnerable and increasingly immigrant pool of labor. This work restructuring, as well as speed up in the workplace now prevalent among meatpackers, also exacerbated the occupational hazards in meat production. While the graphic images of arms and limbs chewed up during the meatpacking process found in Upton Sinclair's memorable descriptions of Packingtown were no longer relevant, there were now a broad array of repetitive motion injuries of muscles, limbs, nerves, and bones. These were injuries associated with "a kind of physical and psychological stress never before imagined," as sociologist Carol Andreas put it.[26]

The middle players, along with their fast-food and supermarket partners, have also had a significant impact on the reduction of crop diversity and related loss of biodiversity, due to their demands for uniformity of growing products and a more homogenous type of raw material for their value-added activity. Take potatoes, for example. Partly to meet the demands of the middle players, only 6 varieties of potatoes are grown today in the United States, compared to 3,000 that are grown and used as dietary staples in the Andes mountains, and 5,000 known varieties that are grown worldwide. Already by 1972, a National Research Council report was warning that North American crops had become "impressively uniform genetically and impressively vulnerable." Such a serious loss of biodiversity, as Joan Gussow has argued, has now become a trend that extends across the food chain.[27]

The growing of potatoes, one of the more notable examples of the loss of genetic diversity, has been largely driven by the development of products like french fries and potato chips. In 1950, prior to the advent of these fast-food and manufactured-food demands, processed potatoes (potato chips, dehydrated potatoes, and frozen potatoes) comprised less than 6 percent of total supplies. By 1983, when the McDonald's french fry had become the single largest demand for potatoes, that number increased to more than 50 percent of all potatoes grown and 60 percent of the potatoes used for food. (Some of the potato crop was utilized for nonfood purposes such as seed and livestock feed.) These changes in the development of the processed potato were further influenced by the demand of the processors and fast-food companies for year-round "consistent product." This meant standardized length (to fit machinery specifications in chopping and dicing), appearance, color, dry matter content, or other growing characteristics. Such requirements invariably influenced the way the potatoes were grown, including growing the same variety of product on the exact same land mass, with increasing reliance on chemical fertilizer and pesticides to facilitate growing a consistent product. Ultimately, this web of relationships between middle player, fast-food outlet, and supplier helped complete the process of transforming food grown into industrial raw material supplied.[28]

The role of the middle players extended to nearly every one of the commodity sectors. Key middle players such as ConAgra and Cargill in fact began to integrate horizontally in several commodity sectors at once, such as soybeans, wheat flour, or corn milling, as well as vertically integrate by dominating some of those same (and other) commodity sectors. Concentration by sector, however, was not a new phenomenon. The rise of the food-processing industries in the first decades of the twentieth century had been characterized by highly concentrated ownership in individual sectors such as meatpacking, flour milling, and sugar refining. However, the trends of the 1970s, 1980s, and 1990s, with their cross-sector ownership and horizontal and vertical integration, had created what in effect amounted to a handful of middle players operating as "supermarkets to the world," as one of the giants, Archer Daniels Midland boasted in its corporate advertising. These shifts were also global in nature. Earlier periods prior to and shortly after World War II had established a global products boom controlled by giant U.S. and European corporations (e.g., bananas and other fruit in central America, sugar in Cuba and Hawaii, and coffee in Brazil, Colombia, and El Salvador). But the new food globalization strategies of the 1970s through the 1990s extended across food sectors and production boundaries.[29]

The activities of the middle players in these commodity sectors also changed the nature of the processing activity itself. Food processing had first emerged as a major industrial activity at the turn of the twentieth century. Companies such as Heinz, Campbell's, and Franco-American became adept at establishing new product lines that involved the substitution of products and services previously produced within the home. By the post–World War II period, these earlier forms of processing such as canning and the related breakdown of food products, such as wheat kernels, into identifiable byproducts like flour, germ, or chaff, were giving way to the development of new types of food products. Value-added became a form of food reconfiguration. This included the vast new use of food additives as preservatives, "smootheners" (to withstand heating or freezing) or food coloring (to provide the appearance of freshness). It included the new technologies that created frozen food, and, subsequently microwavable food products. It also contributed to the shift toward a "fast-food culture," or what *Fortune* called the "relentless pursuit of convenience" in the purchase and consumption of food.[30]

The proliferation of highly processed food products has also been associated with the competition for shelf space in supermarkets and the higher costs of introducing new products. The price charged by supermarkets to cover their costs of stocking a new product, known as "slotting allowances" or slotting fees, can run from $5,000 to $50,000 depending on the chain and the product. The largest of the food manufacturers, including cigarette companies like Philip Morris and R.J. Reynolds who became the dominant players among food manufacturers during the 1970s and 1980s, found that product proliferation contributed to their share of the food product market. With their capacity to spend huge sums on advertising and promotion in order to establish product preference, the large food manufacturers have been able to control what is displayed on the shelf (and reduce or eliminate the possibility of competitors' obtaining such shelf space). In the process, greater and greater numbers of new products are introduced each year, with as much as 90 percent of those products withdrawn by the end of the year.[31]

The contemporary food economy is perhaps most distinguished today by this extraordinarily rapid turnover of products and the effort to differentiate products in stunning and often environmentally wasteful and nutritionally limiting ways. For example, the Kellogg Company introduced in 1998 "Breakfast Mates," a one-stop breakfast concept based on a single-serving package, with the cereal, milk, plastic spoon and bowl all packaged together. Marketed as a way to save time "for today's busy family," the product, according to a Kellogg spokesperson, was "a breakfast with virtually no preparation, and, if you think about it, no cleanup." Although Breakfast Mates had difficulty establishing a market niche, a fate shared by the 90 percent of its new product counterparts, it underlined how the continuous search for value-added products shaped the nature of the food items made available in the super-market aisles.[32]

The shift toward food megastores and superwarehouses in suburban areas also increased the opportunities for processed and packaged foods. These opportunities were particularly seized by the very largest firms, including the cash-rich cigarette companies that had the resources to supply and market a broad line of brand-name consumer foods. A highly processed product-driven food system cycle had been established, associated with continuous product development, intense promotion, and high failure rates. It was highly energy intensive and required the constant introduction of new products. Even as the demand for nonprocessed fresh fruits, vegetables, poultry, and fish expanded and retailers sought to respond to new consumer tastes and interest in health-ier and "whole" foods, the largest of the middle players still continued to dominate what Joan Gussow called "the excessive processing of food." Whole foods, once the primary source of food on the table, had become, in less than a century, little more than a niche market in a system that had changed the very nature of what one meant by food.[33]

Retailers: Selling the Product

The shifts in the food economy in the growing, processing, and manu-facturing of food products also extended to changes in the marketing and selling of food. These changes have included the rise of the super-market format, the differentiation and greater durability of products, and the spectacular emergence of an advertising-driven fast-food culture. Given the consumer-related or community role of food retail, these changes have also had powerful implications in terms of the increasing disconnect between food and place and its related environmental, eco-nomic, and social justice implications.

Though gradual at first, changes in food retailing already began to occur during the first three decades of the twentieth century. During the Progressive Era, food stores, particularly in large urban centers, were diverse in nature and size, and included a number of small immigrant-owners. Many of these stores, despite often charging higher prices, pro-vided crucial neighborhood or community functions, whether product selection (providing culturally specific food items for diverse diets), payment policies (allowing for credit and deferred payments), and as a place for community interaction. The early changes in retailing were primarily associated with the development of grocery chains and product-marketing strategies that facilitated the "Americanizing" or nationalizing of local, regional, and immigrant diets. But the chains, at least through the 1930s, did not entirely alter the structure of food retailing. While there were a few chains (for example, A&P, which controlled as much as 16 percent of the retail business) seeking to control purchasing and warehousing, most were little more than small stores, not significantly different in size and function than their independent counterparts. Partly due to the rise of the chains, however, the independent stores also felt the need to form their own affiliated chains, voluntary groups, and cooperative warehouses and distribution systems. During the 1930s and 1940s, these stores were still able to compete, due in part to the squeeze on capital in the 1930s and building shortages during World War II.[34]

It was during the Depression years that a new kind of food store format—the supermarket—emerged. Although originally introduced by independents, the chains moved aggressively to incorporate this new approach in the late 1940s, 1950s, and 1960s. Supermarkets were defined in part by their sales volume (initially defined as somewhere between $250,000 and $375,000) and selection (which included expanded in-store self-service departments for groceries, meats, produce, and dairy products). From the outset, the bigger companies decided to vertically integrate their operations. They combined warehousing and brand merchandising with traditional retailing. They linked retailing to changes in the food-processing sectors that included product differentiation and the use of slotting fees that reinforced the brand names of the products of the biggest companies capturing shelf space. As a consequence, supermarket chains, which had already abandoned key neighborhood functions such as credit and home delivery, became increasingly divorced from their community role.[35]

From the 1960s through the 1990s, these trends were intensified, due to increases in store size, greater concentration within individual regions, and a link to the car culture with its associated flow of resources, population, and services to the suburbs. In terms of store size, supermarkets expanded from an average size of 15,000 square feet in 1950, to 20,000 square feet in 1963, 38,000 square feet in 1980, and eventually the megastores of 60,000–80,000 square feet of the 1990s. Simultaneously, the rise of the larger stores extended the food retail market share of supermarkets from 23 percent in 1948 to 60 percent by 1960, with gradual increases during the next three decades. Changes in transportation and population growth away from the urban core, with longer trips now required for even routine food shopping, also reinforced the shift toward the larger stores.[36]

The new supermarket formats were increasingly designed to be built on large, often vacant parcels to capture the suburban constituencies that became their prime clientele. The larger size of the suburban supermarkets was characterized by their bigger parking lots, as automobiles assumed the dominant, if not the only mode of transportation. Reinforcing this relationship were the nationwide standards for parking-lot size based on store square footage. Bigger markets required more parking spaces that in turn created the need for larger parcel acquisitions. Suburban land was both cheaper and more available and could be located more conveniently next to freeway exits. With their larger formats and auto-centered transportation focus, supermarkets thus began to promote the idea of weekly, one-stop shopping. You got into your car, exited the freeway where the supermarket was located, and filled your cart with the full range of brand-name products made available from distant locations. Shopping involved not just the purchase of food products (which were themselves increasingly highly processed and "durable" or less perishable) but of household products as well, including those associated with suburban living patterns such as lawn care and auto maintenance.[37]

The shift toward the suburbs and the large format stores with their large parking lots also meant an abandonment of the urban core communities, establishing *lack of access to food* as a critical community food security concern. In low-income areas of Los Angeles, for example, the number of chain supermarkets declined from 44 in 1975 to 31 in 1991. Within this 52-square-mile stretch of low-income residencies could be found the equivalent of one full-service food market (both independents and chains) for every 26,400 people. This compared to an average of one store per 15,200 residents in the more suburban San Fernando Valley. At the national level, similar trends were taking place, with the number of supermarkets declining from 26,815 in 1980 to 24,548 in 1993, while selling areas were increasing from 23,000 to 35,000 people per market.[38]

A new and essentially perverse relationship between reliance on the automobile, parking-lot size, and land-use constraints in urban core communities had been established, extending the problems of food access, quality, and price in those communities. For example, the University of Connecticut's Food Policy Marketing Center conducted a major study of twenty-one metropolitan areas and found that nineteen of them had 30 percent fewer stores per capita in their lowest-income zip-code areas than regionwide averages. At the same time, the zip-coded areas with the fewest supermarkets per capita also had the lowest ratios of vehicle ownership. Moreover, a USDA study found that only 22 percent of food-stamp recipients were able to drive their own cars to purchase groceries compared with 96 percent of non-food-stamp recipients. With land scarce and costs high, supermarkets justified their abandonment of the urban core for the suburbs on the basis of economic performance and industry trends.[39]

As supermarkets located outside the urban core, another kind of food store—the convenience market—filled its place. Located in mini-malls and often more accessible by foot, convenience markets were designed to fill the vacuum due to the lack of large markets and the lower per capita car ownership in inner-city, low-income communities. But unlike specialty food stores or even the small-sized markets that still survived in a few urban areas such as New York City, the convenience markets stocked only a handful of items. These typically included cigarettes, candy, soft drinks, beer, and only a few food staples such as milk, bread, and cereal. The association of convenience stores with liquor outlets in inner-city communities further underlined the community food security concerns for those areas. These included lack of access to fresh foods, more expensive food items, and a growing dependence on highly processed junk foods, whether in the stores or in the fast-food outlets that had also descended on low-income communities during the same period.[40]

Aside from its contribution to the community food security problems of urban core communities, the restructuring of the retail industry in the 1990s led to growing concentration at the national level, in addition to the concentration that had already taken place in regions. Some of the superstores, like Kroger (which combined the Fred Meyer, Ralphs, and Food 4 Less chains), were also leaders in the development of the supermarket private label, extending their influence upstream in the production of food items. This aspect of the business established additional control of the flow of food items from the manufacturing plant to the checkout counter. This expansion by the food chains indicated the difficulty of maintaining any kind of local or community focus in making local products available (whether fresh or processed) or establishing a presence in the community through the services available. Even when the store's customers indicated an interest in local items, such as locally grown fresh produce, the chains, whose buying decisions were centralized and removed from the communities where the stores were located, remained reluctant to introduce items that might otherwise undercut their product mix and formats. Moreover, the huge mass merchandisers or discounters such as Wal-Mart, which had already captured a share of the superstore market, were also poised to extend their food retail market share through both large- and small-scale formats. This included their establishing smaller-sized supermarkets as a more expansive convenience mart, but with discounted prices. Although Wal-Mart was not able initially to erode the market share of the largest of the chains, it nevertheless further weakened the position of the independents. Ultimately, the superstores and the discounters, like the growers and middle players, had become part of a food system where the needs of a community and the local nature of food growing, processing, and selling had become increasingly marginalized.[41]

Food System Outcomes

Is There Enough to Eat? The Problem and Persistence of Hunger
Changes in the growing, manufacture, remaking, and selling of food have had powerful social and environmental consequences, whether in terms of land use and the built environment, diet and health, water and air

quality, or a porous social safety net. Food system outcomes—the winners and losers due to the changes in the production and consumption of food—can in turn represent key indicators about the overall state of society and the environment.

Among those outcomes, the problem of hunger continues to haunt the advocates of a more globalized food system. The problem of not enough to eat amidst surplus crops, green revolution technologies, and highly processed food items has stubbornly and almost inexorably persisted through much of the twentieth century, during periods of presumed economic prosperity as well as decline. This perverse relationship between too much food and too little to eat emerged prominently during the early years of the Depression, when the collapse of markets for farmers and heightened problems of overproduction and dropping prices coincided with the dramatic rise in unemployment and subsistence living. Already by 1932, the presence of large surpluses in wheat began to be viewed in the context of hungry people searching for sources of food in both rural and urban areas. It was in fact the juxtaposition of food surpluses and the rise of hunger—the "breadlines knee deep in wheat" metaphor that so poignantly captured the apparent contradiction of too much food and not enough to eat—that became the crucial political factor influencing the approach that was established.[42]

The key to that approach was the New Deal effort to protect farm income by placing limitations on commodity production and creating price supports, while establishing emergency relief activities. The Agricultural Adjustment Act of 1935, the key legislation at the time, situated the surplus food programs within the USDA, where strong grower constituency interests both shaped and benefited from such programs. Yet the pull of a food assistance or emergency relief approach also remained strong, and the agencies that were established, such as the Federal Surplus Relief Corporation, were mostly focused on getting as much food as quickly as possible into newly established relief programs and channels. Surplus commodities thus provided the basis for government-directed (as opposed to voluntary or charitable) food relief.[43]

The New Deal also represented a period of radical innovation, including new kinds of food strategies. In the area of food relief and food production, for example, a number of programs were developed around the "production-for-use" movements that coalesced during Upton Sinclair's run for governor in California in 1934. Programs such as food production cooperatives (engaged in canning and processing) and community or subsistence gardening programs were examples of this interest in food production experimentation. But the New Deal food surplus programs were more interested in quickly distributing as many surplus commodities to as many people as possible in as short a time frame as feasible.[44]

After the passage of the 1935 legislation and the departure of New Deal liberals such as attorney Jerome Frank from administering the surplus food programs, a subtle yet significant change in the program's name and function occurred. The Federal Surplus Relief Corporation became the Federal Surplus *Commodities* Corporation, and the primary thrust of surplus food programs was clearly established as grower support programs. Even the development of the first food-stamp programs, piloted in the wake of the 1937 recession, was designed by agricultural economists interested in developing new markets for surplus products for constituencies that might not otherwise have purchased such products. This concept of expanding rather than displacing domestic demand for agricultural products cemented the evolution of food aid policy in the New Deal as a temporary adjunct to farm policy rather than as a form of social welfare or income policy.[45]

After World War II, the food assistance/surplus commodity link was significantly eroded. Only a few food assistance programs were maintained after the New Deal, primarily the national school lunch program established through legislation in 1946. Social welfare policy—and the issue of hunger—was disappearing from the policy universe, as the soup lines got shorter and less visible. The surplus commodity programs also went through changes. The PL 480 program, for example, which aggressively established surplus commodity programs as part of the opening of global markets, served to reduce domestic surplus commodities. But the identification of poverty as an issue in the 1960 presidential campaign and heightened awareness of the issues of hunger and poverty in the wake of the publication of Michael Harrington's *The Other America* in 1962 reopened the debate about food assistance as government policy. The Kennedy and Johnson administrations were particularly interested in reviving and expanding surplus food programs, including new food-stamp program initiatives. This political interest coincided with the rise of the civil rights movement, which linked issues of political participation with social and economic rights of poor black constituencies, especially in the rural South. But the Kennedy and Johnson administrations, while seeking to revive food assistance as a potentially popular response to the renewed interest in poverty issues, were also opposed to linking food assistance to a broader championing of equality or income-redistribution approaches. Thus food assistance never fully escaped its characterization as temporary or emergency oriented, albeit government directed, although the rise of social movements and the broad anticorporate sentiment that had also emerged during this decade of activism strengthened the link of food assistance to a social agenda.[46]

Through the late 1960s and 1970s, food assistance programs expanded significantly (from $1.2 billion in 1969 to $8.3 billion in 1977). Higher Congressional authorizations were made possible through a coalition of conservative Farm Belt and liberal urban representatives. Initiatives during this period, such as the creation of the school breakfast and Women, Infants, and Children (WIC) programs and the restructuring of food stamps as an entitlement program, reflected the expanding public policy and civil society focus on hunger and malnutrition. Policy analysts and civic groups argued that hunger, defined as an individual condition requiring government corrective action, could be reduced if food was made available. One hypothesis for this decline was the rapid growth of federal food assistance programs for the poor, with the government serving as provider and social safety net. While the idea of having enough to eat was not quite defined as a right, programs were established on the basis of need as defined by income and thus were established with entitlement status.[47]

However, by the late 1970s, the discourse around food policy, and specifically food assistance, also began to change, undermining the social advocacy that had emerged in the previous ten years. During the late 1960s, the food assistance and hunger debates were framed by the image of the wealthy farmer driving his Cadillac and playing golf in order to not grow crops while people were growing hungry in places like Appalachia and the Deep South. But by the late 1970s, the more prevalent images were about urban (often perceived as African-American) food-stamp recipients engaging in deceptive activities. Food assistance became increasingly embroiled in the antiwelfare and anti-government/anti-entitlement program backlash that fueled the conservative counterattacks that culminated in the election of Ronald Reagan in 1980.

Food assistance programs, already squeezed by inflation, eligibility constraints, and poorly designed criteria for identifying what constituted a minimum food allowance, were seriously undercut by the new Reagan administration. Expenditures for food stamp and child nutrition programs decreased by $12 billion between 1981 and 1984. With federal assistance programs under siege, the Reagan administration, while continuing its efforts to reduce the government's role but wary of the claims of insensitivity, introduced, through the White House press office, the argument that vulnerable populations were still being protected through a well-established safety net despite budget cuts. The figures, however, belied the safety net argument. In 1982, the U.S. Conference of Mayors released the findings from a national survey documenting the rapid rise in hunger in major metropolitan areas as "a most serious emergency." The problem of hunger also came to be defined as not enough to eat *at a given moment in time*, thus extending the problem to the working poor. It was becoming difficult to separate the issue of hunger from the growing income disparities, nutritional deficiencies, supermarket abandonment of the inner city, and a wide range of other food system failures to meet the food needs for a large segment of the population that included both the working and nonworking poor. The problem of not enough to eat had become a condition of food insecurity across communities and vulnerable constituencies caught up in the economic restructuring of the 1980s.[48]

Hunger, like homelessness (another renewed Reagan-era phenomenon) began to assume the status of a "crisis" problem requiring "emergency response," as the mayor's conference put it. And indeed not since the 1930s had there emerged such phenomenal growth and reliance on food pantries, food banks, soup kitchens, and other private charitable food "giveaway" programs to feed the hungry. Rapid increases in emergency food giving coincided with the deep recession of 1981–1982 and the downward pressures on government food assistance programs. During this same period, USDA officials administering government commodity programs began to search for new ways to reduce the huge surpluses in dairy products such as cheese and nonfat dry milk that had resulted from changes in production and a protracted agricultural recession. The Reagan administration was reluctant at first to identify a new government program but was also concerned about the growing public and press attention to the issue of the surpluses and the rediscovery of hunger. As a result, Reagan signed a measure three days before Christmas 1981 to release 30 million pounds of government surplus cheese for emergency food distribution. Two years later the Temporary Emergency Food Assistance Program (TEFAP) was established to handle the distribution of surplus commodities that had initially included rice, cornmeal, and honey, but ultimately extended to several dozen commodities. Coincidentally, supermarkets, partly to counter the charge of inner-city abandonment, established their own food salvage operations to provide emergency food providers with food that was otherwise destined for landfill disposal. Suddenly, emergency food providers, many of whom eventually became dependent on the commodity programs, found themselves with new food sources to distribute.[49]

The establishment of the TEFAP and the availability of salvaged food helped institutionalize a program that had, in its origins, been defined as "temporary" and "emergency-based" but which continued to be renewed because the problem came to be seen as protracted. Thus was created a new type of food system, separate from, yet linked to, the global food system that marginalized low-income communities and developing countries alike, and the new forms of hunger and poverty that engulfed the working poor as well. But expansion of the emergency food system was not simply correlated with economic cycles or food assistance participation levels. The problem of a growing demand for emergency food, insufficient supplies, and an increase in the numbers of the working poor utilizing the emergency food system became particularly pronounced in the period following the passage of the 1996 Welfare Reform Act and the subsequent drop in the welfare rolls. These trends, moreover, all occurred during a period of unprecedented economic "growth." This need for emergency food in the late 1990s even surpassed the high levels that had been established during the severe recession of the early 1980s and subsequent downturn during the early 1990s.[50]

Surveys undertaken of low-income residents in specific neighborhoods identified the extent of these inadequate resources amidst growing demand, independent of either food assistance or emergency food interventions. The large numbers of residents dropping in and out of hunger identified by the *Seeds of Change* study in South Central Los Angeles paralleled the findings in other areas. What the *Seeds of Change* study especially noted was that the emergency food system's web of relationships was also tied to a complex of issues concerning price, quality, and availability of food. After nearly twenty years of emergency food service, it had become increasingly clear that hunger intervention could not be successfully managed when solely defined as a moment of crisis for individuals requiring emergency relief. The problem of not enough to eat ultimately became part of a continuum of food insecurity problems at the community scale.[51]

In terms of discourse, the enlargement and institutionalization of the emergency food system has also led to what Janet Poppendieck has called "the taming of hunger." This has been associated with middle-class participation in emergency food activities and with such events as "Hands across America." These activities taken together have substituted private action for public intervention, a process that intensified with the Welfare Reform Act and the promotion of charitable giving in lieu of public assistance. However, while food banks have represented an enormous institutional presence through the infrastructure established by the TEFAP and other programs, they have not played as direct a political or public advocacy role.[52]

The rise of the emergency food system further obscures the role of the global food system, particularly in relation to low-income communities. Food-system impacts associated with the price, quality, and availability of food in low-income communities have in turn helped precipitate the institutionalization of this dual set of food systems that only intensifies the problem and persistence of hunger. The capacity of the parallel emergency food system to meet food needs, with its focus on the ability to provide more calories, becomes problematic due to the type of food available, as well as the adequacy of the amount of food available. Food security, as Joan Gussow has commented, is simply not achieved through the provision of emergency food. "Outdated canned egg salad," Gussow wryly noted, "does not represent food security." While food security is not achieved based on the kinds of food items offered through the emergency food system, neither, for that matter, does it come with the highly processed or "overprocessed" food items available on the grocery shelves or consumed outside the home. Indeed, the flip side of the global food system's contribution to the persistence of hunger and its contribution to community food insecurity has been the dramatic rise of fast food and its transformation of both diet and the community connection to food source.[53]

The Problems of What We Eat: Fast-Food Culture

"We are what we eat," a slogan for the new century, could also be a statement of how the changes in the food system have overwhelmed the choices and tastes of food consumers, no matter the region, nation-state, or cultural and dietary history. By the 1960s and 1970s, the highly processed, brand-name food item was already accounting for as much as half of all food items purchased. The technicians associated with that change—those employed by the processors and manufacturers to identify the particular strategies for re-engineering each product—were at the forefront of the post–World War II food system restructuring of taste and diet. "Sophisticated processing and storage techniques, fast transport, and a creative variety of formulated convenience-food products have made it possible to ignore regional and seasonal differences in food production," proudly asserted the Institute of Food Technologists, the trade association of the food technicians. But the late 1960s and 1970s also witnessed a counter-trend toward alternative food growing and marketing, battles over chemical food additives, and the search for different methods of shopping and purchasing food (such as the "food conspiracies" associated with cooperative buying approaches). Each of these counter-strategies and approaches sought to identify alternative food products, including the renewed emphasis on whole foods. Food processors and retail outlets in turn sought to accommodate, manipulate, or redirect such alternative perspectives, while at the same time identifying "consumer hot buttons" along with the means "to push those buttons," as a *Fortune* article put it.[54]

Perhaps the most compelling form of this "consumer hot button" approach has been the rise of a fast-food culture and the food system players who have made it a major force in the food industry. The fast-food companies have been able to build on the changes established by the "convenience food product" approach of the processors and have helped to reshape key aspects of food growing and processing as well as the refashioning of food diet and taste. This segment of the industry has also been at the forefront in the shift of food product development and diet globally, promoting what could be considered the antithesis of the local or regional diet and its related local sources of food. Fast-food outlets have been established in city and suburb, domestically and internationally, becoming the same everywhere by destroying the boundaries of local custom and "heralding the new age of Yankee fast-food technology," as one analyst put it in a 1970s celebration of the triumph of fast food.[55]

Fast-food culture has long been associated with the automobile and its impact on daily life in the city and suburbs. Already by the 1920s, restaurants began to experiment with services for drive-in customers. Both the first full-scale "carhop" drive-in (the Pig Stand) and the first drive-in chain (Carpenter's) opened in Los Angeles in the 1930s. The California mode of auto-related, drive-in restaurant service subsequently emerged as a national phenomenon with the development of the Bob's Big Boy franchises in the early 1940s. But the connection of fast-food culture with the automobile was most directly made during the 1950s. With the rapid construction of the interstate highway system, the development of fast-food franchises seemed to literally follow the bulldozers, as they sprang up along the newly completed highways. Built to provide a kind of security and consistency in identity, fast-food outlets were not at all like the earlier railroad-car-like diners and greasy spoon eateries with their wildly diverse designs and signature architecture that had earlier spread along the prefreeway roadways of America. Like the highway system itself, the fast-food restaurant was launched as a suburban-related phenomenon. With its outlets located near suburban traffic intersections and shopping centers and its identical architecture and food products, the fast-food culture's formula for success—e.g., McDonald's famous "quality, service, cleanliness"—were associated with what were presumed to be core suburban values.[56]

The growth of the fast-food franchise during the 1960s and 1970s was simply phenomenal. From its initial base in the suburbs, the fast-food outlet spread to every conceivable setting and in multiple locations, from zoos, office buildings, malls, and hospitals, to even one Arkansas high school.[57] By the 1960s, fast-food outlets were already experiencing a 20 percent annual growth rate. By the 1970s, fast-food chains had become, as one market analyst put it, one of the "truly great economic per-formers" among American service industry sectors. By then, fast-food outlets had entered the cities as well, despite occasional protests in middle-class and upper-class communities. By the 1980s, the restaurant landscape in the inner city, working-class suburb, and outlying suburb alike had been fully transformed, with the "neighborhood bar, ice-cream parlor and restaurant [and even] the vanished neighborhood" itself falling victim to fast food's march through city and suburb, as one *Newsweek* writer lamented. By 1980, McDonald's, the triumphant leader of the fast-food chains, could claim that as many as one of fifteen entry-level workers had his or her start at McDonald's. Moreover, as much as 7 percent of the entire workforce had worked for McDonald's at one point in their lives. At the same time, over 90 percent of the American public would have eaten in a fast-food restaurant within the previous six months, and more than 53 percent would have eaten in a McDonald's during just the previous month. Indeed, by the end of the 1990s, fast food had become a vast economic engine, a $52 billion indus-try that had transformed both diet and unskilled, low-wage work.[58]

In fact, this fast-food phenomenon can be seen as the food industry equivalent of the Taylorist deskilling of work. In Charlie Chaplin's mar-velous portrait of the transformation of work in his film *Modern Times*, an experimental food assembly line is developed to see if a worker could eat his lunch while still working on the line, thus eliminating any loss of down time due to the lunch break. Like Chaplin's metaphor about "efficient" work, the key to the fast-food approach has been the organization of the service itself. "No skilled hands are needed to make hamburgers quickly," one food industry trade publication proudly declared. Instead the hamburgers are simply fed into a conveyer "by unskilled operators who then simply watch the finished product come out the other end." This deskilling and search for industrial efficiency reconfigured the very nature of the service operation. As Theodore Levitt pointed out in his discussion of McDonald's "technocratic hamburger," in a 1972 article in the *Harvard Business Review*, "nowhere in the entire service sector are the possibilities of the manufacturing mode of thinking better illustrated than in fast-food franchising." With its extraordinary attention to detail, ranging from the size of the french fry scoop to the synchronized assembling of the hamburger to the scripted dialogues of the employees, fast-food service had become the Taylorist mode of industrial efficiency writ large. Such "extreme routinization," as Robin Leidner pointed out, could, in fact, "induce rather than prevent idiocy." And, as a number of fast-food observers have also pointed out, that Taylorist vision has been extended to the consumers themselves, by shepherding the food buyers through the lines to "refuel" and be on their way. "The assembly line is stationary," one observer said of the fast-food mode, "and *we*, the consumers, are the moveable parts." "We have worked the business out to a science," McDonald's longtime owner and ideologue Ray Kroc crowed in one early 1968 interview.[59]

The rise of the fast-food culture has also had its academic proponents, perhaps none more prominent than Elizabeth Whelan, a longtime defender of the food and chemical industries, and Dr. Frederick J. Stare, a nutritionist and professor at the Harvard School of Public Health. Stare and Whelan's book, *Panic in the Pantry*, published in the early 1970s, set the tone of those defenders of fast-food culture who argued that the oppositional movements, such as the Center for Science in the Public Interest, constituted a form of "food police." "Eat your additives, they're good for you," they provocatively declared. Whalen subsequently argued that junk food, in fact, could appropriately be designated "fun food." "Fun food" in fact was a term increasingly promoted by the fast fooders to characterize their product.[60]

Aside from their role among fast-food outlets, these fun foods, such as potato chips or other salty, high-fat snacks, emerged as one of the core areas of growth among highly processed food products and a prized "slot" on the shelves in the supermarket. By the late 1990s as many as

5.5 billion salty snacks were consumed on an annual basis, amounting to 22 pounds per capita, or upward of $13 billion spent on snack chips. Fun food or junk food products were also able to rely on heavy advertising and promotion, including for new products. This was not only aimed at establishing particular brand loyalties within the product class of snack foods, but could also influence the "total demand for that product class relative to other food classes with less advertising," as food system analyst Bruce Marion put it. The competitive advantage of the brand name, fun food/snack food heavily influenced what one ate. By the late 1990s, potato chips and french fries, for example, made up more than one-quarter of the vegetable servings eaten by children, and nearly one-third of the vegetables eaten by teenagers, according to a survey undertaken by a biomedical research center at Louisiana State University. The "you are what you eat" concept, influenced and shaped by the fast-food players, had both literal (in nutritional or caloric terms) and symbolic or cultural (in marketing terms) meaning.[61]

These highly lucrative, "value-added" snack food products had also become a significant component of the revenue stream of several of the major food manufacturers and processors, such as Pepsico, whose Frito-Lay subsidiary alone had $6 billion in annual sales. Competition has generally been focused on which value-added product could be dominant in each particular market segment or type of product, such as snack foods. Growth in overseas markets for U.S.-produced snack foods, while reaching more than a $1 billion in export sales by the 1990s, was nevertheless also seen as limited in relation to whether regional and national food preferences could be overcome. What one might eat as well as where such items came from remained heavily contested terrains, even as the European and Asian markets came to be penetrated by both fast-food restaurants and snack items.[62]

With their ability to penetrate global food markets, fast-food chains especially have become highly dependent on their non-U.S. operations in terms of sales growth, while also vulnerable in terms of regional and cultural opposition. McDonald's, for example, was already by 1990 opening more outlets outside the United States than inside. Crossing national borders to establish new outlets in Europe, Asia, Russia, and the old Eastern European bloc, companies like McDonald's were in effect establishing fast-food—and by implication—global food system beach-heads. Previously, the fast-food strategy in the United States had been designed to open stores first in the suburbs, gain sufficient presence to buy television time, and then enter the cities. Outside the United States, the strategy shifted so that urban and highly visible locations like the Champs Elysées in Paris or Red Square in Moscow would be the first sites in that country for a fast-food outlet. To succeed meant in part to achieve a successful outcome associated with a change in discourse as well as eating patterns, since the ability to successfully refashion diet and taste also came through the appeal of symbols and the power of marketing and ultimately the erosion of local food cultures. It was this dimension especially that ignited the opposition to fast food outside the United States. This was demonstrated by the farmers who stormed the McDonald's outlets in France in 1999 and who based their appeal on their regional and national identities in face of the onslaught of the global food system.[63]

Faced with the problem of where and how to expand market share, the fast-food companies have continually sought to develop new value-added strategies. This has included ways to transform product skepticism into a product advantage; that is, how to use the power of fast-food culture to undermine the resistance to it. One notable illustration of this strategy has been the introduction of Olestra products by Procter & Gamble and its marketing partners Frito-Lay and Nabisco. The Olestra story is both revealing and cautionary. In 1968, P&G researchers synthesized a sucrose polyester fat substitute that the company proceeded to name Olestra. Made by heating soybean oil or cottonseed oil and then blending it with sugar, Olestra, whose molecules are larger than ordinary fat and therefore not digested or absorbed by the body, was finally approved after twenty years by the FDA in 1996 for its use in snack foods. This occurred despite the FDA's own warning that Olestra could cause "abdominal cramping and loose stools in some individuals." Doubts had also been raised about Olestra's long-term health effects. But for P&G and its snack food partners, Olestra represented a potential bonanza by its ability to maintain a customer base for snack foods among those concerned about fat content, as well as those customers who might otherwise not be willing to purchase snack foods. "People who left the chip market because of the fat are coming back," the Pringles marketing manager enthusiastically declared about his product's test run in 1997 in Indianapolis. Fun-food advocate Whelan, one of Olestra's core defenders, also suggested that it added "rich taste and smooth texture to food without adding calories." Thus, "fun-food" eaters would no longer have to worry about the calorie content of their snacking. Like other highly processed value-added food products, the Olestra snack food item could be priced higher than its snack food counterparts, presenting the opportunity for a new type of profit center for companies like P&G and Pepsico. Indeed, one stock analyst argued, Olestra could become the "single most important development in the history of the food industry."[64]

Despite its pronounced marketing strategy and rosy projections about new profit centers, Olestra-based products such as Frito-Lay's Wow and Pringles' fat-free chips have had a mixed sales performance since their introduction into national markets in 1998. This relatively weak performance has been compounded by the number of reported incidents of Olestra-induced diarrhea that has also been described by some tongue-in-cheek press coverage. Nevertheless, the transformation of diets associated with the rise of a fast-food culture has continued largely unabated, not only domestically but globally as well. Whether McDonald's reaching 100 billion meals served, or highly processed fun foods grabbing a larger market share and a greater share of store slottings, the fast-food culture now extends throughout the food system.

From Seed to Table: Reengineering Food

Parallel to fast food's influence in reshaping the food system and food choice has been the extraordinarily rapid growth in recent years of genetically modified (GM) food crops or bio-engineered food. From 1996 to 1999, the acreage dedicated to GM or "transgenic" crops increased tenfold, and by 1999, as many as half of the 72 million acres of soybeans in the United States were GM crops. Corn, cotton, and canola among sixty different crops also became targets of opportunities for a new breed of food system player, or what one group called the "gene giants." Most of these changes in production were U.S.-based, which accounted for nearly three-fourths of all the acreage planted to transgenic crops. By 1999, GM food advocates were also predicting that more than 100 million acres would be dedicated to genetically modified plants and crops early in the twenty-first century, and projected double-digit growth rates into the foreseeable future.[65]

While this growth was occurring, a series of corporate takeovers and alliances were aimed at consolidating control in the GM food arena. These mergers and related corporate arrangements involving shared market and technology approaches also led to a "global clustering" of food-related organizations, as William Heffernan put it. These clusters brought together the most powerful of the middle players such as ConAgra, ADM, and Cargill, the large food manufacturing interests such as Philip Morris/Kraft, and the new biotechnology giants such as Novartis and Monsanto. Through a set of linked R&D objectives, market penetration strategies, corporate restructuring, and policy objectives, an even more extensive form of "seed-to-table" throughput control of both the $500 billion U.S. and global food markets was pursued. In the process, these integrated, seed-to-table cluster of firms, in pursuing their goal of carving up the world's food markets, have sought to function much as the seven sister oil cartels did nearly a century earlier.[66]

In pursuing these goals, the gene giants, many of them also agrochemical giants like Monsanto, argued that genetically modified crops represent the next stage in the development of the food system, similar in significance to the green revolution of the 1950s and 1960s. Biotechnology advocates pointed to the prospects of enhanced production capabilities, whether due to growth hormones or newly engineered seed products, or "muscled-up crops," as *Time* magazine characterized the biotech companies' new seed products.[67] The gene giants dismissed any health or environmental concerns for their products, while also arguing that their products should not be subject to biochemical or toxicological tests on the grounds that there is "substantial equivalence" of GM and non-GM foods. Using language reminiscent of the promotion of green revolution technology, the biotechnology advocates further argued that their products could provide the basis for making farmers around the world "more production-oriented rather than remaining subsistence farmers," the better to feed the world. Countering their critics, the biotech advocates also argued that biotech approaches offer important environmental and nutritional benefits by establishing nonchemical means of pest and pathogen control, greater drought and salinity tolerance, and enhanced nutrient content.[68]

But the introduction of genetically modified food products, as in the case of Monsanto, clearly reflected a corporate strategy to reinforce certain types of chemical use, gain further control across different sectors of the food system, and ultimately create a truly global food product. These included the development of herbicide-tolerant crop or seed varieties that could cause, and in fact induce, farmers to utilize more and not fewer pesticides, while also undermining existing opportunities for biological controls. Furthermore, the biotechnology strategy has been designed to produce more of a product (such as through growth-inducing hormones), to reduce rather than expand genetic diversity, and to create new kinds of middle-player dependencies. By pursuing substantial equivalence as a regulatory avoidance strategy, the companies sought to avoid scientific review of the health and environmental impacts of their products, while still claiming that such products are harmless. Given these contrasting views, the debate in the late 1990s over genetically modified foods emerged as a pivotal part of the broader debate over what kind of food system and food system outcomes would emerge in the twenty-first century.

The biotech food revolution of the 1990s sought to influence plant, food, and animal products and production methods through a variety of gene-transplanting or bio-engineering strategies. These included the introduction of herbicide-resistant products (e.g., Monsanto's "Roundup Ready Soybeans"); insect resistant products (such as through the use of the Bt toxin); "identity"-preserving strategies (such as the FlavrSavr tomato, which can delay ripening); and strategies designed to enhance production (such as through the use of growth hormones like bovine somatotropin). For the gene giants, these various genetic interventions have had, as their primary goal, the objective of increased control across various sectors of the food system, extending the trend toward concentration into whole new areas. At the same time, such interventions can and already have had significant consequences for farmers, consumers, communities, and the environment.[69]

This remaking of food and its consequent food system changes through the use of genetic engineering strategies is not entirely a new phenomenon. Some biotechnology advocates have in fact argued that the recent initiatives in this area are simply an extension of what farmers have been doing hundreds of years. But the new genetic engineering strategies, unlike earlier crossbreeding or gene transplant strategies, represent transgenic or gene transplanting across species. What also distinguishes the more recent strategies of food product and plant variety restructuring, including but not limited to the more recent development of transgenic or GM foods, has been their role in the reorganization of the food system, the creation of very different kinds of food products, and what David Goodman and Michael Redclift have called the "refashioning of nature."[70]

One of the more compelling images of the earlier forms of re-engineered food was the development of the "hard tomato." As early as 1947, research at the land grant schools sought to develop a hard tomato product capable of withstanding the stresses of mechanical harvesting for both direct sale and processing. By altering the genetic composition of the tomato, researchers hoped to develop a product with firm walls and a thick flesh, which would be free from cracks and have a square shape for packing and handling. The product could then be gassed with ethylene gas to achieve a red color. This remanufactured tomato was primarily designed to save labor costs and allow for more distant delivery, but it also became a powerful symbol of food system changes. This was made visible in part by Jim Hightower's famous "Hard Tomatoes, Hard Times" tract, revealing how the land grant colleges, in league with the "corporate farmers" and middle players, had turned new growing strategies into a new way of extending market control and remaking food. Subsequently, the hard tomato reemerged as a "biotech" product in the mid-1990s when Calgene (which became a Monsanto subsidiary) introduced its "FlavrSavr" tomato. This genetically engineered product was now designed to also recapture the flavor that had been lost by previous "hard-tomato" interventions, even as such products lost some of their nutritional content. But the "FlavrSavr," which was designed to ripen longer on the vine and maintain hardness through picking, packing, and long-distance transport, never succeeded in establishing its beach-head as a long-distance product.[71]

The identity-preserving and durability goal of the FlavrSavr had been clearly designed to answer the charge that a globalized food product could not compete with a locally grown, seasonal product, where quality remains a significant competitive advantage for local farmers. Local farmers, particularly organic farmers, established growing strategies that served as a counterpoint to the new GM food strategies that also sought to assume the mantle of "sustainability." These contrasting views about sustainability have been a critical aspect of the debate about the intro-duction of a genetically engineered soil bacterium *Bacillus thuringiensis*, or Bt. Organic farmers have used Bt, a naturally occurring pesticide, as a low-cost, low-risk pesticide spray to deal with unanticipated worm or bug infestations. Moreover, they have done so only after a myriad of other nonspray-related strategies have also been employed. Because it has been used so infrequently and in such limited quantities, neither the problem of toxicity nor of pest resistance has become a factor in its use. But, unlike its naturally occurring counterpart, the genetically altered Bt crop was designed to exude the Bt toxin at much higher levels. By intro-ducing the Bt crop widely, organic farmers could, irretrievably over time, become disadvantaged by the genetically modified Bt crop, due to pos-sible insect resistance and other potential impacts such as increased mono-cropping, which is also associated with increased pesticide use. In response, Monsanto, which had purchased the company that held the patent, claimed to have established a "new agricultural paradigm" of sustainability with its genetically engineered Bt product. But "Monsanto's version of sustainable agriculture," as Michael Pollan put it, in fact threatened "precisely those farmers who [have] pioneered sus-tainable farming."[72]

This debate over sustainability can be seen as both a discourse battle as well as a conflict over efforts to control the playing field by over-coming and ultimately devastating potential competitors, including sus-tainable agriculture constituencies. The struggle over growth hormones is a case in point. In states with a strong small dairy farmer constituency such as Wisconsin, strong opposition emerged around the use of the growth hormone (rBGH), also known as rBST (bovine somatrotropin). The introduction of rBGH, designed to significantly increase milk production output, also undermined the position of the small farmers in places like Wisconsin, due to the chronic concerns about overproduction, particularly after the FDA approved the hormone's use in 1993. Monsanto, the leading producer of rBGH in fact claimed that within five years after the FDA ruling, about 30 percent of all U.S. herds were being treated with the hormone, identifying a significant new source of profits for rBGH producers like Monsanto.[73]

But local farm and sustainable agriculture constituencies have been able to challenge the rush to production by Monsanto and other rBGH companies, both within and outside the United States. For example, Health Canada, the Canadian government's equivalent to the U.S. Food and Drug Administration, issued a ruling that turned down Monsanto's licensing application for the use in Canada of rBGH. The Canadian agency based its decision on studies indicating that the growth hormone had potential health impacts on the animals. Moreover, the Canadian scientists, whose findings influenced the decision, cautioned that animal health problems were likely to lead to an increased use of antibiotics, which could in turn end up in milk products and ultimately build up human tolerance for even the strongest antibiotics. While suffering this setback in Canada, Monsanto nevertheless still heavily invested in expanding its global markets, particularly the lucrative European Union market. But while initially declaring rBGH safe in a 1990 ruling, widespread environmental and farmer-led protests also caused the EU to reverse its position, issuing in 1993 a moratorium on rBGH use. Along with the Canadian ruling and the European protests, companies like Monsanto started to become concerned that not only rBGH but other aspects of the biotechnology market were vulnerable.[74]

As the fights over GM products like the Bt toxin intensified, the gene giants undertook a concerted effort to create a favorable climate of public opinion, establish favorable policy initiatives, and fund research. Those efforts at first appeared to pay off, thanks to multimillion-dollar grants to research universities like the University of California, favorable press coverage, and strong USDA support in terms of domestic as well as global trade policy. But opposition to such crops in the United States, and even more extensively outside the United States, remained fierce. Genetically modified food began to assume a position, along with the fast-food culture and, in earlier periods, the green revolution, that symbolized the reach of the global food system in undermining and ultimately destroying regional food economies. Bio-engineered food was also seen as food without a country, creating products that no longer seem to resemble food, captured in the slogan of "frankenfood." In Europe, especially, food retailers began to withdraw GM-related crops from their stores, due to strong consumer pressures.[75]

One of the areas of greatest vulnerability for the gene giants became the issue of whether GM crops should be labeled. During the 1990s, these companies strongly opposed any kind of labeling policy, arguing that it increased costs, and thereby destroyed markets for new products. Such arguments were widely repeated not just by industry spokespeople but government agencies such as the USDA and key government leaders such as Al Gore and Britain's Tony Blair. But despite the political pressure from the United States and the European Union's approval of certain genetically modified foods for sale, labeling approaches and other anti-GM strategies began to be adopted. This was primarily due to the continuing and hardened opposition in Europe to any type of genetically modified foods, with one survey indicating that as many as 85 percent or more would not purchase such products if given a choice. Moreover, a number of European countries such as Norway, Luxembourg, and Austria decided to either ban such products and/or prevent their being grown locally, with mandatory labeling programs instituted as well. As a consequence, the fight over labeling as well as product bans emerged as a core trade policy issue, triggering USDA action to ban European food imports in retaliation over specific labeling requirements adopted by European Common Market countries. A battleground had thus been created, with most of the U.S. press assuming that opposition to genetically modified foods was a foreign rather than domestic phenomena. But with little attention by the press and policymakers, a kind of underground opposition in the United States also began to surface. These diverse groups were able to draw together a range of constituencies that had cut their teeth on support of sustainable agriculture and increasingly focused on the nature and objectives of the big food-system giants and this latest and most sweeping aspect of food-system restructuring.[76]

Perhaps no product came to symbolize this restructuring—and the shifting battleground—more than Monsanto's "terminator seed." Monsanto, which sought to purchase the seed company that had first developed the patent in conjunction with the USDA, argued that its new product would not only be far more productive but could eliminate reliance on all chemical herbicides *except for its own*, its best-selling Roundup. Further, the seed itself, as part of its genetic reengineering, would be rendered infertile, thus requiring the farmer to not only purchase the seed (and the Monsanto pesticide used for it), but to continue purchasing the same seed product afterward. These "suicide seeds" would have genetic traits that could also be turned on and off by an external chemical inducer mixed with Monsanto's patented agrochemical. As part of this approach, Monsanto also established (though later modified due to criticisms from farmer groups) "performance agreements" giving Monsanto the right to inspect the farm in question and hold the farmer liable if the contract was violated, an approach one Monsanto critic characterized as "bio-serfdom." Thus, under the guise of a new production technology, the terminator seed would essentially undermine one of the most crucial of all farming activities—the saving of seeds and thereby the adapting of crops to specific and unique farming environments.[77]

The terminator seed controversy became a key rallying point for critics and activists in the United States, Europe, and elsewhere who focused on such issues as genetic diversity, the monopolization of seeds, and impacts on local farmers. The U.S. press had at first largely ignored the U.S. movements, while characterizing European opposition as an outcome of food safety fears and mistrust of government. But by 1999, opposition had begun to take its toll, not only in Europe but increasingly in the United States as well. The terminator seed became one of the first and most visible casualties. In 1999, Monsanto first reluctantly declared that it would no longer market the terminator seed, and then subsequently dropped its effort to purchase the seed company involved in the development of the terminator technology. But the company, as

Monsanto chairman Robert Shapiro put it, still considered its investment in biotechnology "the most successful product launch in the history of agriculture. What we are doing and what people think of us will eventually come into alignment." Terminator seed technology had reached an impasse, but the role—and influence—of gene technology remained a key battleground regarding the nature and future of the food system itself.[78]

These issues came to a head during the December 1999 Seattle protests linked to the meeting of the World Trade Organization. GM foods had become one of the most visible issues associated with the protests, and what had previously been assumed to be just a marginalized protest movement in the United States suddenly took on more significance as a growing—if not substantial—movement that cut across issues and constituencies. The U.S. press also began to shift in its presentation of the issues, identifying local opposition to GM foods among farmers and food producers as well as among the food movement activists. Meetings of food system–focused groups, such as the Community Food Security Coalition, transformed what had been seen as a more obscure, technical issue, into a campaign that gave prominence to an emerging new politics. By the end of the decade, anti-GM activists had become enormously energized, both by the shift in discourse that was occurring, and their ability to counter what had been seen by some as the inexorable triumph of the gene giants. With U.S. soybean sales overseas beginning to decline for the first time, some of the gene giants such as Novartis decided to spin off some of their GM food activities. By 2000, a consumer backlash against GM foods had intensified in the U.S. as well, forcing a fast-food chain like McDonald's to order its french fry supplier to stop using Monsanto's genetically modified potato. With even Monsanto stockholders concerned that their cash cow had now become an $8 billion investment in jeopardy, it appeared that the global food system players had finally overreached and a battle for the future of the food system had entered a new terrain.[79]

Distance and Durability: Whither the Food System

In a 1992 article, Canadian sociologist Harriet Friedmann raised the concepts of "distance" and "durability" to characterize food system trends, indicating the ways in which food items were capable of traveling anywhere and being marketed anyplace. Products like the Pringles potato chip, first introduced by Procter & Gamble in the early 1970s, had been designed precisely to pursue this distance and durability approach, and overcome the limits to the travel anywhere and be marketed anyplace objective. Due to their short shelf life and fragile nature, potato chips have over the years had one significant marketing constraint: they needed to be processed in locations near the markets to be served. The Pringles potato chip was designed to overcome that barrier in order to become a long-distance and durable product. The chip was made from dehydrated potato mash mixed with a range of chemical preservatives to standardize and lengthen the life of the product. This reengineered product, a perfect circular potato chip, could then be packaged in a hard container and processed in a centralized location in Tennessee to be shipped to more distant parts. Due to its chemical-based remanufacturing, the Pringles chip had also achieved durability status, allowing it to sit on a shelf for a year without tasting stale. As a result, the product could then be advertised nationally, creating a significant marketing advantage. When Olestra was later added to the mix, the reengineering of this overprocessed food product was complete, and any resemblance to the "raw material" of origin had essentially disappeared. The transnational development of the fast-food culture—with its homogenizing of local cultures and communities and reconfiguring and then standardizing products like potato chips and french fries—embodied this durability and distance, or travel anywhere, be marketed anyplace approach.[80]

In her article, Friedmann contrasted the global food system metaphors of distance and durability with the concepts of "locality" and "seasonality," useful metaphors for describing a community food systems approach. "Locality," for Friedmann and for other community food system advocates, can refer to regionally based, a concept that could also be traced back to Lewis Mumford and Benton MacKaye's environmentally grounded "region—to live in." Other analysts have also associated "locality" with "food shed" to describe environmentally appropriate and regionally based resource systems. The concept of seasonality associates the process of food growing—as well as broader food system activities such as marketing and even processing and manufacturing—with the rhythms and resources of a place. Food grown seasonally is food grown regionally. Seasonality also establishes the significance of freshness and quality in the availability and making of food. Seasonal food then is not only regionally grown but regionally made and sold, helping sustain a regional diet in contrast to the "same anywhere" global fast-food diet.[81]

These contrasting metaphors—distance and durability as distinct from locality and seasonality—help identify different kinds of food-system characteristics and potentially contrasting strategies and constituencies. During much of the twentieth century, in the process of becoming more globalized and industrialized, the food system created new meanings about the nature of food products as well as how food is grown, processed, sold, and consumed. The "locality" and "seasonality" metaphors suggest another direction in the twenty-first century for changes in the food system and the role of social movements associated with such change. Both sets of metaphors are useful in turn for helping us understand the barriers and opportunities for change, the nature of the systems that will need to be challenged and transformed, and where and how the new movements and agendas—capable of locating pathways for change—may arise.

6

The Politics of Food: Agendas and Movements for Change

A New Type of Movement?

In August 1994, thirty environmental, community development, sustainable agriculture and antihunger activists and food-system analysts met in a rather barren room in the Hilton Hotel in downtown Chicago. They had been invited by the organizers of the meeting to try to identify a common approach to food-system issues and community food security action, using the venue of the upcoming Farm Bill legislation. The group identified three key objectives: to pass new community food security legislation as part of the Farm Bill; to stimulate new alternative food-system and community food security programs; and to focus public attention and initiate a dialogue among movements about this new community food security approach.[1]

The Farm Bill seemed a useful focus for this new approach. For nearly fifty years this omnibus legislation, introduced approximately every five years, had been structured as a set of subsidy and price support programs for various commodity groups to address their continuing concerns about overproduction. Beginning with the 1981 legislation, three different sets of players—the emerging sustainable and organic agriculture groups, the small-family-farm advocates, and environmentalists focused on soil erosion and pollution concerns—came together to seek revisions and new directions for the legislation. Discussions focused on ways to reduce the intensity of production and link the cost of production to prices, as well as to identify conservation reserves that could take fragile land out of production. These ideas were subsequently introduced as part of a set of "alternative farm bill" amendments during the 1985 and 1990 Farm Bill legislative debates. There were also efforts to link the environmental and rural farm approaches (associated with the growing of food and rural economic development issues) to a fourth crucial "alternative food" player—antihunger groups. These groups were focused on protecting and possibly expanding the Farm Bill's food assistance provisions such as food stamps and the school-lunch program.[2]

The alternative farm bill amendments introduced during the 1981, 1985, and 1990 legislative debates were associated with various coalition-related efforts to influence overall food policy and help consolidate the myriad of alternative or sustainable agriculture groups into a more coherent political force. The coalition initiatives had also provided a forum for mainstream environmental organizations like the Natural Resources Defense Council to raise such issues as nonpoint source pollution from pesticides and fertilizers and to link the problem of agriculture-related toxics with concerns about land and wildlife contamination. Since the sustainable agriculture groups also focused on pesticide use, an environmental/sustainable agriculture alliance seemed possible. A common position regarding rural environments, land use, and, at least for some, farmworker health and safety was also conceivable. While the antihunger groups were not directly a part of this coalition-building process, the sustainable agriculture groups were now including issues of food access and availability for the rural and urban poor as an important dimension of their own advocacy. These linked approaches resembled in some ways the kinds of coalitions that had been established between urban liberals and rural farm interests in the development of food assistance programs during the 1960s and 1970s.[3]

But during this process of Farm Bill–related coalition building, political tensions had also emerged, including unresolved differences in emphasis between environmental groups and rural development advocates. During each of the Farm Bill debates, those differences were never fully resolved by the alternative farm bill initiatives, nor did a fully articulated environmental-rural coalition approach successfully emerge. Rural development advocates remained concerned about the economic squeeze on small farmers and the complex yet critical area of price supports. The concern about economic survival, in turn, was compounded by the continuing economic pressures on farmers to produce, which potentially undercut the environmental initiatives regarding soil conservation, crop rotation, and the land use focus on habitat and open space. The environmentalists, in turn, avoided issues of economic security for farmers and generally ignored broader food-system changes. Farmworker issues also tended to be avoided.[4]

At the same time, the urban-focused antihunger and emergency food system–oriented groups and advocates sought to advance their own agenda of continuing support for the food assistance and emergency food support programs such as food stamps and the TEFAP program. These approaches were developed without directly addressing the powerful changes in food growing, food production, and even food retail. The antihunger and emergency food coalitions became most focused on identifying how hunger had become endemic due to a porous social safety net, and therefore required continuing legislative vigilance in support of the food assistance as well as emergency food provisions. Ultimately, each of these alternative food players had established a set of arguments and carved out an arena for legislative action that addressed only one set of discrete outcomes in relation to food-system changes. It had also become clear by the 1994 Hilton Hotel meeting that a broader, more integrated rural-to-urban approach was needed that could address related social, economic, and environmental problems associated with food-system outcomes.

Was such an approach possible, given the stresses in the coalition process and differing perspectives on agendas and constituencies? One of the research questions posed by the *Seeds of Change* report had been whether and how a rural-to-urban or regional food systems approach might emerge. The report had pointed to the development of groups around the country that promoted such an integrated approach, such as the Hartford Food System (HFS) in Connecticut. First established in the late 1970s through the initiative of a progressive city government, the HFS had initially focused on food access and the abandonment of the inner city by full-service supermarkets. One of the few urban food advocacy groups formed in the 1970s able to survive into the Reagan-Bush era, the group had evolved into an independent nonprofit organization, though it maintained an influential role among municipal and

statewide policymakers. During the 1980s and 1990s, it had expanded
its agenda to include a broader set of food issues, including those impacting farmers as well as urban residents. It promoted and helped stimulate
farmers' markets, food cooperatives, community gardens, and new community-based enterprises. By the early 1990s, this more expansive
approach of the Hartford group had come to be shared by a number of
other food system–oriented groups in places like Austin, Texas, and Minneapolis–St.Paul. What most connected these groups was their desire to
respond to the urgency of the myriad of urban food problems, particularly those faced by low-income communities. This urban-oriented food
advocacy was primarily community focused (and included the use of the
term "community food security"). Several of the community food security groups also sought to address the increasing stresses of small and
local farmers, as well as the goal of achieving a more sustainable food
system. In some ways, these groups articulated the same kind of arguments raised by the food "conspirators" of the late 1960s and early
1970s in their quest for an alternative food system. The community food
security groups of the late 1980s and 1990s, however, were less mercu-
rial and haphazard in their mode of organizing and community outreach.
They were interested in institutional and policy-related changes as well
as an ideological shift. Nevertheless, like their 1970s counterparts, the
community food security groups shared the desire to identify a new way
of talking about food.[5]

Joining their sustainable agriculture, environmental, and antihunger
counterparts, the community food security groups had begun to emerge
as the new players within the food advocacy arena. Partly in the hopes
of cementing this role, the groups meeting in Chicago agreed that the
Farm Bill offered an important opportunity to make more visible their
concepts of community food security and local food systems change.
Farm Bill–related campaigns, it was felt, could aid in eventually coa-
lescing a wide range of groups and constituencies around such concepts.
The challenge, nevertheless, was formidable. Just four months after the
Chicago meeting, with efforts to draft a "Community Food Security Act"
under way, the Newt Gingrich–led Republicans swept to power in the
1994 elections. The Republican ascendancy meant that each of the constituencies previously engaged in Farm Bill initiatives—mainstream envi-

ronmentalists, small farm advocates, and antihunger constituencies—
were placed on the defensive. The Gingrichites were hostile to environmental measures such as soil conservation and crop rotation programs
as well as pesticide use reduction measures. They challenged key small-
farm support programs, such as the various price support programs, and
spoke instead of reducing government intervention and unleashing the
market and creating a "freedom to farm," as the new Farm Bill came to
be known. The Gingrichites were also focused on reducing food assistance programs as part of a broader assault on "welfare" and "entitle-
ments" that would culminate two years later in passage of the welfare
reform legislation.[6]

For the community food security advocates, the triumph of the Gingrichites further underlined the importance of being able to establish a
stronger, more integrated approach and a more inclusive advocacy language about food-system changes. Not only would the conservative
backlash need to be countered, but programs and approaches that
provided an alternative to the antigovernment (and anti-poor people)
promarket rhetoric that was sweeping Washington also needed to be
identified. The language used to advance such programs—community
empowerment, neighborhood and local action, strengthening farmer-to-
consumer links—proved powerful. And the examples available to illus-
trate those approaches—farmers' markets, community gardens, greater
food access, greater food choice, maintaining agriculture in the shadow
of, if not within the city—resonated in terms of the most powerful arguments associated with environmentalism and among other social movements. These included the attachment to place, the search to create
livable communities, and the desire for justice.

To get new legislation passed—a legislative amendment or additional
program to be added on to the Farm Bill—was further compounded by
initial skepticism of the other coalition advocates. Some antihunger
groups feared that the language of community empowerment and neighborhood action associated with community food security would be used
against the traditional antihunger arguments that had relied on a food
assistance and emergency relief framework. Mainstream environmental
groups, while not opposing the initiative, remained aloof and never
directly associated with the community food security campaign. The

mainstream environmentalist position was focused on environmental impacts from the growing of the food, not what happened to the food itself (with the exception of pesticide residues on produce that had been sprayed, which had come to be seen as a winning, middle-class, consumer-related issue). Sustainable agriculture groups were more sympathetic from the outset, and the concept of community food security, beginning with the 1995–1996 Farm Bill debates, became increasingly embraced by these groups as an important urban link to their approach.

The mark-up of the Farm Bill ultimately included a small item—a Community Food Projects program with authorization and funding for the next seven years—that became the successful outcome of the process that had been launched at the Hilton Hotel in August 1994. A new organization—the Community Food Security Coalition (CFSC)—was formed in the process, and hundreds of new programs, organizations, and policies were established at the local level. By 1998, one Web site listing indicated more than fifty major conferences or workshops that were to be hosted in the course of just a single month by local, regional, or statewide community food security groups related to various community food security themes. There continued to be a proliferation of projects and programs, policy initiatives, and new kinds of organizations and enterprises, primarily at the local and regional levels, but which could be directed at national and even global food agencies and issues. In 1999, in response to this outburst of activity, the USDA launched its own Community Food Security Initiative. In developing this program the department had borrowed liberally from some of the core policy recommendations put forth by the CFSC during its 1995–1996 Farm Bill campaign. Community food security had clearly emerged as a new type of political and social language, and a new kind of policy focus. But whether it represented a new type of social and environmental movement—and pathway for change—still needed to be determined.[7]

Exploring the Pathways

Exploring pathways for change—identifying where movements and agendas can emerge and coalesce—flows directly from the discussion in the previous chapter of food-system restructuring and its various outcomes. The remainder of this chapter is in part a discussion about new directions for environmental discourse, identifying how the new food movements can articulate a language that captures the attachment to place, the search to create livable communities, the struggle for justice, and the transformation of a production system.

Food Growing: Searching for Sustainability

During the late 1960s and early 1970s, the term "organic farming," associated with various food-growing practices, began to be identified as a possible alternative approach to the agribusiness market economy. By this period, farmer dependence on off-farm inputs and the near hegemonic role of the middle players had come to be fully embraced by USDA. The 1970 USDA Yearbook, for example, "fairly burst with pride at its description of the transformation of the American farm scene," as USDA critic and food activist Catherine Lerza characterized the USDA perspective, citing the department's own favorable descriptions of changes in food-growing strategies. Other critics complained that the emphasis on off-farm inputs and technological change were undermining the connection of farmers to the land. "Genetically re-designed, mechanically planted, thinned and weeded, chemically readied and mechanically harvested and sorted, food products move out of the field and into the processing and marketing stages—untouched by human hands," wrote Jim Hightower in 1972. Hightower, a leading anticorporate food activist at the time who would subsequently become the Texas Commissioner of Agriculture, characterized the state of agriculture and the growing of food as having less and less to do with the farmer. In challenging this approach, Hightower and others argued, a different type of farming system had to be nurtured, a radical agriculture that not only reduced the use of particular off-farm inputs but promoted an alternative vision of what it actually meant to farm or garden.[8]

For some, though not all of the alternative agriculture proponents of the late 1960s and early 1970s, an alternative vision meant "organic farming" or "biological," "natural," or "low-input" agriculture. The term "organic," popularized by the major advocacy publication for

organic farming at the time, *Organic Gardening and Farming*, was identified as a social as well as ecological approach, including the "decentralization of economy at all levels" and the formation of "communities
small enough to be reasonably self-regulating and self-supporting."
These communities would by their nature be intentional and utopian, an
extension of the counterculture's search for a new way of living, creating in the process a new type of "hip farmer," as the publication *Mother
Earth News* put it.[9]

By the early 1970s, publications such as *The Whole Earth Catalogue*
and *Mother Earth News* would provide the information about how to
construct such an alternative lifestyle, a kind of ideological "how to"
approach. The granddaddy of the organic publications, the Rodale
family's *Organic Gardening and Farming*, had earlier linked gardening
and farming tips with passionate appeals for developing "a simpler,
realer one-to-one relationship with the earth itself." The founder of the
publication, J.I. Rodale, had first used the term "organic" in the 1940s
in relation to his eliminating chemicals on his experimental farm in
Emmaus, Pennsylvania. But, as his son later wrote, the "organic"
concept became for the Rodales the reference "for a revolution in life
styles extending far beyond agriculture." This quest for "moulding lives
in more natural ways" was primarily designed in the 1940s and 1950s
to attract constituencies interested in such issues as health foods and
vitamin therapy. But with the ferment of the late 1960s and early 1970s,
the Rodale publication, which had been previously pegged as a type of
"faddish" publication, grew enormously thanks to the growing interest
in alternative lifestyles *and* reduced chemical use. From its base of 60,000
subscribers in 1958, *Organic Gardening and Farming* grew to as many
as 800,000 subscribers during the early 1970s. In the process, it became
a major force among organic farming advocates while increasingly
appealing to a growing constituency of urban gardeners.[10]

Aside from the followers of *Organic Gardening and Farming* and a
range of other publications that identified organic farming as a new way
of life, there were also critics of industrial agriculture and the agribusiness market economy who were advocates of an alternative agriculture
approach *as a strategy for livelihood*. These advocates were linked to
farm-based movements and small farmer advocacy organizations such as

National Land for People in California and Jim Hightower's Agricultural Accountability Project. They also included groups supportive of the farmworker organizing drives of the 1960s and 1970s that challenged the industrial form of agriculture and the large growers that relied on a cheap and heavily exploited labor force. All of these groups were focused on how to lessen dependence on the off-farm inputs, particularly the nearly universal application of agricultural industrial chemicals. Pesticides and herbicides had in fact emerged as the leading symbol of the dramatic changes on the farm and a lightning rod for protest regarding farmer, farmworker, environmental, and consumer exposures. They were also interested in the survival of the small family farmer and linked the reduced dependence on inputs to the quest to survive economically in the face of food-system changes, or what Ostendorf and Terry called the need to "creatively disengage from the dominant [food] economy." While one wing of this movement spoke of the need for broader social change, other groups focused less on the broad conceptual themes and more on the pressures of the small-farm economy. They were most concerned with the decline of rural communities and their social and economic support systems that had allowed local farmers to survive at the margins of the changing food system.[11]

This search for a more sustainable livelihood as opposed to a lifestyle change was also more directly associated with the range of new organic farming operations and techniques that emerged gradually during the 1970s and early 1980s. These "eco-farmers," as Garth Youngberg characterized this constituency, were small farmers concerned about occupational hazards, environmental impacts, and the growing recognition that pesticide use was turning out to be a mixed blessing. Fears about a "pesticide treadmill" (pest resistance developing in relation to one pesticide creating the need for yet more potent pesticides) created interest in exploring alternative "integrated pest management" approaches. Eco-farmers thus hoped to reduce or eliminate their reliance on chemical inputs through various growing strategies, both new and traditional. At the same time, many of the eco-farmers began to recognize that there might be an emerging market niche of new consumers less focused on alternative social visions than on the desire to shift toward a healthier whole foods diet.[12]

During the 1970s and early 1980s, most of the organic farmers were small growers who fit Youngberg's description of the "eco-farmers," although a variety of large producers and even manufacturers and various middle players began to develop a modest interest in the organic market as well. However, organic farming still remained a small segment of the farm population, given the complexity of the land management and growing skills required and the attention to management details not found in conventional and large-scale farming enterprises. The bigger, more industrialized farming enterprises required greater supervision of part-time and migrant workers and less focus on the actual details of the growing process and subtleties of on-farm management. The numbers of organic farmers in this period remained small—estimates varied from about 5,000 to 20,000 farmers. Differences in estimates were also reflected in the varying definitions of what constituted this new growing practice, or what Youngberg and Fred Buttel called "the changing nature of alternative agriculture as a symbolic phenomenon." The divide between "practitioners" (those who adopted some or all of the alternative, primarily nonchemical, growing techniques) and "romantics" or "visionaries" (who saw in alternative agriculture "not just a different set of techniques but a unique philosophical or ideological lifestyle"), revealed those differences in the symbols and language of organic farming, as Mark Rushevsky put it. The concept of integrated pest management, for example, which identified a variety of pest control techniques that could reduce but not necessarily eliminate the use of chemicals, became one of several terms introduced to emphasize "technique" rather than social vision or political challenge. Similarly, among the visionaries and political advocates were those like Hightower who were interested in broader questions of rural development and the squeeze on small farmers, as well as those who emphasized an environmental focus about food and the rural environment.[13]

While these distinctive perspectives still sought, though not always successfully, to establish a common ground through the concepts of organic as well as low-input food growing, the ability to define what constituted organic or low-input growing remained largely elusive. The problem of definition, including the meaning of such terms as "organic," "low-input," "biological," "natural," and "ecological" farming or growing of

food, was heightened by the lack of interest of the USDA to help facilitate such growing strategies. During the 1970s, when organic growing
appeared to be as much a philosophy as a set of farming practices and
the number of organic growers was relatively small, the USDA failed to
dedicate any direct staff or financial resources to organic farming. By the
1980s, however, pressure had significantly increased on the USDA to
address the organic farming issue through its cooperative extension programs as well as through the land grant colleges and related research
entities, all of whom had long sought to facilitate the use of chemicals
and other nonfarm inputs while also encouraging the rise of the agribusiness market economy. With these growing pressures, a small but noticeable shift in the discourse about organic farming began to occur,
facilitated in part by a renewed sense of crisis about the farm economy,
the growing challenges regarding pesticide use in the policy arena, and
the research that identified the comparative advantages of organic
growing, such as the reduced use of energy per item of food grown. The
market opportunities for organic growers, or "sustainable farming" as
some of its advocates began to call the practices and philosophy of alternative agriculture, also increased during this period, thanks in part to
modest increases in marketing opportunities such as health food stores
and farmers' markets. The growth in sales, to be sure, was still modest
in relation to overall market share for fresh produce. Nevertheless, a
broader sustainable agriculture movement, focused on issues of farm size
and ownership, rural communities and rural culture, and environment
and land management or a land ethic, was clearly on the rise. With the
"alternative farm bill" initiatives in 1981, 1985, and 1990, with new
research funds and programs even at the land grant schools, and with
the continuing battleground over pesticide impacts, sustainable agriculture had become an important political force in the policy arena. It was
also emerging as a recognized and potentially viable alternative for
farmers, particularly small farmers more dependent on local and regional
sales.[14]

As alternative growing practices became more viable and even mainstream, the use of the term "organic" had become, by the late 1980s,
the centerpiece of an important area of conflict and policy dispute. Part
of the concern over the nature and definition of what constituted an

"organic" product or organic farming process had to do with issues of certification and labeling. During the 1980s, various states and independent trade associations had established certification standards, although these varied between states, and most states, including a few with significant agricultural activity, had no certification process at all. Led by the emerging sustainable agriculture movement and a handful of organic producer trade associations, legislation was signed into law in 1990 that sought to establish mechanisms for a federally based, standardized certification process. This legislation, the Organic Foods Production Act, called for the development of national organic standards. As a result of the legislation, the USDA established a National Organic Program and also empowered a National Organic Standards Board (NOSB) to prepare recommendations for a national organic standard to be issued by the secretary of agriculture.[15]

The NOSB process helped mobilize a wide range of constituencies loosely connected to the growing sustainable agriculture movement. Through the NOSB, a set of recommendations was developed between 1992 and 1997 that essentially elaborated the approaches and goals of standardization and certification promoted by the sustainable agriculture advocates. The publication of the recommendations, however, also generated a counter-response from various regulatory and industry interests who wanted to stretch the definition of "organic" to include key elements of the industrial agriculture/global food-system model to also apply to the definition of organic. The most contentious of these efforts, also known as the Big Three, were embraced by USDA in its December 1997 draft set of regulations governing organic food standards. The Big Three incorporated, within the definition of "organic," food irradiation, the use of sewage sludge, and genetically modified crops and hormone-injected beef. Each of those additional items was immediately challenged as undermining core assumptions about what constituted the organic growing of food. Food irradiation techniques, already approved by the FDA, had implications in relation to environmental and health considerations and for the storage of food, which in turn created a potential disconnect between "organic," "locally grown," and food defined as "fresh." The issue of sewage sludge, which also had reference to EPA debates over disposal of the sludge, was seen as potentially undermining core assumptions that organic fertilizer was also non-toxic. The debate over growth hormones and genetically modified food crops cut to the heart of the debates about the nature of the food system itself. These included issues associated with the objectives of the players in the biotechnology field in controlling different aspects of farming and its various inputs. It also raised concerns about the type of food grown and produced that would be labeled "organic." For some critics, the very survival of organic farming, as demonstrated by the concerns over the introduction of genetically altered Bt corn, cotton, and potato crops, was also at stake.[16]

The publication of the December 1997 rules touched off a firestorm of protests and helped consolidate a sustainable agriculture perspective beyond what had been accomplished through the development of the Farm Bill coalitions. More than 275,000 comments were generated, more than for any previous USDA action, with nearly all the comments opposing the incorporation of the Big Three in the proposed regulations. The protests against the Big Three also helped establish a connection between the concerns of the practitioners and the visionaries as well as the political and environmental advocates that had not been previously as clearly articulated. The campaign against the Big Three, similar in some ways to the explosion of interest in community food security and local food-system advocacy, also caused an extraordinary increase in activity around the issues of genetically modified crops specifically and the organic rule issue as a whole. In response, the USDA postponed implementation of the rules and began to retreat from its position around the Big Three. When the proposed regulations were reintroduced two years later, the Big Three had been eliminated.[17]

By the late 1990s, it was clear that sustainable agriculture had become a significant form of advocacy both within and outside the USDA. The USDA's Small Farm Commission issued a report in 1998 that strongly embraced many of the approaches of the sustainable agriculture advocacy groups, and explicitly linked the need for strengthening family farmers and small growers and a renewed commitment to rural development with the issue of growing techniques and strategies. Sales of organic food products were also increasing significantly. Already by the late 1980s and early 1990s, the market for food grown "organically" (which was occurring without a national certification process) was growing at a rate of 20 percent or more annually. By the new century, market analysts were declaring that organic food sales were "not a niche market anymore in terms of consumer interest," with organic production including such diverse categories as vegetable and fruit crops, livestock and poultry, and nursery and flowers. One of the more interesting and intriguing growth areas was in organic dairy products, whose sales increased rapidly in the mid- and late 1990s after the Food and Drug Administration approved the use of Monsanto's genetically engineered growth hormone, rBGH. Those increases extended beyond the sales of plain milk and yogurt products to include specialty items like jalapeño cheddar cheese and chocolate chip ice cream. In terms of specialty markets, the USDA, still reeling from the battles waged over the organic standards rule during the previous year, decided in a January 1999 ruling to allow organic labels on meat. This label was based on the use of organic feed, prohibiting the use of antibiotics and requiring animals to have fresh air and sunlight. Both the USDA and the organic meat growers assumed the ruling would help facilitate the development of a niche or specialty market for small growers or producers, a market also associated with low-fat, lean-meat products such as ostrich and buffalo.[18]

The search for new kinds of organic products extended in a number of directions. McDonald's Swedish subsidiary, for example, began to utilize an organic milk product, while the transnational food companies Nestlé and Sandoz also began to assume positions in the organic food market. *Biofact* magazine even reported that the Mars company, the manufacturers of the Mars candy bar, had decided to develop an organic Mars chocolate product. In Joan Gussow's memorable phrase, the extension of the organics niche market argument could presumably lead to the development of an "organic twinkie" based on the use of organic cream and flour.[19]

For Gussow, a nutritionist who had also emerged as a major advocate linking alternative food growing and food consuming strategies, the power of the concept "organic" was its environmental and land stewardship (and attachment to place) associations.[20] If organic farming meant a different type of connection to the land as well as responsibility for the kind of food grown and marketed, then it also required a form of support that could potentially make it independent of the agribusiness market economy as well. One development along those lines was the Community Supported Agriculture initiative that was introduced in the United States in the mid-1980s. The CSA approach was first developed in Japan in the mid-1960s by the "Seikatsu Club," a group of mothers and wives who wanted to construct a family-based and consumer-related food purchasing strategy that would also sustain local farming operations threatened by food imports and declining farm populations. The "seikatsu" concept, which translates as "putting the farmers' face on food," was subsequently introduced into Europe and then later adapted to the United States and given the name "Community Supported Agriculture" at Indian Line Farm in Massachusetts in 1985. The concept was immediately embraced in this country by pro-organic urban residents and by small farmers worried about survival, several of whom were now obliged to work their farms on only a part-time basis, while seeking off-farm sources of income.[21]

CSAs were organized as a contractual partnership between a grower and a group of households or "shareholders" who agreed to purchase, often on an annual basis, some or all of the crops produced by that grower. The purchase would be made in advance, thus providing a crucial source of secure income for the grower. The shareholders in turn would receive a box or basket of the produce, often on a weekly basis, of the food harvested from the CSA farm. Many of the shareholders would participate in some aspects of the farming activity, particularly the harvesting of the crops. Environmental issues were also paramount, since CSA farms were organic farms. Thus, the attraction for the shareholders was their access to organic produce as well as a shared responsibility for the stewardship of the farm.[22]

The CSA concept was structured around two key objectives. First, it provided the grower with a secure and stable source of income *in advance of the actual cycle of growing and harvesting*. This advance payment, it was hoped, would protect the farmer not just from specific market pressures but allow for their (at least partial) disengagement from the agribusiness market economy itself. Second, through the participation of the shareholders in the life and activities of the farm, including the sharing of the risks and uncertainties of the growing of food, the concept of stewardship could be extended to the consumer end of the farmer-to-consumer relationship. "By entering [into] contracts with local farmers," one CSA group declared, "community agriculture supporters have a say in not only how their food is produced with regard to the environment, but with regard to society." But despite its powerful "throughput" or "food-system" implications, the CSA still suffered from its "niche" status, defined in part by its reliance on consumers who could afford to provide payment for as much as a year in advance. Still, the CSA farm, the number of which increased significantly during the 1990s, emerged as an important component of the sustainable agriculture perspective, providing a potential alternative "direct marketing" or farm-to-consumer dimension to the organic farming approach.[23]

Whether seen as niche market, survival strategy for small farmers, alternative food system approach, or environmental philosophy, it had become clear by the turn of the century that the definition and strategies of organic food growing had become an important battleground. This also extended to the unresolved policy debates that could significantly influence sustainable agriculture/organic farming's market presence and place within the food system. How one defined organic was as much a matter of discourse (David Harvey's "manifestations of power") as semantic distinction, given its importance in identifying alternatives to the agribusiness market economy, as the CSA example seeks to provide. But how viable and widespread the CSA model can become, similar to the debates over whether organic food growing can be more than just a high-priced, niche market at the edges of the food system, raises the broader question of the nature of the alternative food regime itself. Sustainable agriculture, as some of its more radical or system-focused critics have argued, needs to address the question of *who* as well as *what* is to be sustained. It is an argument, for example, that underlines the importance of a system's impacts on farmworkers and rural communities as well as the nature of inputs and environmental outcomes in identifying an alternative form of production as well as distribution. The type of rural-to-urban or grower-to-consumer links that can be established— which includes the issues of food growing and food access *in the city*— parallels the issues associated with the future of farming. While the search for an alternative food regime begins with the search for sustainability at its front end, it then also makes its way into the city where another type of food movement has begun to take shape.[24]

Food in the City

The idea of growing and producing food *in* the cities has been part of the urban and environmental discourse in industrialized societies for more than a century. In western Europe, intensively cultivated gardens, also known as "allotments" or "schrebergartens," were considered part of the cultural as well as physical urban landscape. Lewis Mumford in his classic text, *The City in History*, spoke of the need to bring "life into the city, so that its poorest inhabitant will have not merely sun and air but some chance to touch and feel and cultivate the earth." An urban counterpart to what Aldo Leopold called the need for "land health," the urban garden fulfilled both social needs (for its food-related activities) as well as environmental needs (for its capacity to bring "life into the city").[25]

Despite its role as resource and amenity, the urban garden has had an erratic existence and often shifting set of objectives. There have been periods when urban gardening was seen as a food source for the poor and unemployed ("Pingree's potato patches," which first emerged in the wake of the 1893 economic decline, and the "subsistence" or "Relief Gardens" during the Depression).[26] It has involved initiatives both during World War I (the "United States School Garden Army") and World War II (the "victory gardens") when urban and school garden-related food production were seen as part of the war effort.[27] It has also included community building and value-based initiatives (the school gardens and garden city plots during the Progressive Era and the community garden movements that emerged in the 1970s, 1980s, and 1990s), when urban gardens were seen as contributing to the social and community fabric.

In the Progressive Era, the urban community garden and the school garden were seen by some of its advocates as an extension of the Taylorist interest in work discipline and efficiency of production. One of the leaders of the school garden movement, Henry Parsons, argued that gardens provided a type of early training in the concepts of scientific management, such as "carrying watering cans more efficiently, and by using the proper size tools." "Instructors aimed to make the efficient way a habit by equating moral with material progress," Parsons wrote, equating the role of teachers as equivalent to foremen. Another school garden advocate, Maria Louise Greene, also argued that school gardens instilled in the children "such knowledge of natural forces and their laws as shall develop character and efficiency." The ability of the school garden to be used as an instrument for social and moral education, to reaffirm a connection to the land and agricultural activity, and to prepare young people for entry into industrial life proved enormously appealing to educators during the Progressive Era.[28]

However, the idea of urban gardens serving as a form of food production was strongly opposed during this same period by business interests concerned with potential competitive pressures on labor costs. Thomas Bassett cites the experience of a Philadelphia Cooperative Farm that had been established to provide employment opportunities for unemployed workers. Soon, the success of the farm made it possible to raise the initial baseline wages by more than 30 percent. But the board of directors of the farm, consisting of local businessmen, opposed such increases as capable of "[making] no end of trouble." "If the people [gardening] were to find they could earn as much money as that," the business group concluded, "they would either leave the factories or demand as much pay there." Similarly, opposition from food retailers developed when vacant lot gardens in Minneapolis promoted by the local garden club began to provide a source of local produce for individual gardeners and their families as well as for others. In response, garden club members redefined the goal of the vacant lot gardens as providing neighborhood beautification, which could also then benefit local merchants due to the more attractive surroundings. Other Progressive Era urban garden/urban agriculture advocates took a more radical approach, linking garden development to the single-tax philosophy and cooperative anarchist philosophies of Henry George and Peter Kropotkin. Urban food growing in this context was seen as establishing a community right to the land as well as a cooperative form of organization in contrast to the atomizing, antidemocratic, and socially and environmentally destructive features of industrialism.[29]

During the 1930s, gardening reemerged as a form of economic activity related to the problems of high unemployment. Two models were proposed. One was the individual garden based on a small plot designed to promote the "joy of possession by having as a reward of their labor all the produce from their own particular garden plot." This contrasted with the community or enterprise garden concept, including five- to ten-acre "industrial" plots designed to maximize the production of fresh vegetables ("the greatest amount of food for the greatest number of people"). Gardeners would be paid in cash or in kind, and the success of the garden would be measured by its "output." One study by the Russell Sage Foundation also pointed out that those recruited for the subsistence gardens needed to acquire a broader knowledge of the techniques and purpose of the gardening. This stood in contrast to industrial work where workers would not be accustomed to "take the responsibility of seeing a piece of work through to a finish," but would be used instead to perform "one operation in the company of others who are doing the same thing."[30]

During World War II, the victory garden programs provided a significant source of food for individual families, while serving a community or public purpose to supplement the food rationing that was required. After the war, however, gardens were essentially privatized, both as public spaces and in their role as food source. The backyard garden and front lawn of the individual home replaced the vacant-lot garden and the community sites for neighborhood gardeners. The community garden only reappeared in the late 1960s and 1970s as part of the search for new community identities and green spaces. The 1969 confrontation over People's Park in Berkeley, for example, was primarily a conflict over contending views of appropriate uses of the land—a parking lot for the university versus a more anarchic mix of garden plots as both public space and food source.[31]

During the 1970s, when concerns about energy use were paramount, community gardens were seen as significantly reducing reliance on fossil fuels due to their "grow your own" approach (thus eliminating transport costs) and because they relied on fewer inputs, including energy. Growing one's own food was seen less as a strategy for dealing with food needs than as an expression of self-sufficiency in an energy-intensive, urban-industrial world. As distinct from the concept of the community garden, individual backyard gardening (primarily associated with the concept of the suburban home as an extension of a greener, more Nature-centered landscape) had also become widespread in suburban areas during the 1960s and 1970s. But for those who saw urban gardening as a form of "community open space" or "urban greening," the garden provided a community-building activity in an otherwise bleak, auto- and concrete-dominated urban landscape.[32]

This association of garden with community could also be considered the urban equivalent of the late 1960s, early 1970s "back-to-the-land" communal movements and the related romantic and visionary appeal of organic food growing as a kind of counter-ideology. But the "back-to-the-land" advocates also remained ambivalent about gardening or food producing *in the city*. For example, one of the advocates of the community open-space movement during the 1970s, the Berkeley-based Farallones Institute, debated the wisdom of establishing an urban agriculture or community food approach. Some Institute participants expressed their distaste for the "pollution and constraints of the city as compared with the attractions of beginning a new ecologically harmonious community in a more pristine rural setting."[33] This tension between searching for a "more pristine rural setting" (which paralleled the suburban/exurban search for "natural" landscapes and low-density development) and creating a more livable city also identified the middle-class character of this form of environmental advocacy. Similarly, the desire to create greater ecological design for urban homes—which paralleled the community open-space advocacy—still assumed that the single family home would be the unit for planning and design. Designing the "Eco-Home" to include the growing of food did not directly address the larger dimensions of the central-city housing and land-use dilemmas (not enough housing, contaminated land, vacant derelict lots amid limited open and green spaces). For urban agriculture to become a more significant central-city approach and establish a more expansive agenda, those issues needed to be addressed as well.

The rooftop garden movement in New York City that developed in tandem with housing-related squatters' movements in the 1970s was one of the first of the inner-city community garden initiatives. The popularity of the rooftop gardens paralleled the renewed growth of the vacant lot gardens during the late 1970s and 1980s in several inner-city communities across the country. Together, they provided a major new impetus for developing community gardens as a form of individual food production as well as for their community open-space and urban-greening benefits. In response, a handful of modest funding programs and grant opportunities were established by the USDA and EPA through the 1980s and 1990s. Urban forestry programs were also developed to complement the new interest in urban greening, with tree planting and community gardening technical assistance programs established in nearly every major urban area in the country. Activities for seniors, physical activities programs for children, and environmental education programs associated with school gardens provided a physical fitness and educational backdrop to the development of these programs. A wide range of public and nonprofit entities began to facilitate urban garden programs and help establish some unusual partnerships, such as when the Nebraska national guard was called in to deliver topsoil for an inner-city garden. The American Community Gardening Association, the leading garden advocacy group that formed in 1978, grew substantially in this period, and extended its focus to include advocacy for inner-city community gardens. By the turn of the century, nearly every city in the country had some form of a community gardens program and provided a modest degree of institutional and technical assistance support, much of which was facilitated by the advocacy groups. In the process, urban agriculture emerged as the catch-all phrase for a range of food, environmental, and urban-greening strategies, extending from "worm farms" for composting and fertilizer to the revival of school gardens such as California's "garden in every school" initiative.[34]

What was not as directly addressed in these developments was the purpose and capacity of community gardens and other urban agriculture initiatives providing a significant source of food production in the city. Could community gardens become an effective community food security strategy, and not just a marginal form of urban agriculture? Or were the land use, environmental, and economic constraints so great that joining the concepts "urban" and "agriculture" had to be considered outside the bounds of either an urban or agricultural policy framework? Could urban gardens and urban agriculture programs survive when urban land uses were so thoroughly driven by economic and often speculative real-estate interests, even when such land was publically owned?

One example of this land-use dilemma was the Esperanza community garden in downtown Los Angeles, bordering on the immigrant communities of Pico Union to its west and the high-rise office buildings to its north. The garden originated through a small grant by the Community Redevelopment Agency to a local artist to establish a community art project. The CRA, more than any other agency, had been a key force in restructuring the downtown neighborhoods of Los Angeles in the 1950s, 1960s, and 1970s, bulldozing working-class Latino communities for office towers, monumentalist cultural centers, and high-rise condos for upper-middle-class professionals and mid-level executives. The absence of green space, nightlife, and any kind of community identity in the southern and western edge of downtown contrasted with other areas of downtown largely outside CRA/corporate redevelopment influence, such as the Broadway area to the east with its vibrant Latino street life. The artist decided to use the funds to establish a community garden, and within a year an area that had previously been littered with broken cement and rubble became both green space and a social or community place for its gardener participants. Similar to some farmers' markets, the Esperanza garden provided a cross-class and cross-cultural mix of gardeners, a "magical place," as a *Los Angeles Times* reporter characterized it, "a site where people who might never meet interact and swap gardening tips and share food and land." Similar to other community gardens, however, it was vulnerable to decisions to use the land differently; in this case, a new housing project. As a result, the "magical place" was bulldozed. On a larger scale, another example of that vulnerability involved New York City Mayor Rudy Giuliani's plans to auction seventy-five community gardens on city-owned properties to private bid, an extraordinary move given the popularity and community history associated with New York's community garden initiatives. This fire-sale elimination of the gardens was only prevented when a combination of community garden supporters, including the Trust for Public Land and actress Bette Midler, came up with a sufficient offer of funds to pay for the properties.[35]

In seeking to counter the real-estate market's pressures, community garden advocates sought to demonstrate that urban agriculture could become a key component of any municipal or regional food planning approach. It had become clear that Third World urban agriculture had become a significant food production strategy for geometrically increasing urban populations. But what of the industrial world, including the United States, where urban land costs might be prohibitive for food production and the soils might be contaminated? One interesting historical precedent had been the victory gardens program of World War II. Inaugurated shortly after Pearl Harbor at a national conference hosted by several government agencies, the victory gardens were designed explicitly as a food production initiative to take some of the pressure off the demand for commercial food supplies. A range of programs and services were established or expanded to facilitate garden development. In some counties, the Extension Service emphasized its urban agriculture focus, and provided assistance by making available seeds, fertilizer, and gardening tools. Various government agencies and private landowners were also enlisted to turn over land (including but not limited to vacant lots) to be used as victory garden plots. Literally overnight, a significant level of urban agriculture production was established. By 1943, 20 million victory gardens produced more than 40 percent of the fresh vegetables consumed that year, and the increase in the consumption of fresh vegetables (as well as vitamin C) hit a historic high during the war years.[36]

After the war, the victory gardens approach, designed with public goals and public resources as well as on the basis of national and municipal food planning, was disbanded. Urban gardening and horticulture as forms of *market-based economic activity* targeting individual households as well as private and municipal uses such as golf courses or landscaping catapulted over the community-based garden approach and emerged as a more significant type of business. By the late 1990s, as much as $10 billion was being spent on golf courses, freeway landscape, and cemeteries, and another $1.8 billion on residential yards. Gardening as individual activity had become fully integrated into suburban culture, as newspaper supplements and columns complemented the wide range of gardening publications. The garden became an extension of the private suburban home, rather than a bridge to the community. Food growing, including edible landscapes such as fruit trees, was more afterthought or incidental to the garden as having aesthetic, financial, and therapeutic value. At the same time, the household garden was targeted by the chemical specialty manufacturers as a lucrative and rapidly growing market for pesticide products, becoming a significant contributor to the environmental hazards of the city, such as stormwater pollution.[37]

The rise of the community garden since the 1970s has provided a contrast to the institutional (golf courses, cemeteries) and individual household market-oriented applications and purposes of gardening. Community gardens have been less environmentally hazardous (pesticide applications do occur, but several community gardens have placed restrictions on their use), and they have been explicitly designed to provide new forms of social interaction. Community garden plots for individuals can also become an important source of fresh produce. This is particularly significant for those without access to supermarkets and where cost savings on food, which could amount to as much as several hundred dollars in the course of a year, may be crucial. A widely used estimate has indicated that for each dollar invested in a community garden, a value of six dollars in produce can be generated. But like the earlier debates during the Progressive Era and the New Deal, the question of whether community gardens can be a source of food production for more than just the individual household (as well as a job and skill-training enterprise) remains uncertain and often dismissed.[38]

During the 1990s, a handful of new and innovative "entrepreneurial" or "market" garden programs were developed to address the question of food source and skill-related training in food production. One of the most noteworthy of these programs has been the Crenshaw High School–based "Food from the Hood" project. Food from the Hood was created shortly after the 1992 riots in Los Angeles as a school garden at an abandoned space behind the Crenshaw High School campus in South Central Los Angeles. Organized by Tammy Bird, the school's science and volleyball teacher, as an environmental-related outdoor classroom for her science classes, the program became an entrepreneurial activity when students began to sell their garden produce at a local farmers' market. By the end of the first year, profits from the program were made available for three scholarships for graduating seniors. This initial success led to the idea of a food production enterprise, a salad dressing to be marketed with the name of the project, "Food from the Hood." Start-up funding was provided through Rebuild LA and state antigang funds, technical assistance and management support and participation came from marketing executives and professional food brokers, and advice about distribution and packaging was offered by the owner of Bernstein's Salad Dressing. The project even included a small franchising component after a youth enterprise program in Ithaca, New York, adopted the name for an applesauce product. By linking the school garden to a food business, Food from the Hood became a model for one type of urban agriculture enterprise; an urban food business with a social profile. This included the decision to use the increase in profits from the business to provide more scholarships with higher amounts per scholarship. In this way, Food from the Hood also became a source of identity for the school itself.[39]

The significance of the Food from the Hood model has been its complex mix of social and entrepreneurial as well as food-related goals. "Market garden" initiatives, however—that is, projects that have sought to use the community garden as a source of business activity for selling the items grown—have at times turned out to be more problematic as a type of profitable or even self-sufficient economic activity. The social goals associated with skill training and job development as well as the community functions associated with the use of the land have been more successfully achieved than their capacity to achieve profitability as a small business. Other small-scale food processing enterprises—the alternative "value-added" or alternative middle player approaches—have also had important social but more problematic economic outcomes. Most of these enterprises, such as Home Boy Tortillas in East Los Angeles, or the Riverside Eco-Park in Burlington, Vermont, have been noteworthy for their job creation, community development, or their small business incubator functions, re-creating the connection between producer and consumer in their value-added activities. Yet most of these enterprises have not thrived or been competitive in relation to the larger food-processing or production facilities. A deeper, more policy-driven process, whether in the form of tax policy instruments, social marketing strategies, or the use of institutional purchasing power, is required to turn the small pilot-scale enterprise into a more sustainable production alternative. Ultimately, the strategies of urban agriculture and the alternative food enterprise, whether market garden, community garden, or any other approach toward growing and producing food in the city, have required a far more substantial change in the dynamics and economics of land use, regional, and other forms of alternative production planning for core food and community goals within as well as at the edge of the city.[40]

Access to Food

The difficulties associated with the economics of food growing and alternative food production in the city can also be seen in the problems of maintaining viable small food operations in the retailing of food. Food retail has long been seen as a community enterprise. Food markets, including supermarkets, have remained largely "neighborhood" businesses, even as the definition of the "neighborhood" or the community service area has changed due to the shift to the suburbs and the rise of auto-dependent shopping. Even with the rise of the big-box megastores such as Costco or the entry into the business by operations such as Wal-Mart, most food markets have continued to draw on proximate neighborhoods and communities for their customer base. Despite their changing formats, food stores have also remained vulnerable to organized community pressures and have needed to demonstrate some level of community engagement, even if just minimal support for local schools or civic groups or functions. With the abandonment of supermarkets from the inner cities in the 1970s and 1980s, moreover, supermarkets have become an increasing target of community action. This has included efforts to develop joint ventures linking inner-city supermarkets with community development corporations as well as community-oriented funding sources such as the Local Initiative Support Corporation (LISC). It has involved the development of transportation strategies to increase access to full-service food markets in low-income, transit-dependent communities. It has been associated with community-based campaigns targeting both the quality and availability of food in supermarkets. It has also involved community pressure to make available local produce or other locally grown or produced food items in supermarkets.[41]

Retailers have sought to deflect such community pressures, but have sometimes needed to or been forced to respond, given their potential community vulnerabilities. For example, with the transportation issue, supermarkets have contributed to the dependence on the automobile for food shopping, a trend largely associated with the supermarket shift to the suburbs. Unlike suburbanites, however, inner-city shoppers have become dependent on other forms of transportation due to lower per capita car ownership. This transit dilemma has been compounded by the intercity (i.e., work commute) rather than intra-city (i.e., neighborhood services) focus of the public transit system. Most inner-city supermarket chains have in turn tended to avoid the transportation issue, even though some of the problems—and costs—endemic to inner-city markets, such as shopping cart loss, could be directly traceable to the transportation needs of shoppers.[42]

As the supermarket issue heated up in low-income communities during the 1990s, transportation emerged as one of the critical community demands. A few markets established transportation programs such as a van service that also became a signature program for the market, providing both a community and economic advantage in terms of expanding a customer base and establishing a community identity. Community groups were also successful in reorienting transit services to account for shopping needs, such as the "Grocery Bus" program in Austin, Texas, and new bus routes established in places like Knoxville, Tennessee. By the end of the decade, such community-inspired transportation programs even included a handful of the larger chains above and beyond the ethnic-oriented and smaller independent stores that had taken the lead on the issue. Community food security, environmental, and transportation groups also began to coalesce around these issues. This included an effort by the Surface Transportation Policy Project and the Community Food Security Coalition to include language in the 1998 omnibus transportation bill to fund transit programs to meet community food needs. Though this specific effort failed, for STPP, the food access issue reinforced its argument about transportation programs needing to meet neighborhood and community needs, rather than exclusively for commuter objectives. For the Community Food Security Coalition, it helped expand its food agenda to include a transportation and land-use focus as well.[43]

Community pressures also began to impact supermarket decisions about whether to carry locally grown produce. One large retailer in Dallas, for example, while recognizing the community value and even potential profitability of a local van service that could increase access as well as the store's customer base and sales volume in nearby low-income neighborhoods, decided not to enlist community feedback for such a service. "We can't make decisions on the basis of community interest," the store executive declared. Similarly *Progressive Grocer*, the retail industry's trade publication, in an article about grocer interest in highlighting local produce, cited one chain-store produce manager's ambivalence about that approach. The manager told the trade publication that although the local produce was popular and had a much better taste, he also worried that it could start undermining interest in the produce brought in through their longer distance distribution channels.[44]

Community ties have been more directly associated with various alternative food marketing and selling approaches, such as street vendors, food buying clubs, alternative food stores, and farmers' markets. Street vendors and specialty ethnic food stores, for example, made a significant comeback in the 1980s and 1990s in cities with newly arrived immigrant populations who found that the local supermarkets were not carrying food items popular with various groups. In response, some of the independent markets began to stock ethnic-related specialty items, and a few supermarket chains even experimented with a more ethnic-oriented format, such as Ralph's unsuccessful introduction of the "Viva" chain in Los Angeles that was designed for the Latino market.[45] Food buying clubs and alternative food stores also reappeared in the 1990s, reemerging from earlier, mostly failed experiments of the food cooperatives of the 1970s, and the parallel, though more successful development of the niche-market-oriented health food store of the same period. Unlike the earlier counterculture-oriented "alternative institution" strategies of the 1970s, these programs and businesses began to carve out specific market segments within the food system. Some, though not all such enterprises, targeted higher-income, health-, nutrition-, or environmental-oriented customers who had also become the market for many of the organic growers as well.

The rise of the urban farmers' market in the 1980s and 1990s is perhaps most indicative of the promise and ambiguities of alternative food-system strategies. While the practice of farmers selling food directly to consumers, either through farm stands or at farmers' markets, continued to be a prominent source of income to the farmer and a source of food for the consumer as late as World War II, that system eroded notably after the war. The decline of the central city markets was primarily due to the land-use and transportation changes that included the decline of central business districts and the shift of markets and store shopping to the suburbs.[46] But part of the change in how farm produce was sold was also reflected in the policies governing the packing and grading of produce. During the 1950s and 1960s, national and state policies were established to standardize the size, quality, and packaging of produce. These "standard pack and grade" measures effectively restricted, if not eliminated, the opportunities for farmers' selling directly to consumers outside the immediate vicinity of the farm. However, during the 1960s and early 1970s, a handful of new "direct marketing" opportunities for small local farmers began to reappear, stimulated by the rise of the organics movement and strongly facilitated by the interest of small-farm advocacy groups in identifying farmer-to-consumer programs. Some of these markets were little more than expanded road stands, and several assumed a counterculture format, including barter and crafts activities along with the selling of food. But new policy initiatives at both the national and state levels that sought to bypass the pack and grade restrictions in support of direct marketing were also developed that sought to extend opportunities for small farmers. The interest in direct marketing was aided by what economist Phil LeVeen characterized as the unfulfilled demand for high-quality fresh fruits and vegetables, due in part to the sacrifice of quality and flavor in produce caused by the agribusiness market economy's near exclusive focus on national and international markets. This in turn had opened the way for an alternative urban market network, including the truck farming of organic fresh fruits and vegetables into places like farmers' markets. These opportunities were given an important boost with the passage of the October 1976 Farmer-to-Consumer Direct Marketing Act, which instructed a recalcitrant USDA to support and promote direct marketing activities. Such activities included roadside stands, pick-your-own (PYO) farms, and, most importantly, farmers' markets, the central strategy in the renewal of direct marketing and alternative, locally based urban food approaches.[47]

When a new generation of farmers' markets began to reappear during the mid- and late 1970s, most were farmer driven, organized in states like Michigan, Pennsylvania, Ohio, Indiana, New Jersey, and North Carolina, where direct marketing was seen as an essential component of the small-farm economy. Items sold at the markets included local seasonal staples such as sweet corn, apples, strawberries, and peaches, though a few direct marketing initiatives, such as Pennsylvania's independent producer-retailer "milk juggers," provided for a variety of other products as well. During the late 1970s, a few states also began to allow for exemptions from the pack-and-grade requirements, partly in response to this new *urban* focus on creating greater access for fresh-from-the-farm produce. The pack and grade exemption established in California in 1977, for example, provided for direct farmer-to-consumer sales as long as the seller was a grower, a member of the grower's family, or an employee of the grower. Most of the markets first certified by the California program were also farmer oriented, such as road stands adjacent to a grower's field or public markets in largely rural areas or gateway communities in farm producing areas.[48]

By the late 1970s, the budding popularity of farmers' markets, which extended to as many as 780 markets in thirty-six states, began to attract organizations focused on the needs of low-income residents without access to fresh produce. As one example, the Nutritional Development Services of the Archdiocese in Philadelphia opened three "tailgate markets" on church lots in low-income neighborhoods in 1978 primarily to serve its parishioners. Similarly, the Southern California-based Interfaith Hunger Coalition's "Hunger Organizing Team" (HOT) began to explore in this same period the feasibility of establishing farmers' markets in low-income or "bridge" communities that could attract both middle-income and low-income shoppers. The first of such markets was established in the Gardena area in southwest Los Angeles, with just a handful of farmers participating and uncertainty as to whether a market in such a location would attract a sufficient volume of sales. But, as in Philadelphia, where the tailgate markets had far surpassed initial expectations, opening day at the Gardena market was an unanticipated success, with all the produce brought by the handful of participating farmers sold within two hours. Soon after, another market in a church parking lot was successfully established in South Central Los Angeles that also sought to target low-income shoppers. The early success of these markets, and similar efforts in San Francisco, New York City, and Boston suggested that farmers' markets could become an important community food security initiative.[49]

By the 1990s, farmers' markets had indeed become an important and vibrant community-based program, providing benefits for farmers, consumers, and communities alike. But in large urban communities such as Los Angeles, New York, and Philadelphia, where markets had first been established to improve community food security for low-income communities, new farmers' markets were now located in middle-income communities. One USDA survey indicated that the average household income of farmers' market shoppers by the late 1990s was in the $40,000–$65,000 range. No longer simply focused on affordability, even if prices were comparable on certain items to supermarket prices, farmers' markets began to be promoted as civic enterprises, places where a different type of interaction and relationship between farmer and consumer could be established. Although not all the farmers who sold at farmers' markets were certified organic, the importance of "locality" and "seasonality" or the freshness of the produce did provide a focus on where and when as well as how the produce was grown. Farmers' markets, in fact, had emerged as a critical economic strategy available to small farmers as well as organic farmers who relied on direct marketing for 50 percent or more of their revenue stream. The popularity of the markets—by 1999 as many as 2,500 farmers markets with annual sales of more than $1 billion had come to be located in all fifty states—

offered an important, albeit still limited, opportunity for the small-farm economy and its potential role in an alternative or regionally centered food system.[50]

Farmers' markets had also become a new type of urban place, similar to the community garden and the urban food enterprise, where the concept of food in the city provided both a community and alternative food-system function. Farmers' markets provided a community interchange, a "life in the city" distinct from the functionalist goals (move 'em in, move 'em out) of the fast-food franchise or the retail megastores.[51] But due to their predominant shift to middle-class locations and the "niche market" association, farmers' markets, with a few exceptions, had not been able to extend their initial community food security function in reaching and attracting low-income as well as middle-income constituencies. The promise of the markets—linking the justice dimension in its urban reach to the sustainability of an alternative food approach—still needed to be realized, despite its auspicious start in this direction.[52]

Alternative Outcomes

To develop an alternative food regime requires a long march through the food system, whether in relation to the growing, processing, manufacturing, or selling of food. But changes at each of these steps along the system, difficult as they can be, may not be sufficient. What people eat, whether influenced by the fast-food culture or the lack of choice to meet even basic minimum needs, may remain as problematic a set of outcomes—like issues of food consumption—as changes in the food production system itself. Nevertheless, the dominant food system, embedded as it may be in influencing how food is produced as well as consumed, is not immovable; its outcomes are not inevitable. The remainder of this chapter identifies two possible alternative outcomes based on projects our center helped initiate and evaluate. While descriptive of new opportunities, these alternative approaches also suggest the importance of demonstrating how a change in the discourse—the messages of what is possible—becomes the basis for action, to create the changes in institutions and policies, and in the production and consumption systems overall.

Add the Ketchup and Hold the Tomato: Schools and Food

Perhaps more than any other institution, schools have become a community food security battleground. The high-fat, high-salt, caffeine-laden, fast-food influenced diets that prevail today among school-age children have created what appears at first glance to be a community food security paradox. Whether children are obtaining too few calories or a kind of fast-food fare *or both*, the outcomes are similar: the food that school-age children eat intensifies rather than addresses problems of food insecurity. Moreover, schools have become a marketing opportunity for fast-food vendors as well as the brand name foods of the global food producers. Whether for Pepsi or Pizza Hut, through snack food offerings or à la carte junk food via private food service vendors, by the late 1990s schools had become a target of opportunity.

In the United States today, school food services constitutes a multibillion dollar purchasing and food provision system, with the heart of the school-food universe the federal government's National School Lunch Program. While lunches had been served in school cafeterias for more than a hundred years, the origins of a hot lunch program were first associated with Progressive Era initiatives to provide healthy meals at low prices. These private-based, subsidized programs also sought to "Americanize" the diets of urban, particularly immigrant, schoolchildren by introducing such items as white bread, chicken croquettes, salmon loaf, and scalloped dishes. In the period prior to the Depression, free or subsidized school lunches were also made available on a limited scale by school boards, charitable organizations, and some city governments. These initiatives were also primarily defined as "charitable" or assistance-related programs. The federal government first became involved in the school lunch area during the Depression as part of the government's emergency assistance initiatives tied to availability of surplus commodities. While these efforts were managed through the Works Progress Administration, the establishment of the formal food assistance/surplus commodity link was made possible because of the creation of the USDA's Federal Surplus Relief Corporation.[53]

The National School Lunch Act was passed in 1946 to establish the school lunch program on a continuing basis and to frame it primarily as a nutrition-based program. During and shortly after World War II, the

USDA had been instrumental in publicizing various studies document-ing the health and education problems of military recruits, problems that were traced to Depression-era childhood dietary deficiencies. Of the first one million men called for induction in World War II, 40 percent were rejected for general military service on medical grounds. The school lunch program was therefore conceived as serving national security needs as well as nutrition goals. "If our workers are malnourished," the surgeon general had commented about this national security link, "they cannot be efficient in producing what we need for defense."[54]

Due to this connection between diet and national security, the school lunch program remained more popular among policymakers than other food assistance programs, particularly during the 1940s and 1950s, when most food assistance programs had either disappeared or been reduced in scope. This was largely due to how the program was framed—a social policy with health and nutrition goals rather than strictly welfare-oriented goals. But despite this nutrition mission, the program still served commodity interests, like all the other food assistance programs admin-istered by the USDA, rather than serving as a planning tool for school food service programs. While the program could be financially attractive by potentially expanding the number of students using the program, as well as having a lower cost per meal for school food service providers, it was not always popular with school cafeteria managers. This was due to USDA-associated limits around meal planning, the failure to attract significant numbers of schools and students to the program, and the man-agers' lack of enthusiasm in broadening participation to low-income chil-dren for whom the program was originally designed.[55]

One revealing episode regarding meal planning took place during the winter of 1962, when USDA officials identified orange juice as a surplus item available for the school lunch program. But when a December freeze reversed surplus estimates for orange juice as well as grapefruit, the USDA removed those items from the program. School food service staff bitterly complained about this and other examples of last-minute menu changes caused by the availability or lack of availability of surplus foods. This type of commodity-dominated relationship has continued through-out the history of the program. As a more recent example, in 1998 and 1999, beef supplies made available to the lunch program were significantly increased due to government purchases of $20 million of surplus beef in order to prop up cattle prices for cattle ranchers and the livestock industry as a whole. Similarly, in 1999, the USDA purchased $12 million worth of surplus salmon and then made it available to food services as canned pink and pouched pink salmon. Such purchases are commonly described in the "Commodities Alert" section of the trade publication for food service directors, a section that could be labeled "What's Available." Indeed, for each meal served in the cafeteria, schools obtain about 20 percent of their food through a federal commodities account, which has only minimum transportation costs, to obtain the food.[56]

During the 1960s and early 1970s, when issues of hunger became more visible as part of the policy debates, support for the free and reduced school lunch program for low-income children grew significantly. At the same time, the proportion of free or reduced school lunches increased from 10 percent in 1963 to 40 percent in 1975. However, by the late 1970s, the school lunch program and its companion school breakfast program became more vulnerable, as media and policy-related attention shifted from the concern about hunger to the presumed abuses by poor people of welfare and food assistance programs. Proposed legislation by Ronald Reagan in 1981 sought to significantly restructure the program by reducing the meal subsidy and changing the implementation process through a new lunch meal category. While Reagan's approach stimulated significant opposition, and many of the proposed changes were never implemented, the school lunch program continued to remain a target through the 1980s and 1990s, particularly the subsidies provided for the free or reduced meals for low-income children.[57]

Where the school lunch program has been judged most inadequate has been the quality of the meal itself. During the 1970s, debates were already erupting over what kinds of meals children, teenagers, and adults were eating. Concerns about an obesity epidemic also became pronounced during the 1970s, 1980s, and 1990s. These were underlined by a series of studies that identified 25 percent and more of school-age children as obese, that linked diet-related factors to a growing prevalence of obesity among all age groups and across class and ethnic lines, and that indicated children were obtaining a greater proportion of their nutrient intake from snacks. "This is an epidemic in the U.S.," the director of nutrition at the Centers for Disease Control said of these obesity trends, "the likes of which we have not had before in chronic disease." Nutrition prevalence data indicated high intakes of fat, saturated fat, and sodium in school-age children and that the meals offered in school lunches were above the recommended amounts in those categories and were contributing to, rather than reducing, the problem. Even though a few school lunch programs were designed to change these diet patterns, they were limited in scope and often failed to meet their objectives. This was due to the powerful counter-trends influencing school lunch menus, a problem compounded by the failure to address ways to engage the children in reorienting diet and developing alternative choices.[58]

Part of the explanation is that in nearly all school districts across the country, food services have operated on the basis of the need to generate sufficient revenues to cover costs. As a consequence, another set of trends has also emerged, parallel to the cost-driven commodity pressures. This has included contracting out services to fast-food chains, selling soda and chips to generate additional revenue, and developing exclusive contracts with brand-name junk food producers for the use of their product in vending machines. Faced with cutbacks in funding, many school food services have chosen to squeeze their program, using economic (cost-saving) criteria and menu planning that seeks to attract students by linking offerings to the dominant fast-food culture. Increasingly, there are initiatives to privatize school functions by providing exclusive contracts to companies such as Coke or Pepsi or McDonald's in exchange for income to the school district as well as such donated amenities as sports scoreboards.[59] Fast-food and junk-food companies also often pay for advertising in "educational" videos that are shown in classrooms. In one episode, a Massachusetts school established a relationship with Frito-Lay Products, culminating in a St. Patrick's Day event that had the principal dressed as a bag of popcorn distributing Frito-Lay popcorn to the schoolchildren. In these and numerous other episodes, fast food is legitimized by the schools as much as it is in the larger society. And school food service directors often justify their decision to use fast-food-like products such as chicken nuggets "because that's exactly what they [the students] want," as one director put it.[60]

These are not incidental episodes or marginal trends. Food service trade publications are filled with stories about branding, characterized as the hottest trend today in the school cafeteria. These include a range of initiatives to use "brand" names through vendor or contract relationships. For the fast-food companies that have established these relationships, school lunch has become, as one conference brochure put it, "a unique marketing opportunity" that is able to "reach kids on a daily basis in a $16 billion market."[61] During the 1990s, the use of brand-name foods in schools increased substantially, from 2 percent in 1990–1991 to 13 percent in 1995–1996. Seventy-three percent of the schools used one of four vendors: Pizza Hut (36 percent), Domino's Pizza (27 percent), Taco Bell (22 percent), and Subway (6 percent). Again, cost as well as desire for increased student participation were motivators. To comply with any regulations, such as those related to fat content, some of the brand-name companies have done so by just slightly altering their foods (e.g., slightly less cheese in Domino's Pizza). Meal planning for food service directors remains driven by costs and revenues (with branding providing a source of profits in terms of contracts as well as profit on foods reimbursed). At the same time, food service managers have become concerned with the popularity of the items offered in relation to what they assume the students are willing to eat, given the poor reputation of the quality of the existing cafeteria food and the link of revenues to participation. It becomes clear then why pizza and french fries are often defined as a winner. The concept of "branding" itself—putting a name and a logo on a product that will be familiar with students—ultimately emerges as the governing metaphor in how food services operate.[62]

The importance of branding and the fast-food culture—or what could be considered the penetration of the dominant food system and its products within the confines of the school cafeteria and its captured audience—can also extend into the classroom. "Even nutritional deficiencies of a relatively short-term nature influence children's behavior, ability to concentrate, and to perform complex tasks," argues the Center on Hunger, Poverty, and Nutrition Policy at Tufts University. Take the introduction of caffeinated products into the schools-food universe as one example. In recent years, there has been an increased targeting of caffeine products and beverages for teenagers. Surge citrus soda, a popular teenage drink increasingly found in school vending machines, is designed to appeal, with its "get fully loaded" and "feed the rush" marketing messages, to what the soft drink officials call "the high-energy younger generation." This caffeine explosion has had consequences in terms of learning capacity, significantly reducing attention span. It has ultimately extended the disconnect between food service, nutrition education, and the overall learning process.[63]

By the 1990s, the issue of the diet patterns of school-age children and the focus on obesity preoccupied the public health community. Some nutritionists emphasized the importance of education by "applying knowledge from nutrition science and the relationship between diet and health to their food practices," while others focused on changing behavior (eating foods lower in fat and sodium and higher in fiber). But only a handful of advocates emphasized that what was offered in the cafeteria and what actions the children themselves could take (for example, by participating in school gardens) might offer an alternate route for changing diets.[64]

The rise of the community food security and sustainable agriculture movements and the increased interest in new direct marketing strategies established the school-food universe as an important arena for advocacy for such alternate routes. In 1997, the Community Food Security Coalition adopted a "Healthy Farms, Healthy Kids" initiative that sought to expand farm-to-school purchases in the school cafeteria and "learning-by-doing" strategies for nutrition and environmental education through programs like school gardens. In a number of communities, new projects were established to increase access to fresh and culturally appropriate foods in schools. These included direct marketing strategies, school gardens and farm tours, and even cooking classes (reviving an earlier Progressive Era–"home economics" approach made necessary by the significant decline in knowledge about how to prepare fresh foods). Some programs, such as an initiative in New York to purchase apples from local farmers, were unsuccessful due to purchasing constraints and at times inaccurate assumptions by school food service directors about the logistical barriers and food safety issues involved in purchasing directly from farmers. School gardens could also be considered a major community food security success story but a cautionary example of the difficulties in maintaining and extending new programs. Despite their widespread adoption and even programmatic support, such as California's "Garden in Every School" initiative, which was launched in the late 1990s, many school gardens suffered from lack of resources, overloaded teachers, and a failure to link school gardens to either the curriculum or the cafeteria. Nevertheless, despite the constraints and cautionary examples, the explosion of interest in alternative school-food programs suggested that a social and ecological approach to sustainability in the school-food universe (with implications for the broader food-system issues) could be identified.[65]

The experience of the farmers' market fruit and salad bar at McKinley Elementary School in Santa Monica and its subsequent institutionalization through the school district and expansion into other school districts suggests one type of pathway in this search for a social and ecological food-systems approach. The Santa Monica school district, like a number of school districts worried about the barrage of nutrition-related criticism of food services, had instituted a salad bar option in the early 1990s, thanks in part to the role of local nutrition advocates. A number of the children, particularly in the elementary schools, had been enthusiastic at first, but the numbers of those participating rapidly declined. Complaints were made that the lettuce was brown and wilted, that the choices were limited, and the format unappealing. Purchases in fact were made in bulk and not directly from farmers; in effect, the program was simply an extension of the existing patterns of food selection, preparation, and the assembly-line format in the cafeteria. The number of children selecting the salad bar option had become minimal, to the point that the school food service director considered dropping the option at some of the sites. It appeared to be one of those cautionary tales: healthy foods were not a viable option in the schools-food universe. It also suggested that a more sustainable approach might require insight into how issues of "freshness" and "quality" (where and how the food was grown) and connection to the food (how the food was "presented" and experienced) also came into play.

While the Santa Monica school district explored its options, our center had established a new program, utilizing a community supported agriculture approach to purchase from several farmers through a farmers' market to increase access to fresh produce in low-income communities. Called the "Market Basket," the program eventually evolved into a community-based fresh foods access program, including working with child care centers, community organizations, and low-income women trained to be *promotoras comunitarias*, or health promoters, in their communities. These types of farm-to-consumer programs were also rapidly expanding across the country and the link between "direct marketing," community organizing, and nutrition and environmental education was becoming, for many of the community food security groups, increasingly clear. Working with schools also seemed a logical extension of this fresh-food access approach.

Based on these concepts, we approached the school district to propose a "farm-to-school" purchasing program, through a local farmers' market, with the produce (which would be literally picked the day before on the farm) to be used in the school salad bars. McKinley Elementary School (about half of whose students qualified for free or reduced meals) was selected as a pilot site, with our center initiating and managing the program the first year, while evaluating its cost, participation, and educational functions. Our center staff met with the principal, parents, and cafeteria staff and held focus groups with the students, asking them what they liked and disliked about the current salad bar and what they would do to change it. One suggested a soda machine. But most said what they disliked about the salad bar had to do with the taste and freshness of the produce, with complaints about how the carrot sticks were dried out, the precut lettuce was brown, and the fruit came out of a can. After a series of outreach and educational activities, including a farmers' market tour as well as meetings with the mostly Latino parents at the school, the program was established in the fall of 1997. There was some initial skepticism from the food services staff that the school children would actually select a salad bar as opposed to the usual offerings of pizza, chicken nuggets, and other hot meals available to them as well.

The program was formally launched about two weeks into the school year, competing against pizza on the first day. The results from the outset were impressive and continued to be so through the school year. On average, more than three times the number of children at McKinley were selecting the farmers' market salad bar option than in the previous year, while the cost of the farmers' market salad bar was less expensive than the hot meal option as well as the previous year's salad items. There were links to the school garden (garden items harvested were displayed monthly, also encouraging the children to try new items) and by the end of the year a compost program was established, using the peels and clippings from the salad bar.

Still, there were issues to be addressed, ranging from biases regarding appearance to the need for cultural specificity in the choice and layout of the salad bar (for example, making lemon slices available to be squeezed on the lettuce, a practice suggested by the Latino parents at the school).[66] As the year progressed, the popularity of the program and the children's and parents' enthusiasm and engagement in it generated new support from nearly all of the food service staff, including the food service director, Rodney Taylor. By the end of the year, Taylor was declaring that he had become a convert. "This program," he told one reporter, "has helped me understand that in my job, I have the ability to improve the health of children. That's the real bottom line." By the beginning of the new school year, Taylor had convinced the school board to extend the program to most of the other elementary schools and to one of the middle schools in the district and to change the name of his department to Food and Nutrition Services. The second year was even more successful than the first. The program not only continued to register high levels of participation at McKinley, the original pilot site, but at the seven other schools as well, including, beyond all expectations, the middle school site. At some schools, the average participation was as much as ten times higher than the previous year. Farmer revenues increased significantly as well, including a number of certified organic farmers who were interested in the broad appeal of the program rather than simply assuming that they needed to more exclusively secure a niche market of wealthier consumers in order to survive and prosper. By the third year, the program had become fully institutionalized with its expansion to nearly all the elementary and middle schools, with plans to also develop the program at the two high schools in the district.[67]

The interest in the Santa Monica program grew as the program expanded and became institutionalized. The acting director of the huge

Los Angeles school district agreed to develop several pilot programs in the L.A. district, after visiting the Santa Monica schools and becoming convinced that low-income children could be enthusiastic participants. The Los Angeles school district's farmers' market salad bar program was launched at low-income elementary schools in Chinatown and South Los Angeles and had even more impressive participation among the children than the Santa Monica schools to the surprise of school and food service officials.[68] A sustainable school-food program had proven to be successful, but could that success be translated into a new framework for policy to lay the groundwork for an alternative food systems approach? Like the Santa Monica school district experience, changes needed to be institutionalized. This required not only new policies about food, environment, and schools. It also meant participation by those impacted by new programs and policies, and a change in discourse where the concept of eating healthy also meant enjoying the food, feeling empowered to make choices, and connecting with the farmers and with the process of growing food.

Toward that end, the Community Food Security Coalition established an informal collaboration with the USDA in December 1999, designed to explore at the regional and state level what opportunities could be developed for farm-to-school purchasing programs. While the USDA was most concerned with establishing new markets for small regional farmers, the CFS Coalition focused on the farm-school-community relationships, emphasizing both alternative food system and community food security outcomes. To accomplish such a change required not only pilot programs, but a reorientation of the rules and policies for purchasing, and a shift in how children experienced the food choices made available to them. Such a change, it had thus become clear, required not just a difference in perspective but the availability of the choice itself.[69]

Reexperiencing Food

Building a sustainable food system requires a change in the way food is experienced as well as accessed, whether in the school cafeteria, in community settings, in the supermarket, or in the home. The fast-food culture not only establishes a disconnect between food grown and food consumed but it also changes the way people experience food. The significance of school gardens and community gardens, in this context, resides as much in the area of reexperiencing food (and providing a sense of community) as in identifying an alternative food source.

Another illustration of the new pathways for change around food issues, including the question of how people experience food, has been the development of the "Project Grow" program of the California Department of Health Services, a program that our center first conceived and has helped to develop and evaluate. Project Grow programs have been designed for domestic violence victims, one of the most food insecure populations in the country. These victims are often women and children fleeing from their homes, with limited access to cooking facilities, and dependent on shelters and transitional housing arrangements where the source of food is often limited to donated items or the least expensive store purchases. Some women also have eating disorders as a result of domestic violence, often a consequence of dinner-related arguments and violence. Once victimized and forced to flee their homes, the women and their children may eat poorly and on the run, which in turn compounds their sense of loss and vulnerability.

During the 1990s, as the issues of domestic violence became increasingly visible and prominent as public policy issues, new initiatives were developed in the area of horticultural therapy, the use of gardening as healing activity. The idea that gardening can serve as a form of therapy and as healing activity has paralleled the development of the community garden as a form of social activity and community building. With the rapid increase in the number of domestic violence shelters and other housing strategies for domestic violence victims, the garden as a place to heal began to be explored during the 1990s in several locations. A new literature about horticulture therapy in relation to domestic violence also emerged. The handful of programs that formed in the 1990s, such as the Veteran's Garden in West Los Angeles and the Santa Cruz homeless garden, became the forerunners of this new approach linking physical activity with healing goals.[70]

What is often absent from the horticulture therapy approach has been the question of food source and food needs, an issue particularly pronounced for domestic violence victims. Shelters are one of the core constituents of the emergency food system. With the lack of quality and absence of cultural specificity or security in the availability and source of food, the food experience *at the shelter* does not become integrated into the transition process for developing self-esteem and self-worth. By connecting food security to horticulture therapy, the quality of food—a key issue associated with sustainable agriculture and regional food-system advocacy—can be connected to the sense of place and community purpose often associated with community gardening. Such connections were documented in a study by our center entitled "Gardens for Survivors." The report sought to answer whether a linked horticulture therapy and community food security approach could extend beyond the development of a garden to the experience of food at the shelter, the transitional house, or any other location associated with establishing a transition away from violence in the home.[71]

Following a presentation about community food security at the 1997 Maternal and Children's Health Conference, staff from the domestic violence section of the California Department of Health Services contracted with us to host three workshops in California for domestic violence agency staff and domestic violence victims. The workshops were designed to identify food needs and food experiences at the shelters and transitional houses and explore ways to link community food security strategies with horticulture therapy approaches. For several of the agency staff, the community food security idea provided a kind of epiphany, since food issues and the need for healing activities were so prominent in the daily lives of their constituents. Due to the intensity of the workshop discussions and how they resonated with those participating, the state agency decided to establish a competitive grants program for the agencies to launch gardens and other food security activities. Ten agencies were selected to launch pilot projects that in turn could provide a baseline of information about the healing and food security dimensions of the program. By the end of 1999, most of the ten projects had established active and flourishing programs, creating a suggestive and potentially significant source of information about one constituency's capacity to reexperience food.[72] Food growing as well as food consumption was emerging as a potentially liberating rather than depressing experience, an entry way out of the culture of victimization, also associated with emergency food and the disconnect of the food experience embedded in the fast-food culture.

From Discourse to Action

By the new century, the efforts toward building an alternative food regime, limited in scope, often ad hoc in nature, scattered among constituencies and movements, had nevertheless become a rich and valuable example of the potential for constructing a new pathway for environmental and social change. But the limits involved still seemed formidable in the face of an ever expansive and seemingly impenetrable global food system. What was the actual power and capacity then of the new food movements?

In an analogous vein, my colleague Peter Dreier liked to say, in commenting about the place of progressive social movements in Los Angeles, that the intricate and myriad types of progressive community, institutional, and advocacy groups in the region could literally fill Dodger Stadium. But despite their apparent reach, this would not reflect their capacity to bring about change, given that such groups have remained disconnected, their source of power scattered, their vision often not extending beyond the issue and goal at hand.

The same could be said of the disparate movements and constituencies seeking to build an alternative food regime. Substantial in number, seeking to influence and extend their arguments into the different niches and spaces within the food system, the sum of the new food movements still remains less than their parts. The power of the new food movements resides primarily in the area of discourse, the ability to use the coded messages of power to challenge the system of power—in this case, the global food system. The new food movements have provided a potential pathway for change by challenging the ways we think and talk about food. But the challenge to the movement itself is the need to shift the arguments about discourse to the arena of action where the sum of the different actions, policy initiatives, and movement building activities—whether environmental or socially defined—can become greater than any one of its individual parts. For food issues do reside directly at the intersection of the social and ecological, where social justice and environmental justice movements can meet.

Whether focusing on the food we eat, the clothes we clean, or the places where we live, work, and play, the pathways for change need to be identified, a new language constructed, and strategies for action pursued. In the process, a more expansive and linked set of movements may be able to emerge in this new century, ready to act on the belief that change is not only necessary but possible.

7

Pathways to Change: A Conclusion

Reclaiming the River, Remaking the City

Tucked in a corner of northeast Los Angeles, at the edge of the railroad yards and below the high arch of the Golden State Freeway sits a little pocket park, with a handful of trees, some modest landscaping, and a bench that looks out at the Los Angeles River. To the right of the park where the freeway crosses the river on its way downtown is the river's signature form, the concrete channel that carries the summer trickle of reclaimed water from the upstream treatment plants or the occasional winter storms that can move far quicker than even the most turbulent eastern flood waters. The channel, with its foreboding concrete slanted walls and barren landscaping that argues "keep out," heads to the south and west, past downtown and on through the downstream working-class neighborhoods and largely abandoned industrial sites toward Long Beach harbor.

To the left of the park one can see a quite different, if not unpredictable, scene. This is especially true for those for whom the L.A. River has long represented "danger" and "hostile territory" in the city rather than life on or near the river. The area to the north is one of the "soft-bottom" sections of the river, where the water table is too high to allow for a concrete bed. Nearby is one of the river locations that the locals call "Frogtown," known for its abundance of tadpoles, crawdaddys, and frogs that the kids in the neighborhood have been catching for years. A rich and varied array of vegetation grows in the bed and at the side of the river at this site north of the park, vegetation that changes according to the river flow. Through Frogtown and past the Northeast Trees pocket park there is also now a bike path, a place that has become safer and more inviting after community residents fought to make it a well-lit and available community space. These soft bottom stretches of the river can, of course, be considered "artificial" or even a form of "reinvented Nature," given the river's dependence on the flow of the treated reclaimed water. But is this an environment worth reclaiming? Can the city remake these "unnatural places" to bring nature back into the city in order to make the city more livable?

To reclaim a river that flows through the city, as urban environmental advocates have sought to do with the L.A. River, is tied to the struggle to reclaim the urban environment. Since the 1930s, a consortium of developers, public works engineers, and flood control managers have reconfigured streams, washes, rivers, and creeks in cities nationwide, by channelizing, culverting, riprapping (rocking), and clearing vegetation to turn these areas into storm sewers and concrete straightjackets. By 1972, the last time a study sought to document the extent of urban river reconfiguration, more than 235,000 miles of urban channelization projects had occurred or were about to occur, due to the dominant flood control strategy of agencies like the Army Corps of Engineers. By then, however, a budding movement had also emerged that challenged the management strategies and sought to reclaim the urban stream as community asset.[1]

Along those lines, in 1999, our center developed a year-long program about the prospects for community and ecological revitalization in and around the L.A. River. The program included historical explorations of the prechannelized river. It sought to establish a political dialogue about change along the river and management evaluations of opportunities for ecological restoration, including tearing up some concrete, that could also serve as a new kind of strategy for flood protection. It identified ways to reexperience the river through physical activities such as biking, walking, and kayaking in it and along its edge. It also sought to explore and provide cultural insight into our perceptions, fears, and dreams about the river. We called these multiple events, "Re-envisioning the L.A. River." At one level, this program extended the battles over what the river represented to its various neighboring communities as well as to its varied managers and policymakers. Given that the river cut through the heart of the city, these battles over discourse ("is this indeed a river?") identified a crucial first step in the reenvisioning not just of the river but of Los Angeles itself.[2]

To accomplish such a goal requires more than just a change in discourse, despite the importance of those coded messages that influence agendas and guide action. If the philosophers, filmmakers, poets, water engineers, developers, policymakers, and urban environmentalists have each sought to provide their interpretation of the relationship of the river to the city, the point now is how to change it. The ability to take action, to try to transform the river and the city, requires new kinds of alliances. The impacts from urban river channelization have perhaps most significantly affected low-income or urban core communities, further degrading these areas as livable places. Thus, urban river restoration also becomes an environmental justice issue, with its related community and environmental focus. To reclaim rivers and remake cities requires strategies that can make urban, environmental, social, and industrial change more connected, strategies for action more seamless for the movements that seek to construct those alliances. To build those movements and make change happen, we also need to identify the examples, the pathways, that allow us to begin the process of reenvisioning institutions and systems, while also challenging the structures of power within the contemporary urban, industrial, and global order. This book has offered a snapshot of those possibilities for reenvisioning—and in this concluding chapter, some of the key conceptual arguments associated with that reenvisioning process are elaborated.

An Ethic of Place: Remaking the Global and the Local

Harriet Friedmann's argument about the importance of locality and seasonality in the development of an alternative food regime provides a compelling case of the importance of the "local" in any environmental agenda. However, a strategy for action that fails to address the extent and reach of the global food system runs the risk of appearing marginal, a "small is beautiful" strategy limited to a niche market and appealing to only a limited set of constituencies.

A strategy for action that addresses the global and the local requires on one level what Charles Wilkinson has called an "ethic of place."

Wilkinson's concept refers primarily to those non-urban environments such as forestlands, watersheds, and rural streams that need to be protected or restored for their human or social as well as ecological importance. But an environmental ethic of place also needs to have an urban and industrial and even global form. Bob Marshall's plea to bring the forest to the city also needs to be a plea for community restoration and revitalization. The urban ethic of place needs to focus on neighborhoods, services, jobs, housing, transportation, and green space and open space. It is an argument, and a plan of action, to rebuild and empower communities.[3]

Similarly, the industrial ethic of place stretches the definition of industrial location. It requires what some activists have called an industry "good neighbor policy" that links industrial and commercial facilities and workplaces to neighborhoods and communities. It needs to bring the community interest into the industry and site-specific decision-making processes.[4] It also needs to build and strengthen those industries and enterprises that have by their nature a community purpose or are at least bound to the communities in which they operate. These "sticky industries," as my colleague Peter Dreier has called them, can provide a counterpoint to the global-oriented industries that have no place-based identity other than to maximize their selection of labor markets, materials, transportation, and marketing end points. Often such global-oriented industries seek to operate outside any community or place-based commitments. The global food industry and its products, like the Pringles Olestra ("fat free") potato chip, are a case in point. The facilities, industries, and products of the global food system have been designed to overcome such place-based considerations as locality, seasonality, and community food needs. This locational strategy without borders is pursued in order to turn food grown and processed everywhere, into the same everyday experience of consuming food for everyone.

But can there also be a global "ethic of place"? The respect for place on a global scale, expressed through solidarity and linkages across borders, particularly among NGOs and place-based organizations, is one clear articulation of a global ethic of place. This ethic exists when immigrant communities maintain a connection, through a linked exchange or development process, to strengthen the community of origin and/or recreate the new community, drawing on the identity and needs of the community of origin. A global ethic of place can also be found in the standards and regulations that can be globally applied that at least provide a modest first step in requiring the respect for place. These could include a mandated global minimum wage, a baseline of environmental requirements, or global debt relief. They could also include labor and environmental or human rights stipulations (for example, food security as a right) within trade agreements or even standards like the ISO 9000 and 14000 series, as long as such agreements and standards are enforceable. Such standards and regulations will also work best when they are integrated into public forums for decision making and accountability and are not limited as an exclusive private or "voluntary" activity. Finally, a global ethic of place draws on the respect for nature that captures Raymond Williams's critical insight that we can no longer separate ourselves from nature. The powerful global outcomes associated with such issues as ozone layer depletion, global warming, and loss of biodiversity are each embedded in the actions and choices we make. These are at once local and global, whether they are the transportation systems we develop, the energy sources we rely on, the choice of materials we select, or the types of production systems we design. We can think globally and act locally (as well as think locally and act globally) out of that respect for place. Ultimately, the ethic of place, the conflicts over the global and local, are also political battlefields, requiring changes at both the community and global scale, through the political process, in the management and reinvention of natural and human places, and in relation to public and private decision-making processes. For environmentalism, place does matter, and one can act or think locally as well as globally by focusing, as we speak and act, on all the places we inhabit and share. The enormous significance of the labor, environmental, and food-related protests at the WTO meetings in Seattle in December 1999 provided a compelling example of how this type of linked local and global politics may well stimulate a new set of social movements for the twenty-first century.

New Strategies for Governance, Remaking Politics

When the Clinton administration first came to power in 1993, its focus was on "reinventing government," including its search for the new strategies or paradigms for environmental governance. These approaches, however, mostly signified the narrowness of options rather than the opportunities for change in relation to public activity and government influence or intervention in community, industry, and environmental matters. In terms of community concerns, the issue of brownfields was particularly revealing. Government activity in this area focused on ways to reduce regulation, streamline the development process, and at best identify the need for community participation, rather than explore, as Carl Anthony put it, the strategies for community revitalization *by and on behalf of the community*. The absence of any substantial public goals and public process in land use decisions has meant that a community development process, with its related social and environmental consequences, remains market driven, with government acting as adjunct or partner with market forces. In the area of industry activity, the promise of a public role in addressing the outcomes of industry activity and decision making has by century's end evolved into a defensive and end-point-oriented system of management and mitigation. This much criticized form of environmental management has thus also meant that one part of the government has been forced to clean up the mess that industry or other government agencies have been responsible for. Superfund became the ultimate whipping boy of this end-point system, even as it acted as an indirect constraint on how industry operated. At the same time, the new strategies for environmental governance, beginning with pollution prevention and including the focus on materials, design, producer responsibility, and overall throughput concepts such as industrial ecology, have provided important insights into developing new kinds of public roles and systemwide goals. But the new strategies have also suffered from the absence of a clearly articulated government role, compounded by the unwillingness or inability of government agencies or the legislative process to identify a public role in what continues to be defined as market or proprietary matters.

Today, the contest about control of government—the key to the traditional socialist discourse—seems lost in an era of public hostility to government, the near complete corruption of governance through the influence of money, and the increasing forms of global activity that bypass the nation state. During the 1970s and especially during the Reagan-era 1980s and the Clintonesque downsizing of government initiative (including the call for "reinventing government") during the 1990s, there emerged a growing interest among social movement activists in exploring what has been called "third sector" or "alternative institutions of accumulation." Third sector institutions (for example, food cooperatives or farmers' markets) provide the ability "to amass social surplus for future development," as Christopher and Hazel Gunn have argued. This can occur in a time when most governmental agencies have become caught in a "web of resistance to tax increases, calls for privatization of services, and escalating costs." By focusing on the development of this social surplus, alternatives to traditional corporate and private decision making can be established by placing what the Gunns call "expanded amounts of resources under democratic control." But the development of such alternative institutions and of a "social surplus" suffers from an enormously uneven playing field and often a set of legal and institutional constraints that can only be addressed in the political or public policy sphere.[5]

At the most obvious level, the difficulty in establishing new strategies for governance and these new forms of community, industrial, and environmental decision making and institution building is a matter of politics. Like government activity, politics has become an adjunct of the power of money. The challenge is not just to change government policy, but to change the nature of government and how governance as well as electoral systems are influenced and framed. These are political questions, linked directly to the power of social movements to bring about change.

The rise of green politics in Europe in the late 1970s and 1980s presented an early version of environmentalist efforts to remake politics above and beyond the need for change in government policy or community and industry decision making. But the limits of such a green politics in Europe has also testified to the uneven and often inchoate nature of the new social movements in Europe and even more directly in the United States. However, European green politics has also been able to place on the agenda the search for new kinds of government strategies and the exploration of what some have called the social market as a counterpart to the globalization of capital and industry. In the United States, green politics is far more diffuse and enmeshed in the web of "interest group" politics, rather than establishing a new form of politics and a new agenda. Part of those limits in the United States has to do with the weakness of the labor movement and the absence of any social democratic political traditions since the decline of the Eugene Debs–led Socialist Party during the Progressive Era. But those limits also can be located in the divide within environmentalism itself, between the policy-oriented mainstream groups who lack their own connection to place and community and the local groups who remain fragmented and largely unable to articulate a counter politics. The rise of environmental justice has changed some of the dynamics of environmental politics and provides opportunities to help connect movements and justice-oriented discourses. The limits of environmental justice, similar to the limits of other progressive movements, has to do with narrowness of agendas and the absence of a political form and direction that can embody Hilary Wainwright's "totalizing vision." The ability to provide such a vision could in fact emerge as environmentalism's most important challenge and, if successful, most valuable potential contribution to the development of an integrated alternative politics in the twenty-first century.

Revaluing Work, Remaking Industry
One of the critical, if not crippling, divides impacting environmentalism has been the separation of industry and community and the parallel division between work and environment. This divide can be found in political, administrative, and policy contexts. It emerged full blown in the first great battle over leaded gasoline, when community pressures led to end-point changes in the workplace (better ventilation, more engineering controls) while failing to address the long-term and pernicious effects on the environment. As the automobile penetrated the countryside and, subsequently, reconfigured the inner city, lead emissions became part of the package of social and environmental impacts that changed the face of the landscape and established forms of everyday pollution. Mean-

while, the rise of the automobile, made possible in part by the early contribution of leaded gasoline to its "power and performance," also established a form of workplace management that fundamentally changed the nature of the work experience.

The environmental movement needs to consider and ultimately integrate issues of the workplace. It needs to develop, as central to the environmental agenda, an approach that embraces revaluing and reskilling on the job. It needs to associate an environmental perspective with a more visionary notion of workplace democracy and what the justice-for-janitors movement has identified as the need to assert the dignity of work. Such an approach is also about asserting control over the front end as well as the outcomes of any industrial, agricultural, or service-based industry process. Redesigning work thus becomes an environmental objective, while asserting control over the conditions of work, as well as the outcomes of the work process, becomes a labor issue with strong environmental implications. This intersection of work and environment requires that both the environmental and labor movements advocate "an entirely new relationship between pride of work and pride of place," as Lynn Kaatz Chary has proposed.[6]

Overcoming the work/environment divide is perhaps the most difficult and contentious question facing the future of the environmental and labor movements. So much of the jobs-versus-environment debate or the promotion of labor and environmental alliances suffers from fundamental limitations associated with both labor and environmental discourses. If how we work is an environmental question, it has yet to enter the language of environmentalism. If what is produced at work as well as how it is produced represents a labor question with significant environmental and social consequences, then those considerations too have been largely absent from the language of the labor movement—and of environmentalism as well. A new, integrated language of work and environment needs to be developed. And the alliances that could form as a consequence would likely be more strategic and organic rather than limited and, at best, related to single issues. New kinds of strategies for action would need to be generated, though the basis for that happening still remains a work in progress, and the work and environment action agenda continues to be an elusive goal.

Remaking a Community of Interests

The language of participation and stakeholder-based environmental decision making, partly in response to the rise of environmental justice advocacy, emerged full-blown during the mid- and late 1990s among agencies like the EPA and at the state and local level as well. The occupational and environmental right-to-know movements of the 1980s and 1990s were also important contributions to the idea that environmental decisions needed to have some reference to impacted parties. The right-to-know movement asserted a public role concerning what needed to be known. If an industry released pollutants to the air and land, or created exposures within the workplace, the right to know about those hazards established an implicit connection of workers, communities, and the public to the decision-making process. By knowing, one could act, and such actions, though outside the locations where decision making ordinarily occurred, could nevertheless conceivably influence the process. As a number of chemical and other industry trade publications noted in the early and mid-1990s, right-to-know requirements, such as the 1986 SARA legislation that established the Toxics Release Inventory, had become one of the more significant motivations for pollution prevention among industries. This was particularly noteworthy, given the absence of any direct legislative or regulatory pollution prevention measures.

But the participation and stakeholder-linked programs that the EPA constructed, such as Design for the Environment, the Common Sense Initiative, or Project XL, remained at best advisory and potentially or even likely adversarial. These were programs that ignored or avoided the more encompassing question of whether a community of interests could also be identified and become central to both how and what decisions would be made. As a consequence, most of the stakeholder programs had no real input in the decisions although they might have had a significant stake in the outcomes of those decisions. This became apparent in one of the more visible Project XL efforts involving an Intel facility in Arizona. As Jan Mazurek noted in *Making Microchips*, the Intel effort established an unequal playing field, while favoring the clear industry objective of reducing regulatory burdens rather than establishing a new type of decision-making process addressing where, how, and what was produced. The semiconductor industry, with its rapid product turnover and globalization of manufacturing facilities (particularly wafer fabrication plants, the most polluting of such facilities), had a limited, even negligible, attachment to place. This largely eliminated any long- or even medium-term commitment to the stakeholders and communities where the plants were based; and the Intel example was the rule rather than the exception.[7]

In response to this and other Project XL examples, some of the environmental justice-oriented community groups impacted by the semiconductor industry, such as the Silicon Valley Toxics Coalition, have argued that instead of a dysfunctional stakeholder process like Project XL, another kind of community of interests needs to be established. This could be accomplished by identifying the community, worker, and public role in where, what, and how microchips and other electronics industry products are produced. Such a process in the semiconductor industry could conceivably be structured, the community groups have argued, through the public-private industry consortium, Sematech, which was created in the 1980s to provide government subsidies for research and development to industry to make it more globally competitive. If Sematech were to be the foundation of a true public-private partnership (and, at least implicitly, a more effective community of interests), then the issues of workplace, community impacts, environmental exposures, and product development could be addressed at some level by that partnership.[8]

How that could occur in relation to the more mature and highly concentrated industries such as pulp and paper was addressed by Maureen Smith in a provocative essay in the *Journal of Industrial Ecology*. Smith focused jointly on the importance of regional or place-based perspectives in identifying what constituted appropriate materials choices for the paper industry, and on the crucial relationship between materials choices and production technology, the intersection of which influenced a broad range of social and environmental outcomes. Smith's argument mirrored the position of the Institute for Local Self Reliance regarding what the group called the "carbohydrate economy"—one that moved away from the use of nonrenewable inputs such as fossil fuels and toward the substitution of biomass-based alternatives. Smith specifically argued for the diversification of paper industry feedstocks away from wood dependency and toward greater emphasis on resources including wastepaper and a wide variety of agricultural fiber crops and crop residues. The potential for using this more heterogenous and dispersed resource involves a wide-ranging community of interests, most of which have had little or no prior connection to traditional narrow debates over production technology and resource use in the industry. These interested parties could include farmers concerned about how to address the problem of agricultural waste or seeking alternatives to crops such as tobacco. They could include community residents and policymakers concerned about exposures from incinerating crop residues such as rice straw or corn stalks. They could include a workforce needing to be trained in a new type of work process, as well as innovative pulp manufacturers willing to explore such new technologies. They could also involve a variety of urban recycling interests. The technology implications of such a materials shift, presenting a clear pollution prevention opportunity, could equally involve a change in community impacts and the structure of work, a goal of both environmental justice and workplace democracy advocates. At the same time, both federal and state agencies positioned to support the research and development necessary to identify whether new technologies were viable could help broker the process and thereby facilitate bringing together this community of interests.[9]

The community of interests concept, elaborated earlier in the discussion of wet cleaning and dry cleaning, may be most compelling when applied at the community scale. This would be particularly true among those "sticky industries," like garment care and food retail, that require a presence in the community and a degree of goodwill in their operations. New kinds of cooperative and joint venture arrangements, like the Pathmark supermarket in Newark, New Jersey, or the CSA farm-to-community model, suggest the forms that could be constructed to institutionalize the arrangements established and the decisions that get made. Ultimately, the community of interests concept can extend and transform the notion of public-private partnership and how we view the social and environmental goals and the public obligations of an industry or an individual facility operating in a real place in real time.

Mapping All Our Assets: The Joining of the Social and the Ecological
During the 1990s, a new approach to community organizing called
"asset mapping" began to take root among a handful of community
organizations and "social entrepreneurship" advocates. In some respects,
asset mapping appeared to be similar to the mapping exercises of the
antitoxics groups during the 1980s and 1990s who had sought to locate
key community institutions and players in relation to the *location* of a
hazardous facility, the better to challenge that facility. The antitoxics
groups, like other social movement groups, wanted to identify the
problem areas the community needed to confront. Where asset mapping
differed was its primary focus on identifying the positive resources and
potential contributions of the constituencies and institutions for empow-
ering a community and making it more livable. Without using that term,
asset mapping became essentially an exercise in identifying a potential
community of interests.[10]

Like the social movements they have drawn upon and sought to guide,
the asset mappers often remain focused on particular issues and needs,
even as they have set out to identify and bring together a broad array of
players and interests. The asset mappers have also focused on the human
or social resources available within a community—community-based
organizations, schools, churches, and other institutions linked to civil
society. Absent from the approach has been the idea that physical
resources—the ecology of particular places—should also be mapped and
thus become central to any initiative to create more livable and empowered communities. Moreover, asset mappers are community- and con-
sumer-oriented (retail, for example, has been an important asset to be
considered), but not production-centered, where what is produced and
how it is produced could be counted as an asset or at least a potential
asset, depending on prospects for change.

Despite these limits, asset mapping as a community-building strategy
did begin to resonate among social movements seeking to locate new
ways to talk about community and place, particularly in urban areas
where community, like "environment," needs to be reconstructed. In
doing so, asset mapping has also facilitated a process of linking movements and perspectives, allowing the community building exercise to

focus simultaneously on such concerns as food, education, health, cultural life, and, more broadly, civil society itself.

While most of the asset mapping of the 1990s focused directly on civic institutions, the rise of an urban environmental movement during this same period also led to the beginnings of what could be considered an environmental asset mapping approach. The growth of a wide range of urban greening initiatives such as community gardens, urban reforestation, urban stream restoration, and even efforts to turn some brownfield sites into parkland reinforced and extended these efforts to essentially assert that nature in the city was not an oxymoron. It is a view of the city where, as Sam Bass Warner put it, the "health, liveliness, creativity and sociability of its citizens" is also connected to "the richness of the supporting nonhuman environments among which they live."[11] But environmental asset mapping, whether in the city or in rural areas, involving a one-acre community garden plot or the vast grasslands of the prairies or the forests of the Pacific Northwest, still needs to recognize that all those assets, as Benton MacKaye had argued, are human connected as well.

Mapping all our assets still requires a perspective, a discourse if you will, where the social and the ecological are not just meeting halfway from their separate spheres but have become joined as part of a common exercise and through the construction of a common vision. It is this common vision that can help liberate environmentalism from its confines as a bounded movement, where it has largely been defined on the basis of a separation of the social and the ecological. Part of the historical dilemma for environmentalism has been its lack, as Fred Buttel argues, of a " 'natural' or enduring constituency that anchors its base of support."[12] But that "absence of constituency" may also reveal a potential strength of environmentalism in its capacity to pursue this broader view in joining the social and the ecological. If "environmental issues are ubiquitous in [the sense] that there is scarcely a social relationship that does not involve some implication for resource use, pollution, ecosystem processes, or the biosphere," as Buttel argues, then environmentalism needs to see itself as a social movement with a broad view and a broad agenda. Similarly, when any environmental issue can be seen as socially determined, whether resource use, pollution, ecosystem processes, the biosphere, or any other environmental focus, then environmentalism's great task will also be to see itself as a primary agent of social change.

When the social and the ecological are joined together, movements for change have the capacity to become more powerful actors in the struggles to come. An environmentalism unbound can help point the way.

Notes

Introduction

1. Andy Fisher, *Hot Peppers and Parking Lot Peaches: Evaluating Farmers' Markets in Low Income Communities* (Venice, Cal.: Community Food Security Coalition, January 1999); personal communication with Ted Galvan, market manager, Pico Farmers' Market, Santa Monica, August 13, 1997.

2. Denny N. Johnson and Errol R. Bragg, *National Directory of Farmers' Markets* (Washington, D.C.: U.S. Department of Agriculture, November 1998); *Southland Farmers' Market 1998 Annual Report* (Los Angeles: SFMA, 1999).

3. Kelly Lamkin, "1996 Certified Farmers' Market Survey: Data and Findings," Southland Farmers' Market Association and Pollution Prevention and Education Center, Los Angeles, 1996.

4. Robert Gottlieb, *Forcing the Spring: The Transformation of the American Environmental Movement* (Washington, D.C.: Island Press, 1993); *Reducing Toxics: A New Approach to Policy and Industry Decision-Making*, edited by Robert Gottlieb (Washington, D.C.: Island Press, 1995).

1 Environmentalism Bounded: Discourse and Action

1. See the discussion about the "reinventing nature" concept in *Uncommon Ground: Toward Reinventing Nature* (New York: W.W. Norton & Co., 1995); also, William Cronon, "The Trouble with Wilderness; or Getting Back to the Wrong Nature" (and comments in response), *Environmental History* 1, no. 1 (January 1996): 7–55.

2. See Martin Melosi's presidential address to the Environmental History Association, reprinted as "Equity, Eco-racism, and Environmental History," in *Out of the Woods: Essays in Environmental History*, edited by Char Miller and Hal Rothman (Pittsburgh: University of Pittsburgh Press, 1997), 194–211; J. Donald Hughes, "Whither Environmental History," *ASEH News* 8, no. 3 (Autumn

1997): 1ff.; Martin V. Melosi, "The Place of the City in Environmental History," *Environmental History Review* 17, no. 1 (Spring 1993): 1–23.

3. The coded messages comment is in David Harvey, *Justice, Nature, and the Geography of Difference* (Blackwell: Oxford, England, 1996), 78; see also John S. Dryzek, *The Politics of the Earth: Environmental Discourses* (New York: Oxford University Press, 1997), 38; Marteen A. Hajer, *The Politics of Environmental Discourse: Ecological Modernization and the Policy Process* (Oxford: Clarendon Press, 1995), 15. Harvey also refers to Raymond Williams's insight that "the idea of nature contains, though often unnoticed, an extraordinary amount of human history. Like some other fundamental ideas which express mankind's vision of itself and its place in the world, 'nature' has a nominal continuity, over many centuries, but can be seen, in analysis, to be both complicated and changing, as other ideas and experiences change." Cited in Raymond Williams, "The Idea of Nature," in *Problems in Materialism and Culture: Selected Essays* (London: Verso Books, 1980), 67. See also David Harvey, "The Nature of Environment: Dialectics of Social and Environmental Change," in *Real Problems, False Solutions: Socialist Register 1993*, edited by Ralph Miliband and Leo Panitch (London: Merlin Press, 1993), 39.

4. Harvey, "The Nature of Environment," 3.

5. "A Big Splash: Moments That Changed the Fox Valley," *The Courier News*, Elgin, Illinois. December 26, 1999.

6. Both quotes are from *Forest Outings* (by thirty foresters), edited by Russell Lord (Washington, D.C.: U.S. Department of Agriculture, U.S. Forest Service, 1940). The first quote is from a Marshall essay in *Nature Magazine*, 1938 (p. 73); the second quote is derived from Marshall's perspectives on the need for greater access of low-income groups in relation to U.S. Forest Service lands and activities.

7. *Forest Outings*, 282–283.

8. Ibid., 79.

9. Robert Marshall, *Arctic Village* (New York: Harrison Smith and Robert Haas, 1933), 378; Benton MacKaye, "Outdoor Culture—the Philosophy of Through Trails," *Landscape Architecture* 17 (April 1927): 163–171.

10. Robert Marshall, "Recreational Limits to Silviculture in the Adirondacks," *Journal of Forestry* 23, no. 2 (February 1925): 173–178; "Wilderness as Minority Right," *Service Bulletin*, U.S. Forest Service, August 27, 1928, cited in James M. Glover, *A Wilderness Original: The Life of Bob Marshall* (Seattle: The Mountaineers, 1986); Robert Gottlieb, *Forcing the Spring: The Transformation of the American Environmental Movement* (Washington, D.C.: Island Press, 1993), 15–19.

11. Samuel P. Hays, *Beauty, Health, and Permanence: Environmental Politics in the United States, 1955–1985* (Cambridge: Cambridge University Press, 1987); Samuel P. Hays, "From Conservation to Environment: Environmental Politics in the United States Since World War II," in *Out of the Woods*, 107; see also David

Brower, "Introduction," in Eliot Porter, *The Place No One Knew: Glen Canyon on the Colorado* (San Francisco: Sierra Club, 1963); the "tourist gaze" concept is elaborated in John Urry, *The Tourist Gaze: Leisure and Travel in Contemporary Societies* (London: Sage Publication, 1990), 11.

12. Brower's quote is from *The Sierra Club Wilderness Handbook*, edited by David Brower (New York: Ballantine Books, 1967), 17. The National Park Service, under its director, Stephen Mather, established a strong emphasis both on recreational tourism and national identity, characterizing the parks under Park Service jurisdiction as "national playgrounds." Cited in Richard West Sellars, *Preserving Nature in the National Parks: A History* (New Haven: Yale University Press, 1997), 244; see also Alfred Runte, *National Parks: The American Experience*, 2d ed. (Lincoln: University of Nebraska Press, 1987).

13. Sellars suggests that the Park Service saw this tourism approach establishing "the highest and best use of the national parks' scenic resources." Such a "use" concept, as Sellars pointed out, also paralleled the highest and best use arguments associated with resource management and exploitation. With this conception of Nature as commodity, preservationist and conservationist perspectives find common ground in their distinctive though related views about commodity development. *Preserving Nature in the National Parks*, 88; see also Robert W. Righter's review of the Sellars book in *Environmental History* 3, no. 3 (July 1998): 389–390; Louis Warren, *The Hunter's Game* (New Haven: Yale University Press, 1997), 129.

14. Robert Marshall, *The People's Forests* (New York: Harrison Smith and Robert Haas, 1933).

15. See, for example, Gifford Pinchot, "The Lines Are Drawn," *Journal of Forestry* 17, no. 8 (December 1919): 900; George T. Morgan Jr., *William B. Greeley: A Practical Forester, 1879–1955* (St. Paul, Minn.: Forest Historical Society, 1961).

16. The Grand Plan scenario is described in Peter Wiley and Robert Gottlieb, *Empires in the Sun: The Rise of the New American West* (New York: Putnam, 1982), 42–46.

17. Harry N. Scheiber, "From Science to Law to Politics: An Historical View of the Ecosystem Idea and Its Effect on Resource Management," *Ecology Law Quarterly* 24 (1994): 631–651.

18. In a little noted passage in *The People's Forests*, Marshall argued that the way resources were used needed to be addressed, due to "the unmitigated evils of the present high pressure salesmanship which the forest industries find necessary in order to retain their markets in competition with other materials." In the search for expanding markets, Marshall speculated, forest companies and their resource manager allies had relied on massive and destructive timber cutting, "regardless of whether other products might not be better suited for the purpose than wood." Robert Marshall, *The People's Forests*, 130. A contemporary perspective on materials use is elaborated by John E. Young and Aaron Sachs, *The*

Next Efficiency Revolution: Creating a Sustainable Materials Economy, World-watch Paper no. 21 (Washington, D.C.: Worldwatch Institute, 1995).

19. Blake Gumprecht, *The Los Angeles River: Its Life, Death, and Possible Rebirth* (Baltimore: Johns Hopkins University Press, 1999).

20. Dana Plays, "River Madness," A film montage presented at the "Hollywood Looks at the River" event in the Occidental College "Re-envisioning the L.A. River" program, CBS Studio Center, Los Angeles, April 6, 2000. In the panel that followed the screening of Dana Plays's montage, filmmaker Wim Wenders commented that "landscapes ask for their own stories to be told. The L.A. River, as it now exists as a cemented river, has a story of aggression to tell."

21. Jill Leovy, "Pumping New Life into the L.A. River," *Los Angeles Times*, November 29, 1998, B1ff.; Judith Coburn, "Whose River Is It Anyway? More Concrete versus More Nature—The Battle over Flood Control on the Los Angeles River Is Really a Fight for Its Soul," *Los Angeles Times*, November 20, 1994.

22. "While floods in other lands are wholly evil in their effects," Southern California chronicler J.M. Guinn wrote in 1890, "ours, although causing temporary damage, are greatly beneficial to the country. They fill up the springs and mountain lakes and reservoirs that feed our creeks and rivers, and supply water for irrigation during the long dry season. A flood year is always followed by a fruitful year." See J.M. Guinn, "Exceptional Years: A History of California Floods and Drought," *Historical Society of Southern California Annual* 1, no. 5 (1890): 38; the Crespi diary quote is cited in Gumprecht, *The Los Angeles River*, 38; see also Bernice Eastman Johnston, *California's Gabrielino Indians* (Los Angeles: Southwest Museum, 1962).

23. In one interesting exchange of letters in the *Los Angeles Times*, a land developer building on the banks of the river dismissed the danger of floods, but also argued that "if there is any danger from an overflow at any time in the future, the authorities should see to it in time to prevent any disaster." "A Warning," *Los Angeles Times*, August 2, 1882; The *Times*, as the leading promoter of the 1880's boom, also weighed in on the river as impediment to development. This "treacherous stream," the paper proclaimed, "cannot be trusted. It needs to be restrained within its banks." *Los Angeles Times*, January 20, 1886.

24. By the 1950s it was estimated that more than two million people lived along the L.A. River floodplain. The work of the Corps in channelizing the river continued to encourage the belief that the river was immaterial in decisions about the nature and direction of development. Moreover, any concern about water resources in the region, as the Corps's 1940 survey report on the L.A. project declared, would best be addressed by importing new sources of water into the region. See Richard Bigger, *Flood Control in Metropolitan Los Angeles* (Berkeley: University of California Press, 1959), 3, 89.

25. Part of the Corps's single purpose flood control mandate in Los Angeles was established as a consequence of the defeat of New Deal efforts to establish mul-

tipurpose water projects as a form of social planning, a debate that had its origins in the conflicts between the Corps and Reclamation Service during the Progressive Era. By decoupling the concept of flood control from river-basin development in the legislation that authorized funding for Los Angeles flood control activities, it provided the legislative underpinning for the idea that the flow of the L.A. River presented an obstacle to development and thus needed to be rendered harmless. Aside from its work on the Mississippi River, the Corps's Los Angeles project became its largest single authorized flood control program. See *Flood Control in Metropolitan Los Angeles*, 27–28. The "rebuilding this fractious stream" comment is from Andrew R. Boone, "River Rebuilt to Control Floods," *Scientific American* 161, no. 5 (November 1939): 265; see also "Flood Control Program for Los Angeles," *Western Construction News* 14 (April 1939): 148; "Concrete Flood Control Channel for L.A. River," *Western Construction News* 23 (February 1948).

26. The "killer encased in a concrete straightjacket" quote is from the Spring 1978 issue of *Aqueduct*, the publication of the Metropolitan Water District of Southern California. The title of the article summed up the perspective of the water agency and its engineering counterparts: "The Los Angeles River is: * a big joke, * a killer in concrete, * a 50-mile-long ditch, * all of the above." See also Christopher Kroll, "Changing Views of the River," *California Coast and Ocean* 9, no. 3. (Summer 1993): 32; Richard Lester, "New Deal for Los Angeles County, Raw Deal for Its Environment? The Environmental Impact of Work Relief," paper presented at the 1997 Southern California Environment and History Conference, September 18, 1997; Kevin Starr, *Endangered Dreams: The Great Depression in California* (New York: Oxford University Press, 1996), 320–321.

27. Lee R. Henning, "Concrete Lining for a River Channel," *Western Construction News* 33 (February 1958): 31.

28. "All of us are in service to an idea," MacAdams wrote, "creating a Los Angeles River Greenway from the mountains to the sea." Lewis MacAdams, "Restoring the Los Angeles River: A Forty-Year Art Project," *Whole Earth Review* no. 85 (Spring 1995): 63.

29. In a revealing moment, in the midst of a contentious meeting over proposals to further enlarge and extend the river's concrete barriers and clear the channel, MacAdams and a county official engaged in this battleground over language. Each time the county official sought to refer to the river as a flood control channel, MacAdams interrupted, and said "River!" The interruptions were continuous; and the language for MacAdams was more than symbolic. For this poet-activist and the FoLAR activists, this was "a battle over the definition of the river, and what the river is going to be." "Friends of the L.A. River: Improve It," See *California Coast and Ocean* 9, no. 3 (Summer 1993): 26; Gumprecht, *The Los Angeles River*, 298.

30. Lewis Mumford, *The City in History: Its Origins, Its Transformations, and Its Prospects* (New York: Harcourt, Brace & World, 1961), 462, 470; see also

David Schuyler, *The New Urban Landscape: The Redefinition of City Form in Nineteenth-Century America* (Baltimore: Johns Hopkins University Press, 1986), 183.

31. Robert Hersh, "Race and Industrial Hazards: An Historical Geography of the Pittsburgh Region, 1900–1990," Discussion Paper 95–18, Washington, D.C., Resources for the Future, March 1995; Sam B. Warner Jr., *Streetcar Suburbs: The Process of Growth in Boston, 1870–1900* (Cambridge: Harvard University Press and the MIT Press, 1962), 32; Schuyler, *The New Urban Landscape*.

32. The garden advocates also promoted the concept of school gardens for rural school children in order "to make children better satisfied with country life, and to induce them to stay on the farm in peace and contentment instead of drifting to the city." Louise Klein Miller, *Children's Gardens for School and Home: A Manual of Cooperative Gardening* (New York: D. Appleton & Co., 1904), 5; the Sam Bass Warner quote is from *Streetcar Suburbs*, 14.

33. Florence Kelley, *Notes from Sixty Years: The Autobiography of Florence Kelley*, edited and with an introduction by Kathryn Kish Sklar (Chicago: Charles H. Kerr, 1986); Florence Kelley, "The Sweating System," in *Hull House Maps and Papers* (New York: Thomas Y. Crowell, 1895); Dominick Cavallo, *Muscles and Morals: Organized Playgrounds and Urban Reform, 1880–1920* (Philadelphia: University of Pennsylvania Press, 1981); Galen Cranz, *The Politics of Park Design: A History of Urban Parks in America* (Cambridge: MIT Press, 1982).

34. The suburb as an anti-urban environment was underlined by the Levitts' argument that it was no longer "socially desirable to build rental housing—or to live in the cities." "Levitt's Progress: Two Sons and a Father Who Revolutionized Home Building," *Fortune*, October 1955, 154.

35. "The ultimate solution," Ford declared in his series of articles for the *Dearborn Independent*, "will be the abandonment of the city." See also the articles in the same collection: "We Shall Solve the City Problem by Leaving the City," "The Modern City—A Pestiferous Growth," and "The Exodus from the City," in Henry Ford, *Ford Ideals: Being a Selection from "Mr. Ford's Page" in the* Dearborn Independent (Dearborn, Michigan: The Dearborn Publishing Company, 1922), 154–158 and 425–428; the "middle realm" quote is from Evan Eisenberg, *The Ecology of Eden* (New York: Alfred A. Knopf, 1998), xvi; the anticity perspectives of John Muir are elaborated in Stephen Fox, *John Muir and His Legacy: The American Conservation Movement* (Boston: Little Brown, 1981).

36. Hays, *Beauty, Health, and Permanence*, 72–73; see also Peter J. Schmitt, *Back to Nature: The Arcadian Myth in Urban America* (New York: Oxford University Press, 1969); Carol A. Christensen, *The American Garden City Movement and the New Towns Movement* (Ann Arbor: UMI Research Press, 1986).

37. Benton MacKaye, "The Townless Highway," *The New Republic* 62 no. 797 (March 12, 1930): 93–95.

38. MacKaye, "The Townless Highway," 94, 95.

39. James J. Flink, *America Adopts the Automobile, 1895–1910* (Cambridge: MIT Press, 1970), 112; C.J. Galpin, "Better Highways to Relieve City Congestion," *American City* 28, no. 2 (February 1923): 187; Henry Ford, in his attack on city life as "unnatural" also argued that the rise of an auto-centered transportation system would render "confinement within the City unnecessary for large numbers of people," allowing the process of suburbanization to extend to the working class as well as the well-to-do. In Ford, *Ford Ideals*, 156.

40. "The automobile," Kenneth Jackson argued in his seminal study on suburbanization and the rise of the automobile suburbs, "had a greater spatial and social impact on cities than any technological innovation since the development of the wheel." Kenneth T. Jackson, *Crabgrass Frontier: The Suburbanization of the United States* (New York: Oxford University Press, 1983), 188; Studebaker President Paul G. Hoffman is cited in Mark H. Rose, *Interstate: Express Highway Politics, 1941–1956* (Lawrence, Kansas: The Regents Press of Kansas, 1979), 1; the Urban Land Institute figure is cited in Mark S. Foster, *From Streetcar to Superhighway: American City Planners and Urban Transportation, 1900–1940* (Philadelphia: Temple University Press, 1981). 146.

41. Tom Lewis, *Divided Highways: Building the Interstate Highways, Transforming American Life* (New York: Viking Penguin, 1997).

42. For example, in St. Paul, Minnesota, developers worked closely with the highway engineers so that the freeways could also "provide a buffer between the blighted areas proposed for renewal and those not so designated." Alan A. Altschuler, "The Intercity Freeway," in *Introduction to Planning History in the United States*, edited by Donald A. Krueckeberg, Center for Urban Policy Research (Rutgers: The State University of New Jersey, 1983), 207; see also Tom Lewis, *Divided Highways*, 120.

43. Joan Didion, *The White Album* (New York: Pocket Books, 1979), 83; the car commute numbers are cited by Donald C. Shoup, "Congress Okays Cash Out," *Access*, no. 13 (fall 1998): 2.

44. Benton MacKaye, *The New Exploration: A Philosophy of Regional Planning* (Urbana: University of Illinois Press, 1962).

45. MacKaye, *The New Exploration*, 66; see also Paul Thompson Bryant, "The Quality of the Day: The Achievement of Benton MacKaye" (Ph.D. diss., University of Illinois, 1965).

46. Christensen, *The American Garden City Movement*, 44; "Lewis Mumford, "Regions to Live In," *The Survey Graphic* 54 (May 1, 1925): 152; Catherine Bauer, *Modern Housing* (New York: Arno Press, 1974), 114.

47. Benton MacKaye, "The Challenge of Muscle Shoals," *The Nation* 136, no. 3537 (April 19, 1933): 445.; Benton MacKaye, "Outdoor Culture: The Philosophy of Through Trails," *Landscape Architecture* 17, no. 3 (April 1927): 164–165.

48. MacKaye was also an early advocate of public ownership or control of forest resources as a means to facilitate conservation-oriented management or "timber cropping," as opposed to the anti-conservation strategies of "timber mining" pursued by the private logging interests. Similar to Bob Marshall, MacKaye developed radical perspectives on issues of labor and community development. He also identified with some of the Socialist ideas and political circles that emerged in the post–World War I period at a time when the Forest Service was itself shifting toward a more conservative, pro-industry approach. Benton MacKaye, "Some Social Aspects of Forest Management," *Journal of Forestry*, 16, no. 2, February 1918, p. 212; Paul Sutter, "'A Retreat from Profit:' Colonization, the Appalachian Trail, and the Social Roots of Benton MacKaye's Wilderness Advocacy," *Environmental History*, 4, no. 4, October 1999: 553–577.

49. Though MacKaye's vision of the Appalachian Trail was primarily a notion of regional reordering between city, countryside, and undeveloped or primeval lands, it was established more as recreation than "re-creation," a 2,000-mile trail that extended from Maine to Georgia. Bryant, "The Quality of the Day," 126; Benton MacKaye, "The Appalachian Trail: A Project in Regional Planning," *American Institute of Architects*, October 1921, 325–330. In language similar to that used by environmental justice advocates seventy years later, MacKaye also described the trail as a "place to live, work and play on a non-profit basis." Cited by Paul Sutter, "'A Retreat from Profit,'"563.

50. The "basic communism" quote is cited by Gail Radford, *Modern Housing for America: Policy Struggles in the New Deal Era* (Chicago: University of Chicago Press, 1996), 181; see also *Planning the Fourth Migration: The Neglected Vision of the Regional Planning Association of America*, edited by Carl Sussman (Cambridge: MIT Press, 1976); Mumford, *The City in History*, 527; Lewis Mumford, "Megalopolis as Anti-City," *Architectural Record*, December 1962, cited in Lewis Mumford, *The Urban Prospect* (New York: Harcourt, Brace & World, 1968), 140; Ramachandra Guha, "Lewis Mumford, the Forgotten American Environmentalist: An Essay in Rehabilitation," *Capitalism, Nature, Socialism* 2, no. 3 (October 1991): 67–91.

51. Catherine Bauer, "Slums Aren't Necessary," *American Mercury* 21, no. 123 (March 1934): 297; see also Catherine Bauer, *Modern Housing* (New York: Arno Press, 1974); Radford, *Modern Housing*, 76.

52. Catherine Bauer Wurster, "The Social Front of Architecture in the 1930s," *Journal of the Society of Architectural Historians* 24, no. 1 (March 1965): 48.

53. Catherine Bauer, "The Dreary Deadlock of Public Housing," *Architectural Forum* 106 (May 1957): 141–142.

54. Robert Fogelson, *The Fragmented Metropolis: Los Angeles, 1850–1930* (Cambridge: Harvard University Press, 1967).

55. By 1950 the urban land mass of the 168 major metropolitan areas had reached more than 200,000 square miles and a population of 84 million people. Yet the central cities within this stretch of urban expansion still accounted for nearly 60 percent of the population in these areas and thus still tended to dominate urban America. But the period between 1950 and 1990, as David Rusk has argued, witnessed an extraordinary explosion of outlying development, which both qualitatively and quantitatively changed the urban and regional landscape. The escape from the city phenomena now emerged as a model of exurban dispersion—the search for semi-urban spaces at the edge of the city. By 1990, the metropolitan land mass had extended to nearly 350,000 square miles and its population had nearly doubled to 159 million people, but the central cities now accounted for only one-third of that population. David Rusk, "The Exploding Metropolis: Why Growth Management Makes Sense," *The Brookings Review* 16, no. 4 (Fall 1998): 13–15; see also David Rusk, *Cities without Suburbs* (Washington, D.C.: Woodrow Wilson Center Press, 1995), distributed by Johns Hopkins University Press; William Whyte's quote is from his book *The Exploding Metropolis* (Berkeley: University of California Press, 1993), 7, and is cited by Jane Holtz Kay in *Asphalt Nation* (Berkeley: University of California Press, 1997), 244; see also Sam Bass Warner Jr., *The Urban Wilderness: A History of the American City* (New York: Harper & Row, 1972).

56. Josephine Goldmark, *Fatigue and Efficiency: A Study in Industry* (New York: Russell Sage Foundation, 1912), 203; see also Barbara Sicherman, *Alice Hamilton: A Life in Letters* (Cambridge: Harvard University Press, 1984); *Notes of Sixty Years: The Autobiography of Florence Kelley*, edited by Kathryn Kish Sklar (Chicago: Charles H. Kerr Publishing Co., 1986).

57. The Brandeis quote is from his *Brief on Behalf of the Traffic Committee of Commercial Organizations of Atlantic Seaboard*, Interstate Commerce Commission, Docket no. 3400, 1911, cited in Goldmark, *Fatigue and Efficiency*, 192. On the "gospel of efficiency" concept, see, for example, the exchange regarding scientific management between Upton Sinclair and Frederick Taylor in "The Gospel of Efficiency," *The American Magazine* 72 (June 1911): 243–245; Samuel D. Hays, in his invaluable study, *Conservation and the Gospel of Efficiency*, constructed his central thesis about the rise of conservationist discourse in the Progressive Era in the context of how key concepts such as the use of expertise and "science" as a management tool were related to the cult of efficiency.

58. Testimony of Frederick W. Taylor in *Hearings before Special Committee of the House of Representatives to Investigate the Taylor and Other Systems of Shop Management Under Authority of House Resolution 90*, January 25, 1912, reprinted in Frederick Winslow Taylor, *Scientific Management* (New York: Harper & Brothers, 1947), 40.

59. Taylor's "deskill" quote is from a 1906 talk and is cited in Robert Kanigel's biography of Taylor, *The One Best Way: Frederick Winslow Taylor and the Enigma of Efficiency* (Oxford: Blackwell, 1997), 169; the "monotonous, tiring,

and uninteresting" quote is also cited in Kanigel, *The One Best Way*, 139; the reform-minded socialist is William English Walling, and his quote is from *Progressivism—And After* (New York: The MacMillan Co., 1914), 58; Taylor's "maximum efficiency" quote is from his book, *The Principles of Scientific Management*, 1911, reprinted in *Scientific Management*, 9; see also Samuel Haber, *Efficiency and Uplift: Scientific Management in the Progressive Era* (Chicago: University of Chicago Press, 1964), 24.

60. The "forests vanishing" quote is part of Taylor's more elaborate argument about his association with the Progressive Era promotion of efficiency in addressing "the waste of material things" and its relation to the sphere of work. *The Principles of Scientific Management*, 5; the comparison of Fordist and Taylorist approaches is explored in *Scientific Management: Frederick Winslow Taylor's Gift to the World?* edited by J.-C. Spender and Hugo J. Kijne (Boston: Kluwer Academic Publishers, 1996), xiv–xvi; in a 1915 article on "Fordism," Horace Arnold and Fay Faurote wrote, "The Ford Motor Company has no use for experience, in the working ranks anyway. It desires and prefers machine tool operators who have nothing to unlearn, who have no theories of correct surface speeds for metal finishing, and will simply do what they are told to do, over and over again from bell-time to bell-time." "Ford Methods and the Ford Shops," *Engineering Magazine*, 1915, 41–42, cited in James J. Flink, *The Automobile Age* (Cambridge: MIT Press, 1988), 117. Ford himself, in his 1926 autobiography, defined mass production as "the reduction of the necessity for thought on the part of the worker and the reduction of his movements to a minimum. He does as nearly as possible only one thing with only one movement." Henry Ford, *My Life and Work*, in collaboration with Samuel Crouther (Garden City, New York: Doubleday, Page & Co., 1926), 80; however, Peter Drucker, in a 1976 essay for the *Conference Board Record*, argued that Taylor would have been highly critical of Fordist notions of production and assembly line work because it violated his basic principles such as the "finding, training, and developing of the individual for the job he is fitted for." Drucker associates Taylorism with the principle of labor *productivity*, and not assembly line production. Though recognizing that the two concepts became largely interchangeable in subsequent years, Drucker's revisionist portrait of Taylor was designed to make his theories of functional foremanship and workplace efficiency more consonant with the emerging concepts of flexible production strategies taking shape in the period in which Drucker was writing. Drucker's essay, "The Coming Rediscovery of Scientific Management," is reprinted in *Toward the Next Economics, and other Essays* (New York: Harper & Row, 1981), 96–106.

61. Jane Addams, *The Spirit of Youth and the City Streets* (New York: The Macmillan Company, 1914), 131; labor's equity right concept can be found in John P. Frey, "The Relation of Scientific Management to Labor," *Federationist* 20 (April 1913): 296–302, cited in Ny, Milton J. Nadworny, *Scientific Management and the Unions 1900–1932, A Historical Analysis*, (Cambridge: Harvard University Press, 1955), 88.

62. The "new suave unionism" quote is in Arthur W. Calhoun, "Labor's New Economic Policy," in *American Labor Dynamics In the Light of Post-War Developments: An Inquiry by Thirty-Two Labor Men, Teachers, Editors, and Technicians,* edited by J.B.S. Hardman (New York: Harcourt, Brace & Co, 1928), 320; see also Nadworny, 112; Haber, *Efficiency and Uplift,* 129–131. Taylorism had both its radical and conservative interpreters. In the United States, several members of the Taylor Society (the leading organization of scientific management advocates) began to assume a more progressive stance after Taylor's death in 1915. This included recognition of the role of unions in furthering efficiency and production goals and even such potentially radical concepts as joint councils of workers and managers to discuss production problems or appealing to the "creative motive" of the worker "by giving him an understanding of his particular contribution of a finished product and making him conscious of his place on the production team." In the Soviet Union, the Taylorist approach to work discipline and scientific management became a lightning rod for debates over the evolving economy and the role of workers and managers. While Lenin in 1913 and 1914 condemned Taylorism as the "scientific method of extortion of sweat," by 1918, with the Bolshevik Party consolidating its power over the state while in the midst of civil war, Lenin revised his assessment. In this context, Lenin assumed that piece work and a Taylorist approach to work discipline and production efficiency were essential in enabling the new Soviet Union to "set its people the task of learning to work." The Stachanovite worker, celebrated by Stalin's Russia for his ability to maximize his productivity, follow orders, and work fully on behalf of his factory and the state, could be considered, as one analyst put it, the model outcome of a Taylorist system. Ultimately, this new industrial work culture came to be most associated with a Fordist, modernist, and productionist philosophy (the goal of a production system associated with its outputs, rather than its inputs, its structure, and its processes). The "scientific method of extortion" and "the task of learning to work" are from Lenin's *Works,* vol. 27 (Moscow: Progress Publishers, 1972), and are cited in Rainer Traub, "Lenin and Taylor: The Fate of 'Scientific Management' in the (Early) Soviet Union," *Telos,* no. 37 (Fall 1978): 82–92. The Stachanovite analogy is made by Hugo Kijne and J.-C Spender, xvi; James J. Flink, in *The Automobile Age* (112–113), also describes how Fordist mass production methods, which were widely chronicled in Pravda, came to be widely emulated in the Soviet Union during the 1920s. "In workers' processions," Flink writes, "Ford's name was emblazoned on banners emblematic of a new industrial era" See also Haber, *Efficiency and Uplift,* 164; Daniel Nelson, *Frederick W. Taylor and the Rise of Scientific Management* (Madison: University of Wisconsin Press, 1980). Nelson argues that by the 1920s the earlier mistrust "between corporate and shop management" that concerned Taylor and especially Thorsten Veblen had been superseded by the rise of the efficiency and production-oriented middle managers and the triumph of the modern factory system and the centrally controlled bureaucratic/corporate enterprise, 201. Taylorism itself, as labor historian Richard Edwards has argued, also became less significant in terms of its influence on the

actual production strategies adopted by various industrial operations (which ultimately tended to be minimal) than the significance of the Taylorist *idea* that became critical to the evolving structure of power relations in the workplace, the continuing emphasis on deskilling, and the effort to establish management control at each stage of production. Richard Edwards, *Contested Terrain: The Transformation of the Workplace in the 20ᵗʰ Century* (New York: Basic Books, 1979), 97–104.

63. Hoover's talk was delivered on February 14, 1921, before the executive board of the American Engineering Council, Federated American Engineering Societies, of which Hoover was president. The abstract of the talk was reprinted under the title "Industrial Waste" in the *Bulletin of the Taylor Society*, April 1921, 77–79; the Ida Tarbell quote is from her autobiography, where she approvingly compares Taylor's concepts with Henry Ford's. Ida Tarbell, *All in the Day's Work: An Autobiography* (New York: Macmillan, 1937), 293.

64. By the 1990s, one major global company, Heineken, would come to characterize itself as "a marketing company, with a production facility." Cited by Richard J. Barnet and John Cavanagh in *Global Dreams: Imperial Corporations and the New World Order* (New York: Simon and Schuster, 1994), 168.

65. *Dying for Work: Workers' Safety and Health in Twentieth-Century America*, edited by David Rosner and Gerald Markowitz (Bloomington: Indiana University Press, 1989); Alice Hamilton, *Exploring the Dangerous Trades: The Autobiography of Alice Hamilton, M.D.* (Boston: Little, Brown, 1943). See also *Reducing Toxics*, pp. 170–176; Toxic Circles; "The Political Economy of Occupational Disease," Charles Levenstein and Dominick J. Tuminaro, in *Work, Health, and Environment: Old Problems, New Solutions*, edited by Charles Levenstein and John Wooding (New York: The Guilford Press, 1997); Alice Hamilton, *Industrial Poisons in the United States* (New York: Macmillan, 1925).

66. The "nightmare" quote is from Charles Kettering, *The New Necessity* (Baltimore: Williams and Wilkins, 1932), 75; the "one of the greatest steps" quote is from *Prophet of Progress: Selections from the Speeches of Charles F. Kettering*, edited by T.A. Boyd (New York: E.P. Dutton & Co., 1961), 52 and 115.

67. Fittingly, to accomplish this production breakthrough, the Ethyl Corporation was established as a partnership between General Motors and Standard Oil of New Jersey. The DuPont company, which itself had a major stock ownership in GM, in turn contracted with the Ethyl Corporation to manufacture the lead additive required for the production of ethyl gasoline. For each of the parties, the introduction of leaded gasoline represented an extraordinary financial and marketing windfall. "GM, in effect, made money on almost every gallon of gasoline sold anyplace by anyone," as Peter Drucker would later put it. *Management*, Peter Drucker (New York: Harper & Row, 1973), 712. High profits were generated by the Ethyl Corporation (which recorded net sales of $1 billion between its incorporation in 1924 and when ethyl gas patents expired in 1947). General

Motors' patent royalties in that period amounted to $43.3 million, while profit on sales due to partial ownership in the company were $82.6 million. DuPont, as the manufacturer of tel, had profits of $86 million. David A. Hounshell and John Kenley Smith, *Science and Corporate Strategy: DuPont R&D, 1902–1980* (Cambridge: Cambridge University Press, 1980). Moreover, the arrangement was crucial for GM in that it required on its part only minimal managerial effort, while still maintaining partial ownership in the product through its ownership share of the Ethyl Corporation. At the same time, GM's interlocking relationship with DuPont meant that the chemical manufacturer would also become the primary supplier for this new product. GM became the major beneficiary (as co-owner) as well as end-point user. This co-producer/-supplier/user web of relationships further meant that GM could define itself more strictly as "mechanical people dealing with metal processing," while DuPont and the Ethyl Corporation would constitute a separate management structure of "chemical engineers and chemical marketers," as Sloan later defined these relationships. See *U.S. v DuPont & Co.*, U.S. Supreme Court briefs (1956, p. 1248) cited in Arthur J. Kuhn, *GM Passes Ford, 1918–1938: Designing the General Motors Performance-Control System* (University Park, Pa.: Pennsylvania State University Press, 1986), 98; Alfred P. Sloan Jr., *My Years with General Motors*, edited by John McDonald with Catharine Stevens (Garden City, NY: Doubleday & Co., 1964), 150, 220–221; see also Ed Cray, *Chrome Colossus: General Motors and Its Times* (New York: McGraw-Hill, 1980), 241–242.

68. Midgley argued that there was no specific reason to believe that any potential occupational health concerns were warranted, despite indications from other scientists that tetraethyl lead likely represented a significant occupational hazard. The amount of lead that would be released into the environment, Midgley declared in a 1924 interview, was minute and thus inconsequential. Stuart Leslie, "Thomas Midgley and the Politics of Industrial Research," *Business History Review* 54, no. 4 (Winter 1980): 480–502; see also Alice Hamilton, "What the American Woman Thinks Concerning Motor Car Gasoline," *The Woman Citizen* 10 (July 11, 1925): 14ff.

69. Flink, *The Automobile Age*, 214.

70. The "one newspaper account" is from the *Bridgeton Evening News* and is cited in "Insanity Gas," *Literary Digest* 83, no. 8. (November 22, 1924): 18. The revelations about acute occupational health hazards also led to public concern about environmental exposure to lead and the call for government intervention. This problem had been anticipated by the producers when they named their product "ethyl gasoline" rather than leaded gasoline. "However efficient this new compound may be in eliminating the 'knock,'" one publication commented, ". . . public safety and public health are paramount. The history of poisoning in occupational diseases in this country and abroad proves abundantly the need for Government supervision." "Tetraethyl-lead on Trial," *The Outlook* 140 (June 3, 1925): 174; see also "Ethyl Gasoline," *Science* 60, November 21, 1924, supplement, xii; see also "Employees Stricken; 1 Dead, 4 Insane," *New York Times*, October 27, 1924, 1; *New York Times*, November 27, 1924, 14;

Silas Bent, "Deep Water Runs Still," *The Nation* 121, no. 3131 (July 8, 1925): 62–64.

71. The Bureau of Mines' research had been financed by the tetraethyl lead companies and was released in 1924 just as the tel controversies began to reach fever pitch. Rosner and Markowitz, *Dying for Work*, 123–124; the Midgley quote is from *Literary Digest*, November 22, 1924, 18; the Hamilton quote is in *Woman Citizen*, 14; see also "Possible Effects of Production on Industry," *New York Times*, May 8, 1925, 21; "Sober Facts about Tetraethyl Lead," *Literary Digest* 83, no. 13 (December 27, 1924): 25–26.

72. "Ethyl has stood her trial, and the jury have returned a Scottish verdict of 'Not Proven,'" the *London Times* underlined this interpretation of the surgeon general's report and its role in allowing leaded gasoline to return to the market. Cited in "Use of Lead Tetraethyl," *Science* 69 (March 1, 1929): 239–240.

73. Despite eventual regulatory action that eventually eliminated the lead content in gasoline in the 1980s, DuPont continued to manufacture tetraethyl lead until as late as 1991 for uses other than leaded gasoline. This production continued even as additional studies indicated that, despite the occupational controls on worker exposure to tel, the workers were nevertheless experiencing serious long-term health risks, including colorectal cancer. William Fayerweather, M.E. Karns, I.A. Nuwayhid, and T.J. Nelson, "Case-Control Study of Cancer Risk in Tetraethyl Lead Manufacturing," *American Journal of Industrial Medicine* 31 (1997): 28–35. The regulatory actions during the 1970s and 1980s are described in Albert L. Nichols, "Lead in Gasoline," in *Economic Analyses at EPA: Assessing Regulatory Impact*, edited by Richard D. Morgenstern, (Washington, D.C.: Resources for the Future, 1997), 50. The sixty refiners' numbers are in Graham Edgar, "Defeating the Gasoline Knock," *Scientific American* 142 (January 1930): 60–61. Interestingly, this article, which touted the technical breakthroughs associated with tetraethyl lead, failed to mention any of the intense occupational and environmental controversies that had taken place just a few years earlier. The average lead content numbers are identified in Paul P. Craig and Edward Berlin, "The Air of Poverty," *Environment* 13, no. 5 (June 1971): 2–9.

74. Charles Kettering, "Keep the Consumer Dissatisfied," *Nation's Business* 17, no. 1 (January 1929): 30; David Hounshell argues that while Taylorism focused primarily on work processes and Fordism on machinery and production, Sloanism could be considered the triumph of marketing over "pure production." David A. Hounshell, *From the American System to Mass Production, 1800–1932: Development of Manufacturing Technology in the U.S.* (Baltimore: Johns Hopkins University Press, 1984), 267.

75. Joseph Romm, who ran the Office of Energy Efficiency and Renewable Energy at the Department of Energy, has argued that the new management concepts lend themselves to a post-Taylorist view of reskilling the workplace. In the 1950s and 1960s, Romm writes, Shigeo Shingo and Taiichi Ohno, who developed the Japanese "fast-cycling" and "just-in-time" type production strategies, extended the Fordist/Taylorist notions of efficiency to include worker engage-

ment in the linking of more efficiency *processes* with more efficient *operations.* The new style systems approach that the Japanese advocated sought to overcome the distinction between process and operations issues and establish a more efficient total production system, including such core productivity outcomes as energy efficiency as well as worker productivity. By reengaging workers—both production line workers as well as Taylor's engineers—the argument went, new opportunities for pollution prevention and technological innovation could be identified. Unlike the Taylor system, thinking (associated with research and development) could thus be connected back with doing (associated with hands-on experience). However, the global dispersion of production and therefore of labor through these same systems has limited whatever reskilling potential the new management approaches afforded. Worker vulnerability becomes the flip side of flexible production. See Joseph J. Romm, *Cool Companies: How the Best Businesses Boost Profits and Productivity by Cutting Greenhouse Gas Emissions* (Washington, D.C.: Island Press, 1999), 28–45; see also Robert H. Hayes, Steven C. Wheelwright, and Kim B. Clark, *Dynamic Manufacturing Creating the Learning Organization* (New York: The Free Press, 1988), 228–229; Keith D. Denton, *Enviro-Management: How Smart Companies Turn Environmental Costs into Profits* (Englewood Cliffs, N.J.: Prentice Hall, 1994); *Industry, Technology, and the Environment* (Washington D.C.: U.S. Congress, Office of Technology Assessment, OTA, 1994).

76. The Saturn plant's issues were discussed in a presentation by William R. Miller III, manager, Environmental Affairs, Saturn Corporation, at the Public Interest Workshop on Clean Production and Clean Products, University of Tennessee, Knoxville, May 28, 1998; the "poster boy" quote is from Davan Maharaj and Stuart Silverstein, "A Series of Setbacks for Labor, Employers," *Los Angeles Times*, September 15, 1998; see also *Organizing the Landscape: Geographical Perspectives on Labor Unionism*, edited by Andrew Herod (Minneapolis: University of Minnesota Press, 1998).

77. Williams, "The Idea of Nature," 81.

78. Ibid., 83.

79. Michael Pollan, *Second Nature: A Gardener's Education* (New York: Dell, 1991), 198; Mindy Thompson Fullilove and Robert E. Fullilove III, "Place Matters," in *Reclaiming the Environmental Debate: The Politics of Health in a Toxic Culture*, edited by Richard Hofrichter (Cambridge: MIT Press, 2000).

80. Hilary Wainright, *Arguments for a New Left: Answering the Free-Market Right* (Oxford: Blackwell, 1994), 212.

2   Livable Regions and Cleaner Production: Linking Environmental Justice and Pollution Prevention

1. Personal communication with Lois Gibbs, October 10, 1995.

2. On the changes in environmentalism during the 1980s, see Michael McClosky, "Twenty Years of Change in the Environmental Movement: An

304   Notes to Chapter 2

Insider's View," *Society and Natural Resources* 4 (1991): 273–284; on the "Superfund for Workers" concept, see Michael Merrill, "No Pollution Prevention Without Income Protection: A Challenge to Environmentalists," *New Solutions* 1, no. 3 (Winter 1991): 9–11; the Superfund for Workers concept was subsequently renamed "Just Transition" partly to avoid Superfund's negative associations.

3. Gary Cohen, "It's Too Easy Being Green: The Gains and Losses of Earth Day," *New Solutions* 1 no. 2 (Summer 1990): 9–12; Gottlieb, *Forcing the Spring*, 201–204.

4. The "new paradigm" comment was made by Gus Speth at the preinauguration December 1992 Economic Summit and is cited in Gottlieb, *Forcing the Spring*, 307.

5. See, for example, Stephan Schmidheiny, *Changing Course: A Global Business Perspective on Development and the Environment*, with the Business Council for Sustainable Development (Cambridge: Mit Press, 1992).

6. Mitchell Bernard et al., *Breach of Faith: How the Contract's Fine Print Undermines America's Environmental Success* (New York: Natural Resources Defense Council, February 1995); *The Worst Congress Ever: A Review of the Environmental Record of the 104th Congress at Mid-point and the Outlook for 1996* (Washington, D.C.: The Wilderness Society, December 27, 1995).

7. The Hatcher quote is in "The Rise of Anti-Ecology," *Time*, August 3, 1970, 42; Whitney Young is cited in the *Trenton Evening Times*, February 16, 1970, in Harold Sprout, "The Environmental Crisis in the Context of American Politics," in *The Politics of Ecosuicide*, edited by Leslie L. Roos Jr. (New York: Holt, Rinehart & Winston, 1971), 46. Interestingly, Hatcher developed an alliance with various environmental constituencies in Gary on issues involving U.S. Steel. Those shifts are explored by Andrew Hurley in *Environmental Inequalities: Class, Race, and Industrial Pollution in Gary, Indiana, 1945–1980* (Chapel Hill: University of North Carolina Press, 1995). The National Urban League, which had also been at the forefront in criticizing environmentalists as hostile to the civil rights agenda, also eventually reversed its position with the rise of the environmental justice movement. In its 1992 *State of Black America* report, for example, it included for the first time since the report had been issued a discussion of environmental issues faced by African-American communities. See Robert D. Bullard, "Anatomy of Environmental Racism and the Environmental Justice Movement," in *Confronting Environmental Racism: Voices from the Grassroots*, edited by Robert D. Bullard (Boston: South End Press, 1993), 24.

8. Willard R. Johnson, "Should the Poor Buy No Growth," *Daedalus* 102, no. 4 (Fall 1973): 170–171; Norman Faramelli's comments are in "Ecological Responsibility and Economic Justice," in *Western Man and Environmental Ethics: Attitudes Toward Nature and Technology*, edited by Ian G. Barbour (Reading, Mass.: Addison-Wesley Publishing Co., 1973), 192.

9. The "slum rats" and other urban environmental protests are described by George Berg in "Environmental Pollution in the Inner City," *Science and Citizen*,

June–July 1968, 123–125; the issue of lead in the ambient environment from leaded gasoline is discussed in Daniel T. Magidson, "Half Step Forward," *Environment* 13, no. 5 (June 1971): 11–13; see also Gottlieb, *Forcing the Spring*, for the discussion of lead paint and housing, 244–250; see also *Environmental Quality and Social Justice in Urban America*, edited by James Noel Smith (Washington, D.C.: Conservation Foundation, 1972). Sydney Howe, the executive director of the Conservation Foundation, had sought during this period to develop an environmental justice–oriented direction for his organization. As a consequence, he was subsequently fired by his board and replaced by William Reilly (later, EPA director under George Bush), who steered his organization away from Howe's more urban–oriented focus. Personal communication with Sydney Howe, February 13, 1992.

10. See, for example, Daniel Zwerdling, "Poverty and Pollution," *Progressive* 37, no. 1 (January 1973): 25; Julian McCaull, "Discriminatory Air Pollution: If Poor, Don't Breathe," *Environment* 18, no. 2 (March 1976): 26–31. In an interesting commentary written in 1976, two political scientists argued that clear divisions existed between a black political and "social justice" or civil rights agenda and an environmental agenda. In analyzing the voting records of black congressmen, however, the authors were confronted with a high degree of support among the black delegation for "pro-environment" votes, as identified by the League of Conservation Voters' ratings of congressional votes. That discrepancy was explained by distinguishing between issues that the analysts considered "environmental concerns *in their own right*" (author's italics) such as a vote on a dam construction bill that could be construed as a "wilderness-conservation" issue. These contrasted with other votes, such as one that would have weakened provisions of the Occupational Safety and Health Act, that the authors contended were more specifically "workplace environment" or "urban problems" votes than purely environmental considerations. John M. Ostheimer and Leonard G. Ritt, *Environment, Energy, and Black Americans* (Beverly Hills: Sage Publications, 1976), 23–24.

11. Ken Geiser and G. Waneck, "PCBs and Warren County," *Science for the People* 15, no. 4 (1983): 13–17; Charles Lee, *Toxic Wastes and Race in the United States, A National Report on the Racial and Socio-Economic Characteristics of Communities with Hazardous Waste Sites* (New York: Commission for Racial Justice, United Church of Christ, 1987).

12. Andrew Szasz and Michael Meuser make the point that by reconceptualizing the discourse around environmental justice to place it within the context of a broader shift in social forces and political economy as well as what Szasz and Meuser call "the larger problem of modernity," then some of the key texts in environmental history and historical sociology could be read again as commentaries on environmental inequality as well. Andrew Szasz and Michael Meuser, "Environmental Inequalities: Literature Review and Proposal for New Directions in Research and Theory," *Current Sociology* 45, no. 3 (July 1997): 99–120.

13. Presentation by Mike Miller and Drew Sones, Los Angeles Bureau of Sanitation, UCLA, February 4, 1987; *Solid Waste Program in the City of Los Angeles: Issues and Recommendations, Overview*, Delwin A. Biaggi, director, Bureau of Sanitation, City of Los Angeles, January 1987. During 1986–1987 a group of my students undertook a year-long evaluation of the 1,600 ton per day incinerator project, part of the overall Los Angeles City Energy Recovery project or LANCER. The study documented and evaluated the approach of the Bureau of Sanitation, including the efforts around facility siting. Louis Blumberg et al., *The Dilemma of Municipal Solid Waste Management*, Graduate School of Architecture and Urban Planning, UCLA, 1987; these issues were further elaborated by Louis Blumberg and Robert Gottlieb in *War on Waste: Can America Win Its Battle With Garbage?* (Washington, D.C.: Island Press, 1989).

14. Cerell Associates, "Political Difficulties Facing Waste-to-Energy Conversion Plant Siting," J. Stephen Powell, senior associate, Waste-to-Energy Technical Information Series, chapter 3a (Los Angeles: California Waste Management Board, 1984); see also Lillie Craig Trimble, "What Do Citizens Want in Siting of Waste Management Facilities?" *Risk Analysis* 8, no. 3 (1988).

15. *Superfund Amendments and Reauthorization Act of 1986* (SARA), Pub. L. no. 99–499, 100 Stat. 1613 (codified as amended at §§ 9601–9675, 1988 and Supp. V 1993); Gottlieb, *Forcing the Spring*, 191.

16. See, for example, Citizen's Clearinghouse for Hazardous Wastes, *History of the Grassroots Movement for Environmental Justice, 1981–1986* (Arlington, Va.: CCHW, 1987); Penny Newman, "We Are the Power," *Everyone's Backyard* 7, no. 4 (Winter 1989): 3; Richard Moore, "Toxics, Race, and Class," paper presented at the Annual Briefing of the Interfaith Impact for Justice and Peace, Washington, D.C., March 18, 1991; see also *Toxic Struggles: The Theory and Practice of Environmental Justice*, edited by Richard Hofrichter (Philadelphia: New Society Publishers, 1993).

17. Tom Estabrook argues that the dissolution of the National Toxics Campaign had several contributing factors. These included an overdependence on foundation support, the failure to establish effective accountability mechanisms exacerbated by the rapid growth of the organization, and the perception of white male dominance in decision making, although, as Estabrook argues "in actuality it had become much more inclusive." Racial issues, however, pervaded much of the debate over the organization's role and future. See Thomas H. Estabrook, "Labor/Community/Environment: The Spatial Politics of Collective Identity in Louisiana" (Ph.D. diss., Clark University, 1996), 302–308; personal communication with Ted Smith, October 15, 1992.

18. Gottlieb, *Forcing the Spring*, 3–5.

19. The ability of the mainstream environmental groups to tap funding sources on the basis of their environmental justice claims, from foundations, through government grants, and through court settlements such as the Los Angeles settlement, rankled the local environmental justice groups who saw such fundraising strategies as competitive with their own efforts to survive. "After struggling

to attain a new funding category for environmental justice activities," one leading environmental justice advocate wrote, "low-budget, community-based environmental justice groups find themselves shut out once again." Peggy M. Shepard, "Issues of Community Empowerment," *Fordham Urban Law Journal* 21 (1994): 754; see also, Presentation by Robert Garcia, Environmental Defense, Occidental College, Los Angeles, February 16, 2000; John Adams, "Environmentalism and Justice at NRDC," *The Amicus Journal*, Spring 1994, 2. The environmental justice orientation of the Los Angeles office of Environmental Defense was made possible through a settlement from a court case involving the construction of the Century Freeway through a number of low-income communities in Los Angeles. To ensure that Environmental Defense maintained an environmental justice approach, a trustees' group was formed. The group included leading social justice advocates and was established to oversee the expenditure of the $1.5 million earmarked for the office and ensure that Robert Garcia, the head of the local organization and a one-time civil rights attorney, would be able to focus on environmental justice concerns.

20. See, for example, Carl Anthony's commentary in *Greenpeace*, July–August 1991, 19–20.

21. The EPA definition is in *United States Environmental Protection Agency Draft Guidance: Environmental Justice in EPA's NEPA Compliance Analyses* (Washington, D.C.: U.S. EPA, 1995); see also statement by Dr. Clarice Gaylord, director of Office of Environmental Equity, U.S. Environmental Protection Agency, in *Hearings before the Subcommittee on Civil and Constitutional Rights of the Committee on the Judiciary*, U.S. House of Representatives, 103rd Cong., 1st sess., March 3–4, 1994, 19–21.

22. The comparative risk approach, despite EPA efforts to establish an "environmental justice" dimension to the process of conducting such assessments, became a concern to environmental justice advocates. A number of groups and individuals, including those who participated in the EPA's state-by-state "Comparative Risk Project" during the early 1990s, warned that the structure of contemporary risk assessment understated the hazards faced by the most vulnerable populations. See "An Environmental Justice Perspective on Comparative Risk," in *Toward the Twenty-first Century: Planning for the Protection of California's Environment*, Sacramento California Comparative Risk Project, California Environmental Protection Agency, May 1994.

23. Along those lines was a widely noted comment by Robert Wolcott, the chair of the EPA's Environmental Equity Workgroup, who bluntly sought to separate the issue of measuring environmental burdens from the question of decision-making intent. "Surprise, surprise. We have these [hazardous] facilities near poor people," Wolcott was quoted in *U.S. News and World Report*. "Look back 500 years and you'll find the same thing. Some like to picture industrial executives sitting down in the most venal way and siting these pernicious facilities in the backyards of minorities as some kind of social vengeance. I have to be very skeptical of this." Wolcott's comments are in Michael Satchell, "A Whiff of

Discrimination?" *U.S. News and World Report*, May 4, 1992, 34. As part of the EPA "equity of burden" process, the agency had established its Environmental Equity Workgroup to focus explicitly on distributional issues through a risk-based approach. See William Reilly, "Environmental Equity: EPA's Position," *EPA Journal*, March–April 1992, 19; *Environmental Equity: Reducing Risk for all Communities* (Washington, D.C.: U.S. EPA, 1992); *Worst Things First? The Debate Over Risk-Based National Environmental Priorities*, edited by Adam M. Finkel and Dominic Golding (Washington, D.C.: Resources for the Future, 1994); E. Donald Elliott, "Superfund: EPA Success, National Debacle?" *Natural Resources and Environment*, Winter 1992, 48; William Stevens, "What Really Threatens the Environment?" *New York Times*, January 29, 1991.

24. Dan Tarlock points out how a few early National Environmental Policy Act (NEPA) related lawsuits even sought to use the requirement of an environmental impact statement required by NEPA to oppose, "as a form of pollution," the potential movement of minorities into white neighborhoods. Dan Tarlock, "City versus Countryside: Environmental Equity in Context," *Fordham Urban Law Journal* 21 (1994): 471.

25. During the 1970s, the civil rights/distribution-related criticism of the EPA concerned the failure to provide Clean Water Act funds for communities of color to build sewage treatment facilities. Interestingly, the concerns about community impacts subsequently shifted to a focus on negative impacts from such facilities, with one of the more prominent environmental justice/discriminatory intent lawsuits filed over the decision to site a sewage treatment facility in West Harlem in New York. The lawsuit, filed by the Natural Resources Defense Council in conjunction with the community group West Harlem Environmental Action, also resulted in a court settlement of $1.1 million, to be administered by the NRDC and WHEA and directed toward community environmental and public health issues. See Vernice D. Miller, "Planning, Power & Politics: A Case Study of the Land Use and Siting History of the North River Water Pollution Control Plant," *Fordham Urban Law Journal* 21 (Spring 1994); see also Christopher H. Foreman Jr., *The Promise and Peril of Environmental Justice* (Washington, D.C.: The Brookings Institution, 1998), 41–42; James H. Colopy, "The Road Less Traveled: Pursuing Environmental Justice through Title VI of the Civil Rights Act of 1964," *Stanford Environmental Law Journal* 13 (1994): 180–185.

26. See, for example, Philip Shabecoff, "Environmental Groups Are Told They Are Racist in Hiring," *New York Times*, February 1, 1990; see also the letter from Richard Moore et al., to John O' Connor et al., May 20, 1990, in author's possession. Several of the letters are also reprinted in "The Whiteness of the Green Movement," *Not Man Apart*, April–May 1990, 14–17.

27. The use of civil rights discourse as applied to the environmental arena emerged for some as a purely distributive focus, or what Iris Young called "the distributive paradigm of justice." Such an approach tended to ignore "the institutional context that determines material distribution," as Young put it. In terms of environmental justice, this argument distinguishes between the distributive

model, with its concerns about the patterns of distribution (the discriminatory intent argument), in contrast with the position that focuses on the cause of the environmental problem and the decision making associated with it. Iris Young, *Justice and the Politics of Difference* (Princeton: Princeton University Press, 1990), 18; Sheila Foster draws on John Rawls's concept of the "difference principle" that would allocate social goods to result in the greatest benefit or least burden to the least-advantaged social classes. However, a number of other analysts have argued that what is new in terms of the environmental justice movement has been its ability to transform "the possibilities for fundamental social and environmental change through processes of redefinition, reinvention, and construction of innovative political and cultural discourses and practices." See Giovanna di Chiro's "Reframing Environmental History: Nature, Environment, and Community in the Environmental Justice Movement" cited in Sheila Foster, "Justice from the Ground Up: Distributive Inequities, Grassroots Resistance, and the Transformative Politics of the Environmental Justice Movement," in *California Law Review* 86 (July 1998): 840, 790–791.

28. See, for example, Vicki Been, "What's Fairness Got to Do with It? Environmental Justice and the Siting of Locally Undesirable Land Uses," *Cornell Law Review* 78 (1993), and "Locally Undesirable Land Uses in Minority Neighborhoods: Disproportionate Siting or Market Dynamics?" *Yale Law Journal* 103 (1994); Michael Fisher, "Environmental Racism Claims Brought Under Title VI of the Civil Rights Act," *Environmental Law* 25 (Spring 1995): 285–333; Ann Bowman, "Environmental (in)Equity: Race, Class, and the Distribution of Environmental Bads," in *Flashpoints in Environmental Policymaking: Controversies in Achieving Sustainability*, edited by Sheldon Kamieniecki, George A. Gonzalez, and Robert O. Vos (Albany: State University of New York Press, 1997), 155–175; "Environmental Justice in the United States—A Primer," *Michigan Bar Journal* 76, no. 1 (January 1997): 62–69.

29. Alice Kaswan, "Environmental Justice: Bridging the Gap Between Environmental Laws and Justice," *The American University Law Review* 47 (December 1997): 221–300.

30. In an important commentary that sought to address this potential distinction, Bunyan Bryant, a leading academic analyst who had been engaged in the development of this new civil rights discourse, sought to distinguish between a movement concerned with environmental racism and environmental justice. Bryant defined environmental racism as "the unequal protection against toxic and hazardous waste exposure and the systematic exclusion of people of color from environmental decisions affecting their communities" (the discriminatory intent argument). Environmental justice, on the other hand, references "decent paying and safe jobs; quality schools and recreation; decent housing and adequate health care; democratic decision-making and personal empowerment; and communities free of violence, drugs, and poverty." Bunyan Bryant, Introduction to *Environmental Justice: Issues, Policies, and Solutions*, edited by Bunyan Bryant (Washington, D.C.: Island Press, 1995), 5–6.

31. Deeohn Ferris and David Hahn-Baker, "Environmentalists and Environmental Justice Policy," in *Environmental Justice*, edited by Bryant, 67. Among environmental justice advocates, there have been efforts to broaden the concept of "justice" beyond the civil rights focus on discriminatory siting of facilities like landfills or incinerators. Robert Bullard, for example, has long emphasized the *political* character of environmental justice by arguing that the lack of decision-making capacity by communities of color is a central concern of the environmental justice movement. Daniel Faber, identifying an "eco-socialist perspective" as distinct from a liberal framework, seeks to extend the "justice" discourse by arguing that environmental justice activity needs to distinguish between a "distributional justice" perspective addressing what are essentially procedural issues such as siting to a "productive justice" framework that can address such issues as capital investment and production choices. Daniel Faber, "The Struggle for Ecological Democracy and Environmental Justice," in *The Struggle for Ecological Democracy: Environmental Justice Movements in the United States*, edited by Daniel Faber (New York: The Guilford Press, 1998), 15; Sheila Foster, "Race(ial) Matters: The Quest for Environmental Justice," *Ecology Law Quarterly* 20 (1993): 721–753; on the comparative risk argument, see "An Environmental Justice Perspective" in *Toward the 21ˢᵗ Century*; Margaret Kriz, "The Color of Poison," *National Journal*, July 11, 1998.

32. In that case, *Coalition of Concerned Citizens Against I-670 v. Damian*, 608 F. Supp. (S.D. Ohio 1984), the Supreme Court ruled against the plaintiffs because the court felt that although the siting of the highway might have had differential impacts, the defendant was able to demonstrate "adequate justification" for the action. The "adequate justification," ironically, was based on the fact that alternative sites identified also included communities of color. More significantly, the court was clearly reluctant to stop a project where significant planning, property acquisition, and overall project development had already occurred, and where the broader issue of alternatives to a highway construction project itself was not central to the argument. See Michael Fisher, "Environmental Racism Claims," 325–326; *Civil Rights and the Environment: Bridging the Disciplines*, legal compendium, Washington, D.C., Environmental Justice Project of the Lawyers' Committee for Civil Rights Under Law, March 1993; see also Sidney D. Watson, "Reinvigorating Title VI: Defending Health Care Discrimination—It Shouldn't Be So Easy," *Fordham Law Review* 58 (1990): 939–978.

33. See, for example, Luke Cole, "Civil Rights, Environmental Justice and the EPA: The Brief History of Administrative Complaints Under Title VI of the Civil Rights Act of 1964," *Journal of Environmental Law and Litigation* 9, no. 2 (Fall 1994): 309–398; David Oedel, "The Legacy of Jim Crow in Macon, Georgia" in *Just Transportation: Dismantling Race and Class Barriers to Mobility*, edited by Robert S. Bullard and Glenn S. Johnson (Gabriola Island, British Columbia: New Society Publishers, 1997); Sam Howe Verhovek, "Racial Tensions in Suit Slowing Drive for 'Environmental Justice,'" *New York Times*, September 7, 1997, 1; Michael Fisher, "Environmental Racism Claims," 288.

34. Sheila Foster, "Justice from the Ground Up," 779–786; Michael de Courcy Hinds, "Pennsylvania City Hopes It's Bouncing Back from the Bottom," *New York Times*, January 5, 1992; see also Wesley D. Few, "Recent Development in Case Law: The Wake of Discriminatory Intent and the Rise of Title VI in Environmental Justice Law Suits," *South Carolina Environmental Law Journal* 6 (Summer 1997): 108–120; "Industry's Role in Achieving Environmental Justice" (presentation by Dominique R. Shelton, program on "Environmental Justice in California," Los Angeles, May 20, 1999).

35. Interview with Nancy Nadel, April 9, 1999.

36. Of the first fifty administrative complaints filed with the agency after 1993, twenty-eight were denied outright, and the remainder still remained under review or investigation, despite the length of time that had elapsed, which in some cases had been more than five years. Allan Zabel, senior counsel of the EPA (presentation at U.C. Berkeley, Boalt Law School, November 16, 1998), cited in "The Promise of Title VI of the Civil Rights Act for Environmental Justice Claims," Julia Larkin, University of California School of Public Policy, December 1998.

37. Luke W. Cole, "Civil Rights, Environmental Justice and the EPA," 311.

38. Some conservative civil rights figures, such as Arthur Fletcher, the chairman of the National Black Chamber of Commerce, have argued against certain environmental regulations, such as the particulate standards developed through the Clean Air Act Amendments as "pitting clean air against jobs and physical health against economic well-being in depressed neighborhoods." These critics make a particular point that small businesses such as dry cleaners, bakeries, and auto body shops, who might employ minority workers in these low-wage sectors, are especially vulnerable in terms of air quality standards. See Arthur Fletcher, "Clean Air or Jobs: What a Choice," *Los Angeles Times*, November 1, 1998; see also David Friedman, "A Darker Shade of Green: The Inner City's Latest Foe is a Mainstream Environmentalist," *Los Angeles Times*, June 20, 1999; the "distributive outcomes" quote is from Sheila Foster, "Justice from the Ground Up," 791.

39. Interestingly, the term "brownfield" has also appeared in the business organization literature as referring to traditional Fordist or Taylorist production facilities "with mass production methods and systems of social organization." See James P. Womack and Daniel T. Jones, *Lean Thinking: Banish Wealth and Create Wealth in Your Corporation* (New York: Simon & Schuster, 1996), 305; EPA's brownfields definition is in *Brownfields National Partnership Action Agenda* (Washington, D.C.: U.S. EPA, n.d.).

40. The change in status is discussed in James G. Wright, *Risks and Rewards of Brownfield Redevelopment* (Cambridge, Mass.: Lincoln Institute of Land Policy, 1997), 21; U.S. Environmental Protection Agency, *Guidance on Settlements with Prospective Purchasers of Contaminated Property* (Washington, D.C.: U.S. EPA, May 1995); U.S. Environmental Protection Agency, May 1997, EPA 600-F-97-090; GAO report, *Superfund: Progress, Problems and Reauthorization Issues*, by Richard L. Hembra (Washington, D.C.: U.S. GAO, April 21, 1993);

the "number one issue" quote is from Charles Bartsch and Richard Munson, "Restoring Contaminated Industrial Sites," *Issues in Science and Technology* 10, no. 3 (Spring 1994): 74; see also U.S. General Accounting Office, *Community Development Reuse of Contaminated Sites* (Washington, D.C.: U.S. GAO, June 30, 1995).

41. Carl Anthony, "Making Brownfields Bloom," *Land and People*, Fall 1996, 26.

42. The "eco-risk exaggeration" term is used by David Friedman in "A Darker Shade of Green," June 20, 1999; Carol Browner's quote is from her testimony before the House Committee on Transportation and Infrastructure, Water Resources and Environment Subcommittee, June 27, 1995, and is cited in Charles Bartsch and Elizabeth Collaton, *Brownfields: Claiming and Reusing Contaminated Properties* (Westport, Conn.: Praeger, 1997), 41; see also U.S. General Accounting Office, *Environmental Protection: Agencies Have Made Progress in Implementing the Federal Brownfield Partnership Initiative*, GAO/RCED-99-86 (Washington, D.C.: U.S. GAO, April 1999); Mark S. Dennison, *Brownfields Redevelopment: Programs and Strategies for Rehabilitating Contaminated Real Estate* (Rockville, Maryland: Government Institutes, 1998); Kirk Johnson, "Washington Steps Back, and Cities Recover," *New York Times*, November 16, 1997; Dave Gatton, "Mayors Seek Bipartisan Action on Brownfields/Superfund Bill," *U.S. Mayor* 66, no. 9 (May 17, 1999): 1ff.; J. Thomas Black, "Brownfields Cleanup," *Urban Land* 54, no. 6 (June 1995): 47–51.

43. Carl Anthony's discussion of environmental justice in a regional context is in National Environmental Justice Advisory Council, *Environmental Justice, Urban Revitalization, and Brownfields: The Search for Authentic Signs of Hope— A Report on the "Public Dialogues on Urban Revitalization and Brownfields: Envisioning Healthy and Sustainable Communities"* (Washington, D.C.: U.S. EPA, revised September 1997), 11 (www.epa.gov/swerosps/ej/pdf/nejacjpdf/htm); see also *Building Upon Our Strengths: A Community Guide to Brownfields Redevelopment in the San Francisco Bay Area* (San Francisco: Urban Habitat, 1999); "Brownfields Redevelopment: A New Era of Urban Revitalization," in *Working for Sustainable Communities: A Resource Guide for the San Francisco Bay Area* (San Francisco: Environmental Careers Organization and Urban Habitat Program, 1998), 12–15; Paul Stanton Kibel, "The Urban Nexus: Open Space, Brownfields, and Justice," *Boston College Environmental Affairs Law Review* 25 (Spring 1998): 565–613; Raymond Hernandez, "Rethinking the Cleanup Rules for Polluted Sites," *New York Times*, August 29, 1999.

44. An article in *Governing* magazine cited one Chicago-based environmental justice advocate complaining that the limited brownfields cleanup strategy simply "perpetuated the pollution that communities like the Southeast Side of Chicago have been living with for 100 years." Tom Arrandale, "Lots of Trouble," *Governing* 10 (July 1997): 24; see also Lynn Grayson, "The Brownfields Phenomenon: An Analysis of Environmental, Economic, and Community Concerns,"

Environmental Law Reporter 25 (July 1995): 10340; Samara F. Swanston, "An Environmental Justice Perspective on Superfund Reauthorization," *St. John's Journal of Legal Commentary* 9, no. 2 (Spring 1994): 565–572.

45. See Lenny Siegel, "Subsidies and Gentrification," *Citizen's Report on Brownfields* 1, no. 2 (April 1999): 2, and Lenny Siegel, "Resolving Title VI Uncertainty," *Citizens' Report on Brownfields* 1, no. 1 (November 1998): 3–4; see also Charles Bartsch and Richard Munson, "Restoring Contaminated Industrial Sites," *Issues in Science and Technology* 10, no. 3 (Spring 1994): 74–78. Even more disturbing to some of the environmental justice groups was the possibility that the sudden availability of brownfields funds could be used to promote land uses and industrial projects that undercut opportunities for more community influenced or directed development initiatives. For example, during the late 1990s, community and environmental groups began to explore opportunities for transforming the Union Pacific railroad yards that bordered the Chinatown district of Los Angeles, just north of downtown. The groups sought to identify community needs (e.g., housing, schools) and opportunities for open space and recreation (given that Chinatown was one of the densest neighborhoods in the city). As these plans were being explored, a developer group, with ties to the mayor's office and the Union Pacific, moved quickly to close a deal that would turn the Chinatown Yards into tilt-up warehouses and some manufacturing. The key for the developers was a pledge of $11.5 million in federal brownfields grants and loan guarantees. The developer characterized this brownfields money as "designed to help the developer" close the deal as opposed to most other inner-city funding mechanisms that were designed to help the tenants of a development rather than the development itself. The use of brownfields funding in this case disregarded community and environmental input and potentially threatened the residential character of the Chinatown community itself, which had previously been forced to relocate due to the intervention of outside developers. See, George Ramos, "Seeds of Discord Grown over Downtown's 'Cornfield,' " *Los Angeles Times*, October 2, 1999; Stephen Siciliano, "Brownfields Funding for Cornfields," *Los Angeles Downtown News* 28, no. 35 (August 30, 1999): 1ff.; Lewis MacAdams and Robert Gottlieb, "Changing a River's Course: A Greenbelt vs. Warehouses," *Los Angeles Times*, October 3, 1999; *The River Through Downtown: Proceedings* (Los Angeles: Friends of the Los Angeles River, 1998); Mark Haefele, "It's Chinatown, Jake," *L.A. Weekly*, September 3–9, 1999, 15.

46. Statement of Angela Rooney cited in *Environmental Quality and Social Justice in Urban America*, edited by James Noel Smith (Washington, D.C.: The Conservation Foundation, 1974), 85; see also Deanna Welch, "Freeway Construction Connected, Divided City," *Los Angeles Times*, December 5, 1999.

47. Like the term "sustainability," which was used by some to refer to "perpetual growth" and became a much-argued and open-ended term, the concept of "livability" or "livable communities," first popularized by environmental justice–oriented neighborhood development and environmentally oriented land

use advocates in the 1990s, was seized upon as a broader, less-focused phrase designed to enhance Al Gore's presidential aspirations in 2000 through initiatives that addressed middle-class concerns over green space, traffic congestion and sprawl, crime, and educational needs. See Judy Pasternak, "2000 Budget to Focus on Urban Communities," *Los Angeles Times*, January 11, 1999; Don Chen, "Transportation Equity and Smart Growth," *Progress*, 10, no. 1, Feb.–March, 2000, 1.

48. Urban Habitat defined transit-oriented development, a broad concept that became increasingly popular after the passage of the 1991 Intermodal Surface Transportation and Efficiency Act (ISTEA), as "combining the goals of improving transportation access and mobility with affordable housing [and] community-based economic development." By the late 1990s, Urban Habitat, which had originated as a project of the Earth Island Institute but subsequently became an independent organization, had emerged as the environmental justice group most focused on a broader, place-based and land use concept of urban and regional renewal along environmental and social justice pathways. Henry Holmes, "Just and Sustainable Communities," in *Just Transportation*, 30. See also *What If We Shared: Findings from the San Francisco Bay Area Metropolitics* (San Francisco: Urban Habitat, 1998); F. Kaid Benfield, Matthew D. Raimi, and Donald T. Chen, *Once There Were Greenfields: How Urban Sprawl Is Undermining America's Environment, Economy, and Social Fabric* (Washington, D.C.: Natural Resources Defense Council and Surface Transportation Policy Project, 1999); Robert Bullard, Glenn S. Johnson, and April O. Torres, "Dismantling Transportation Apartheid Through Environmental Justice," *Progress*, 10, no. 1, Feb.–March, 2000, 4–5.

49. See, for example, Todd Campbell and Lesley Dobalian, *Failing the Grade: How Diesel School Buses Threaten Our Children's Health* (Los Angeles: Coalition for Clean Air, 1999); West Harlem Environmental Action represents an important example of a group that emerged through a classic, facility-siting-related struggle around a sewage treatment plant that situated the civil rights–oriented, risk discrimination-focused origins of the group, but which, in the context of its community or place-based roots, subsequently broadened its agenda to include a focus on transportation and land use, health, and food issues, as well as air quality and organizational and community capacity building. Presentation by Peggy Shepard, West Harlem Environmental Action, at Workshop on Clean Production and Clean Products, Knoxville, University/Public Interest Partnership, May 28, 1998; Peggy Shepard, "Issues of Community Empowerment," 745.

50. The Rogers quote is cited in Marc Cooper, "The New Oakland Raider," *The Nation* 266, no. 19 (May 25, 1998): 20; Myron Orfield, *Metropolitics: A Regional Agenda for Community and Stability* (Washington, D.C.: Brookings Institution Press, and Cambridge, Mass.: Lincoln Institute of Land Policy, 1997).

51. Ken Geiser presentation at the "Environmental Education and Learning for the 21st Century" session at the 6th International Conference of the Greening

of Industry Network, "Developing Sustainability: New Dialogue, New Approaches," University of California at Santa Barbara, November 16, 1997; Barry Commoner, *Making Peace with the Planet* (New York: Alfred A. Knopf, 1990).

52. Cheryl Kennan and Joshua Kanner, "Taking Stock: An Evaluation of Massachusetts' Toxics Use Reduction Act," and Michael J. DeVito, "An Alternative to TURA," *Pollution Prevention Review* 8, no. 1 (Winter 1998): 85–94, 95–105; Manik Roy, "Toxics Use Reduction in Massachusetts: The Whole Facility Approach," *Pollution Prevention News*, U.S. Environmental Protection Agency, February 1990.

53. Joseph Ling, "Industry: Making Cleanup Pay," *Environment* 22, no. 3 (1980): 42–43; Thomas Zosel, "How 3M Makes Pollution Prevention Pay," *Pollution Prevention Review* 1, no. 1 (1990–1991): 67–72; Michele Ochsner, "Pollution Prevention: An Overview of Regulatory Incentives and Barriers," *New York University Environmental Law Journal* 6, no. 3 (1998): 605–610; Peter Sinsheimer and Robert Gottlieb, "Pollution Prevention Voluntarism: The Example of 3M," in *Reducing Toxics*, 389–420.

54. Laura Beck, et al., *Toxics Reduction: Policy Tools, Strategies, and State Initiatives* (Los Angeles: Pollution Prevention Education and Research Center, 1992); U.S. Environmental Protection Agency, *Pollution Prevention 1991: Progress on Reducing Industrial Pollutants* (Washington, D.C.: Office of Pollution Prevention, 1991), EPA 21P-3003; Manik Roy and Eric V. Schaeffer, "Waste Solution Must Involve Water, Air, Land," *Forum for Applied Research and Public Policy* 8, no. 1 (Spring 1993): 116–124; David K. Rozell and Roy Brower, "Pollution Prevention Facility Planning: The Oregon Experience," *Pollution Prevention Review* 3, no. 3 (Summer 1993): 277–283.

55. The definition of pollution prevention is found in Sec. 6603 of the *Pollution Prevention Act of 1990*, PL 101–508, Title 6, 104 STAT. 1388 (1990); on the background to the debates over how to interpret pollution prevention, see the discussion in *Reducing Toxics*, 124–131; see also Robert F. Blomquist, "Government's Role Regarding Industrial Pollution Prevention in the United States," *Georgia Law Review* 29 (1995): 349–448; Harry Freeman et al., "Industrial Pollution Prevention: A Critical Review," *Journal of the Air and Waste Management Association* 42, no. 5 (May 1992): 618–656.

56. Joel S. Hirschhorn, "Why the Pollution Prevention Revolution Failed—And Why It Ultimately Will Succeed," *Pollution Prevention Review* 5, no. 2 (Winter 1997): 11–31; the OTA study was entitled *Serious Reduction of Hazardous Waste: For Pollution Prevention and Industrial Efficiency*, Office of Technology Assessment, U.S. Congress, Pub. No. OTA-ITE-317, 1986.

57. Warren Muir, "Facing Facts," in *Pollution Prevention 1997: A National Progress Report*, U.S. Environmental Protection Agency, Office of Pollution Prevention and Toxics (Washington, D.C.: U.S. EPA, June 1997); EPA 742-R-97-00; "Companies Don't Always Carry Out Pollution Prevention—Even When It's Good for Business," *The Gallon Environmental Letter* 1, no. 20, Canadian

Institute for Business and the Environment, November 18, 1997); U.S. Environmental Protection Agency, *Searching for the Profit in Pollution Prevention: Case Studies in the Corporate Evaluation of Environmental Opportunities*, by James Boyd (Washington, D.C.: U.S. EPA, Office of Policy, Planning, and Evaluation, April 1998), EPA 742-R-98-005.

58. John Cross's arguments are in "You Say You Want a P2 Revolution?" *Pollution Prevention Review* 8, no. 2 (Spring 1998): 18; Ken Geiser had also weighed in concerning the debate between Hirschhorn and Cross in the *Pollution Prevention Review*. Geiser, unlike Hirschhorn, sought to keep the focus on particular policy gaps that hindered pollution prevention implementation, even if actions to overcome such gaps bordered on the "incrementalist" approach that Hirschhorn had so passionately criticized. But underlying Geiser's focus on policy gaps was the interest among a number of pollution prevention advocates that pollution prevention provided the basis for the discussion of "clean products" and a "materials policy" approach, signifying a new type of "industrial policy." Ken Geiser, "Can the Pollution Prevention Revolution Be Restarted?" *Pollution Prevention Review* 8, no. 3 (Summer 1998): 71–80.

59. U.S. Environmental Protection Agency, *Pollution Prevention 1997*, 26, 47; see also Eric Jay Dolin, "Case Studies: EPA's Voluntary P2 Programs Pay Off," *Pollution Prevention Review* 8, no. 1 (Winter 1998): 15–26.

60. See U.S. Environmental Protection Agency, *Industrial Pollution Prevention: Incentives and Disincentives*, August 1994 (Washington, D.C.: GPO, 1994), EPA-820-R-94-004; U.S. Environmental Protection Agency, *Industrial Pollution Prevention Project (IP3): Summary Report*, July 1995 (Washington, D.C.: GPO, 1995), EPA-820-R-95-007.

61. Iris Marion Young, *Justice and the Politics of Difference* (Princeton, N.J.: Princeton University Press, 1990), 83.

62. *Reducing Toxics*, 95–108, 26–33; R. Shep Melnick, *Regulation and the Courts: The Case of the Clean Air Act* (Washington, D.C.: The Brookings Institution, 1983).

63. *Forcing the Spring*, 133–143.

64. Nixon's statement can be found in his "Message of the President Relative to Reorganization Plans Nos, 3 and 4 of 1970," H.R. Doc. 366, 91st Cong., 2nd sess. (1970), cited in "On Integrated Pollution Control," James E. Krier and Mark Brownstein, *Environmental Law* 22 (1991): 121; see also John C. Whitaker, *Striking a Balance: Environment and Natural Resources Policy in the Nixon-Ford Years* (Washington, D.C.: American Enterprise Institute for Public Policy Research, 1976).

65. J. Clarence Davies and Jan Mazurek, *Pollution Control in the United States: Evaluating the System* (Washington, D.C.: Resources for the Future, 1998); Gregg Easterbrook, *A Moment on the Earth: The Coming Age of Environmental Optimism* (New York: Viking, 1995).

66. For example, the *New York Times*, citing the EPA's own inspector general, noted the "widespread failures by Federal and local officials in several states to police even the most basic requirements of the nation's clean air and water laws." John H. Cushman Jr. "EPA and States Found to Be Lax in Pollution Law," *New York Times*, June 7, 1998.

67. U.S. Environmental Protection Agency, *Study of Industry Motivation for Pollution Prevention*, by Manik Roy and Ohad Jehassi, Pollution Prevention Policy Staff, Office of Prevention, Pesticides and Toxic Substances, draft paper, April 23, 1997; on the attention-focusing aspect of Superfund, see Mary E.S. Raivel, "CERCLA Liability as a Pollution Prevention Strategy," *Maryland Journal of Contemporary Legal Issues* 4, no. 1 (Fall/Winter 1992–1993): 131–151; on the attention-getting role of regulatory enforcement, see Thomas L. Eggert, "Moving Forward by Looking Back: The Role of Enforcement in Promoting P2," *Pollution Prevention Review* 9, no. 2 (Spring 1999): 1–5.

68. DeWitt John, *Civic Environmentalism: Alternatives to Regulation in States and Communities* (Washington, D.C.: CQ Press, 1994); see also Public Policy Institute, *Civic Environmentalism in Action: A Field Guide to Local and Regional Initiatives* (Washington, D.C.: Public Policy Institute, 1999); *Setting Priorities, Getting Results: A New Direction for EPA* (Washington, D.C.: National Academy for Public Administration, 1995); on the criticism of Superfund see Daniel Mazmanian and David Morell, *Beyond Superfailure: America's Toxics Policy for the 1990s* (Boulder: Westview Press, 1992).

69. *The Alternative Path: A Cleaner, Cheaper Way to Protect and Enhance the Environment*, The Aspen Institute Series on the Environment in the 21st Century, The Aspen Institute Program on Energy, the Environment, and the Economy (Washington, D.C.: The Aspen Institute, 1996), 27.

70. *Reinventing Environmental Regulations*, National Performance Review, President Bill Clinton and Vice President Al Gore, Washington, D.C., March 16, 1995; U.S. Environmental Protection Agency, *The Common Sense Initiative: A New Generation of Environmental Protection* (Washington, D.C.: U.S. EPA, June 1995), EPA 400-F-95-001; U.S. Environmental Protection Agency, *Project XL: Proposals for Facilities, Sectors, and Government* (Washington, D.C.: Office of Policy, Planning, and Evaluation, 1995).

71. Sanford Lewis, "Greening of the Electronics Industry: Prospects and Dilemmas for Stakeholder Participation," in *Pollution Prevention Strategies and Decision-Making in the Electronics Industry Sector: An Evaluation of Project XL and the Stakeholder Input/Regulatory Relief Model* (Los Angeles: Pollution Prevention Education and Research Center, 1999); Ken Geiser, "Why Public Interest Groups Should Focus on Clean Products and Clean Production," Workshop on Clean Production and Clean Products, Knoxville, University/Public Interest Partnership, May 1998.

72. John H. Cushman, "Congressional Republicans Take Aim at an Extensive List of Environmental Statutes," *New York Times*, February 22, 1995.

73. On the background to the next generation approach see Marian R. Chertow and Daniel C. Esty, *Thinking Ecologically: The Next Generation of Environmental Policy* (New Haven: Yale University Press, 1997); *EHE: The Environmental Protection System in Transition: Toward a More Desirable Future*, The Enterprise for the Environment, available at http://www.csis.org/e4e/; *Towards Sustainable America: Advancing Prosperity, Opportunity, and a Healthy Environment for the 21ˢᵗ Century* (Washington, D.C.: The President's Council on Sustainable Development, May 1999), 113–119.

74. Members of both the National Academy of Engineering and the National Academy of Science played a formative role in the development of the industrial ecology approach. Robert White, the president of the National Academy of Engineering, provided one of the first definitions of industrial ecology, which became a frequently cited and influential guidepost to the core arguments in industrial ecology. White's definition emphasized the "flow of materials and energy in industrial and consumer activities" as well as the importance of the economics, political, regulatory and social factors on the flow, use, and transformation of resources." Cited in Charles W. Powers and Marian R. Chertow, "Industrial Ecology: Overcoming Policy Fragmentation," in *Thinking Ecologically*, 27. For the proceedings of the 1991 symposium see L.W. Jelinski et al., *Proceedings of the National Academy of Sciences, USA* 89, no. 3 (February 1, 1992); see also, S. Erkman, "Industrial Ecology: An Historical View," *Journal of Cleaner Production*, 5, no. 1–2, 1997, 1–10.

75. Robert A. Frosch, "Toward the End of Waste: Reflections on a New Ecology of Industry," in *Technological Trajectories and the Human Environment*, edited by Jesse H. Ausubel and H. Dale Langford (Washington, D.C.: National Academy Press, 1997), 158; Marc Ross, "Efficient Energy Use in Manufacturing," *Proceedings of the National Academy of Sciences, USA* 89, no. 3 (February 1, 1992): 827–831; Jesse H. Ausubel, "Can Technology Spare the Earth?" *American Scientist* 84 (March–April 1996): 166–178.

76. The "Copernican turn" concept is developed by Jesse Ausubel in "The Liberation of the Environment," in *Technological Trajectories*, 1–13; see also Jesse H. Ausubel, "Industrial Ecology: Reflections on a Colloquim," *Proceedings of the National Academy of Sciences, USA* 89, no. 3 (February 1, 1992): 879– 884.

77. Allenby makes the argument that certain domestic industries, such as the electronics industry, have played a "critical leadership role in shifting environmental considerations from a marginal to a strategic position in their operations." Braden Allenby, "Industrial Ecology in the United States," *Journal of Industrial Ecology* 1, no. 3 (Summer 1997): 6; see also Chauncey Starr, "Sustaining the Human Environment: The Next Two Hundred Years," in *Technological Trajectories*, 198; In an interesting passage in his textbook on industrial ecology, Allenby notes that aside from the more traditional efficiency approaches such as water reclamation, a "mitigating technology" for an underdeveloped country such as Mexico would be bio-engineered agricultural technologies that

could overcome the inefficiencies of traditional agricultural practices. Branden R. Allenby, *Industrial Ecology: Policy Framework and Implementation* (Englewood Cliffs, New Jersey: Prentice-Hall, 1999), 298.

78. Valerie Thomas and Thomas Spiro, "Emissions and Exposures to Metals: Cadmium and Lead," and Wayne France and Valerie Thomas, "Industrial Ecology in the Manufacturing of Consumer Products," in *Industrial Ecology and Global Change*, edited by Robert Socolow et al. (Cambridge: Cambridge University Press, 1994); on the single-use camera issue, see the discussion in "Pollution Prevention Voluntarism: The Example of 3M," Peter Sinsheimer and Robert Gottlieb, in *Reducing Toxics*, 407–409.

79. Franz Berkhout, "Nuclear Power: An Industrial Ecology That Failed," in *Industrial Ecology and Global Change*, 319, 324; Nicholas Ashford, "Industrial Safety: The Neglected Issue in Industrial Ecology," *Journal of Cleaner Production 5*, no. 1–2 (1997): 115–121; Kirsten U. Oldenburg and Kenneth Geiser, "Pollution Prevention and . . . or Industrial Ecology?" *Journal of Cleaner Production 5*, no. 1–2 (1997): 103–108.

80. Tachi Kiuchi, invitation letter for "Industrial Ecology IV—The Profit in Sustainability" conference, November 30, 1998, in author's possession.

81. Bette K. Fishbein, "EPR: What Does It Mean, Where Is It Headed," *Pollution Prevention Review 8*, no. 4 (Autumn 1998): 43–44; Bette K. Fishbein, *Germany, Garbage, and the Green Dot: Challenging the Throwaway Society* (Washington, D.C.: Inform, 1994); Beth Rogers, "European Packaging Laws: Can It Happen Here?" *Waste Age*, June 1996, 61–68; Reid J. Lifset, "Take It Back: Extended Producer Responsibility as a Form of Incentive-Based Environmental Policy," *Journal of Resource Management and Technology 21*, no. 4 (December 1993): 163–175.

82. President's Council on Sustainable Development, *Sustainable America: A New Consensus for Prosperity, Opportunity, and a Healthy Environment for the Future* (Washington, D.C.: GPO, February 1996), 27; see also "Product Stewardship: Tackling Takeback and Moving Beyond It," *Business and the Environment 9*, no. 3 (March 1998): 2–5.

83. National Research Council, *Improving Engineering Design: Designing for Competitive Advantage* (Washington, D.C.: National Academy Press, 1991), 59–60; American Electronics Association, *The Hows and Whys of Design for Environment* (Washington, D.C.: November 1992); D.A. Gateby and G. Foo, "Design for X: Key to Competitive, Profitable Markets," *A.T.&T. Technical Journal 63*, no. 3 (1990): 2–13, cited in Braden R. Allenby, "Integrating Environment and Technology: Design for Environment," in *The Greening of Industrial Ecosystems*, edited by Braden R. Allenby and Deanna J. Richards, National Academy of Engineering (Washington, D.C.: National Academy Press, 1994), 135.

84. The OTA report is U.S. Congress, *Green Products by Design: Choices for a Cleaner Environment* (Washington, D.C.: Office of Technology Assessment,

September 1992), 11, OTA-E-542; Joseph Fiksel, *Design for Environment: Creating Eco-Efficient Products and Processes* (New York: McGraw Hill, 1996), 54, 95; *Implementing ISO 14000: A Practical, Comprehensive Guide to the ISO 14000 Environmental Management Standards*, edited by Tom Tibor and Ira Feldman (Chicago: Irwin Professional Publishing, 1997).

85. Deborah L. Boger and Claudia M. O'Brien, "The U.S. EPA's DfE Program," in *Design for Environment*, 375–409. In a striking comment during a conference call involving EPA officials and environmental advocates, one DfE EPA official characterized the EPA's design for the environment objective (with respect to dry cleaning's use of the hazardous chemical solvent perchloroethylene) in these terms: "EPA doesn't want to reduce or eliminate perc; it wants to reduce exposures to perc." Comments by Cindy Stroup, U.S. Environmental Protection Agency Design for the Environment, January 27, 1999; see also, Katherine M. Hart, Deborah L. Boger, and Michael A. Kerr, "Design for the Environment: A Partnership for a cleaner future, *Printed Circuit Fabrication* 18, no. 4, April 1995, 16–18.

86. Paul Hawken, "Natural Capitalism," *Mother Jones*, March–April 1997, 40–62; see also Paul Hawken, Amory Lovins, and L. Hunter Lovins, *Natural Capitalism: Creating the Next Industrial Revolution* (Boston: Little Brown & Co., 1999).

87. Ernest A. Lowe, John L. Warren, and Stephen R. Moran, *Discovering Industrial Ecology: An Executive Briefing and Sourcebook* (Columbus, Ohio: Battelle Press, 1997), 3.

88. Frederick Anderson, "From Voluntary to Regulatory Pollution Prevention," in *The Greening of Industrial Ecosystems*, 102.

89. Stuart Hart, "Beyond Greening: Strategies for a Sustainable World," *Harvard Business Review*, January–February 1997, 71; "Toward a New Conception of the Environment-Competitiveness Relationship," Michael E. Porter and Claas van der Linde, *Journal of Economic Perspectives* 9, no. 4 (Fall 1995): 97–118.

90. Maarten Hajer, *The Politics of Environmental Discourse*, 32–33.

91. "Principles of Clean Production," Lowell and Knoxville: Lowell Center for Sustainable Production and the Center for Clean Products and Clean Technologies, October 21, 1999; see also Beverly Thorpe, *Citizen's Guide to Clean Production* (Lowell, Mass.: Clean Production Network, August 1999).

3   Dry Cleaning's Dilemma and Opportunity: Overcoming Chemical Dependencies and Creating a Community of Interests

1. See, for example, U.S. Environmental Protection Agency, *Small Business and the Environmental Protection Agency: Building a Common Sense Approach to Environmental Protection*, (Washington, D.C.: Office of Policy, Planning, and Evaluation, June 1995), EPA 230-K-95-002.

2. In 1925 a heavy petroleum cleaning agent with a higher flash point known as Stoddard solvent was introduced as a way to lessen petroleum solvent's fire hazards. Three years after the introduction of Stoddard solvent, the U.S. Commerce Department, working directly with the National Assocation of Dyers and Cleaners, established Commercial Standard CS3-28. The standard specified a minimum flash point based on what the Stoddard solvent was capable of meeting, linking the rule to the availability and continued use of the product. Stoddard solvent was also named after William J. Stoddard, the president of the National Association of Dyers and Cleaners, for his role in encouraging the research that contributed to the development of this petroleum solvent. See Albert R. Martin and George P. Fulton, *Drycleaning: Technology and Theory*, a report of the National Institute of Drycleaning (New York: Textile Book Publishers, 1958), 109; Ben and Marie Pearse, "Ever Get Spots on Your Clothes?" *Saturday Evening Post* 228, no. 21 (November 19, 1955): 36ff.

3. While centralized plants maintained a significant share of the dry-clean market in the immediate post–World War II period, the number of "agency" or "drop shops" associated with the centralized facility also increased in number, reflecting the expansion of dry cleaning into multiple locations. One of the first franchising operations that emphasized multiple locations and quick service was "One Hour Martinizing." This company also introduced the use of perc as its cleaning solvent as early as 1949 to facilitate that shift. The growth of perc use in turn began to erode the share of the market for petroleum, a trend that already began to occur during the 1950s. By the 1950s, in fact, the dry-cleaning "method" had come to be fully characterized as either synthetic-based (associated with perchloroethylene, carbon tetrachloride, and trichloroethylene use) or petroleum-based (primarily associated with Stoddard solvent). See Dorothy Siegert Lyle, *Focus on Fabrics: Selection and Behavior in Drycleaning* (Silver Spring, Md.: National Institute of Drycleaning, 1958); "Good Clean Profits," *Franchising Opportunities* 22 (August 1990): 65–68. Bernard Landis, "New Methods Help Cleaners Set Record," *Nation's Business* 42 (February 1954): 31; Morton Sontheimer, "How to Get on with Your Dry Cleaner," *Woman's Home Companion*, February 1950, 40–41ff.; "Dry Cleaning: Large Larger, Small Smaller," *Economist* 206:234, January 19, 1963; The "little businessmen" quote is in "Dirty Work at the Cleaners," *Fortune* 37 no. 5 (May 1948): 126.

4. "Danger from Cleaning-Fluid Vapor," *Scientific Digest* 27 (May 1950): 49–50. "When It Comes to Cleaning Agents, Fire Isn't the Only Hazard," *Good Housekeeping* 136 (February 1953): 25.

5. While the head of the Drycleaning Institute's Consumer Education Service was still arguing in 1955 that carbon tetrachloride was "the most practical, general-purpose, dry cleaning solvent" then available, concerns about significant acute health effects were already dominating industry discussions about carbon tet and its future in the cleaning industry. Carbon tet, in fact, became the first of the dry-cleaning chemical solvents to be subject to human health-related guidelines or voluntary standards. During the 1950s, the American Conference of Government Industrial Hygienists (ACGIH) developed an advisory or voluntary standard of

a maximum allowable concentration in the air of fifty parts per million for those exposed over an eight-hour period. These nonenforceable ACGIH standards also understated possible occupational health hazards and were only developed after considerable evidence had been established that health risks were associated with the use of a particular substance such as carbon tet. In fact, in carbon tet's case, its eventual decline and elimination as a dry-cleaning solvent occurred prior to the development of formal government (and enforceable) standards established in 1971. By then, carbon tetrachloride in dry cleaning already had been displaced by other chemical solvents. See the discussion in the November 19, 1955, *Saturday Evening Post* article referenced above (p. 162) and the response on December 31, 1955, "Dangerous Cleaner," Dr. T.B. Magath, p. 4, on the debates regarding carbon tet's risks; see also Dennis J. Paustenbach, *"Risk in Perspective," Harvard Center for Risk Analysis* 5 (January 1997): 1.

6. Martin and Fulton, *Drycleaning: Technology and Theory*, 120.

7. "Cleaners: $85-Million Chemical Customers," *Chemical Week* 83 (August 30, 1958): 61–65; "Drycleaning Due for Do-It-Yourself Boom?" *Chemical Week*, March 19, 1960, 105–108.

8. More than 240,000 people were employed in dry cleaning in 1969, a growth of 18 percent since 1948. Elinor W. Abramson, "Employment in Laundry, Drycleaning, and Valet Services," *Monthly Labor Review* 93 (November 1970): 43–47.

9. These declines in the late 1990s were primarily in the volume of use but did not substantially alter the percentage of sales to dry cleaners by the chemical suppliers. Thus, the chemical solvent manufacturers and distributors' continued to rely heavily on the dry-clean market to maintain production of perc. In public forums, however, both dry-cleaning trade officials and chemical industry spokespeople sought to emphasize that perc use had declined to a level that reduced its importance as an environmental hazard. However, evaluations of perc ambient levels still registered high numbers in urban areas such as Los Angeles. California Air Resources Board, *Toxic Air Contaminant Fact Sheets: Perchloroethylene* (Sacramento: California Air Resources Board, 1997); comments by Manfred Wentz, in U.S. Environmental Protection Agency, *Garment and Textile Care Program: An Eye to the Future*, 1998 Design for the Environment Conference, January 1998; For earlier perc use figures see, "Will Perchlor Clean Up?" *Oil, Paint and Drug Reporter* 185 (May 25, 1964): 9. A subsequent analysis in the same publication three years later placed dry cleaning's market share at 88 percent. "Chemical Profile: Perchloroethylene," *Oil, Paint and Drug Reporter* 192, no. 11 (September 11, 1967): 9.

10. Margaret Allen, "Changing a Habit in Great Britain," *Printers' Ink* 293, no. 3 (February 10, 1967): 27–29. The ICI campaign paralleled an effort during this same period by a dozen apparel manufacturers in the United States who also sought to promote dry cleaning through an advertising campaign (in this case in conjunction with a major dry-cleaning chain). These clothing manufacturers, such as Haggar, Worsted-Tex, and Catalina had themselves incorporated dry-

cleaning garment care as part of their own marketing framework. "Apparel Makers Join in Sanitone Magazine Effort," *Advertising Age* 37, no. 41 (October 10, 1966): 3.

11. Perc manufacturers hoped that dry-cleaning "expertise" (associated with the use of perc) combined with a continuing increase in the coin-op market (particularly as dry cleaners expanded into both business lines) would become "the answer to a stable business," as one chemical industry publication put it. And although the coin-ops were responsible for dry-cleaning establishments declining in overall numbers (as much as 10 percent in the late 1960s), coin-operated laundries, the fastest-growing segment of the commercial cleaning industry, were also installing their own do-it-yourself dry-cleaning equipment, thereby increasing overall perc sales. As late as 1977, one industry analyst called the growth of the coin-ops, which also paralleled the rise of the budget or discount cleaners, "the greatest breakthrough in clothing maintenance from the consumer's standpoint ..." Dorothy Siegert Lyle, *Performance of Textiles* (New York: John Wiley & Sons, 1977), 338; see also "Dollar-saving Dry Cleaning," *Better Homes and Gardens*, October 1973, 32; "Coin-Operated Dry Cleaning: Birth of an Industry," *Sales Management* 86 (January 6, 1961): 49–51; "US Perchloroethylene Market in '71 Expected to Top 750 Million Pounds," *Oil, Paint and Drug Reporter* 192 (September 11, 1967): 3ff.

12. "The experience of small drycleaning shops" quote is from "Drycleaning Due for Do-It-Yourself Boom?" *Chemical Week*, 105–106; see also "Will Perchlor Clean Up?" *Oil, Paint and Drug Reporter* 179 (January 23, 1961): 5ff.

13. During the 1970s and 1980s, perc output remained somewhat volatile, due in part to the continuing uncertainties in the dry-cleaning market, as well as other contributing factors such as shifts in fluorocarbon production. With fluctuations in the market, the industry also experienced a process of shakeout and consolidation, with essentially four major producers (Dow Chemical, Vulcan Chemicals, Diamond Shamrock, and PPG) dominating the industry by the late 1980s. See "Chemical Profile: Perchloroethylene," *Chemical Marketing Reporter* 210 (August 9, 1976): 9; "Drycleaning Spruces Up Perchlor Sales," *Chemical Week* 111 (November 1, 1972): 35; "Perchloroethylene Boosted by Dow; Trade Studying Half-Cent Advance," *Oil, Paint and Drug Reporter* 195 (March 3, 1969): 25; "PERC Output Takes Big Dip as CFC Phase-Out Is Felt," *Chemical Marketing Reporter*, January 20, 1992, 14–16; "DuPont Pulls Out of 'Perc,'" *Chemical Week* 139 (July 16, 1986): 9.

14. Betty Furness, "Clothing-Care Labels That Last," *McCall's*, October 1970, p. 46.

15. "This Is No Way to Wash the Clothes," *Consumer Reports* 33, no. 2 (February 1968): 66–67; "At Last! Permanent Care-Labels," *Consumer Reports* 37, no. 3 (March 1972): 170–173.

16. "Cleaning Instructions Labeling of Imported Fabrics," *Hearing on H.R. 6143 (A Bill to Require that Certain Textile Products Bear a Label Containing*

Cleaning Instructions) before the Subcommittee on Interstate and Foreign Commerce, House of Representatives, 92nd Cong., 1st sess., May 10, 1971; the ordering of three pairs of identical garments was undertaken for the "repeat clean" technical performance evaluation for the report, *Pollution Prevention in the Garment Care Industry: Assessing the Viability of Professional Wet Cleaning,* that was prepared for the U.S. Environmental Protection Agency, California Air Resources Board, and South Coast Air Quality Management District (Los Angeles: Pollution Prevention Education and Research Center, December 1997).

17. The care labeling rule identified the use "of any common organic solvent" as constituting the accepted "dry clean only" process in the care label, specifically identifying perc, petroleum, and fluorocarbons as examples of such a solvent (FTC 16 CFR 423); see Connie Vecellio, "FTC Care Labeling Revision" in *Apparel Care and the Environment: Alternative Technology and Labeling Procedures, Proceedings,* U.S. Environmental Protection Agency, Office of Pollution Prevention and Toxics, (Washington, D.C.: GPO, September 1996) EPA 744-R-96-002; see also *Performance of Textiles,* 389–390; The 50 percent figure about garment manufacture problems is from the 1969 National Institute of Dry Cleaning laboratory evaluation of problem garments, cited in Betty Furness, *McCall's,* October 1970.

18. Richard B. Carnes, "Laundry and Cleaning Services Pressed to Post Productivity Gains," *Monthly Labor Review* 101 (February 1978): 38–42. William Seitz, "The State of the Industry," *American Drycleaner,* February 1986; Carol Lynn Goedert, "Dry Cleaners Don New Image," *Venture* 5 (May 1983): 101–102. Lawrence B. Katz and Dev Strischek, "Lending to Dry Cleaners," *The Journal of Commercial Bank Lending* 70 (November 1970): 26–34.

19. Between 1965 and 1975, more than 140,000 Koreans entered the United States, with more than 80 percent of that total arriving between 1971 and 1975. Between 1975 and 1985, 318,000 Korean immigrants arrived, with the numbers increasing slightly during the mid-1980s. Kayung Park, *The Korean-American Dream: Ideology and Small Business in Queens, New York* (Ph.D. diss., City University of New York, 1990). Ivan Light and Edna Bonacich, *Immigrant Entrepreneurs: Koreans in Los Angeles, 1965–1982* (Berkeley: University of California Press, 1988).

20. The rapid entry of the Koreans into dry cleaning had led to the belief among some non-Koreans in the business that the Koreans had opted to become discount cleaners with poor environmental practices, often located in low-income, high-crime areas. However, a survey of cleaners in the southern California area indicated that Korean cleaners were no different than their non-Korean counterparts in terms of their locations and marketing strategies (ranging from discount to higher end). Korean Youth and Community Center (KYCC), Analysis of South Coast Air Quality Management District Dry Cleaner Data Base, 1996; see also Pyong Gap Min, "Filipino and Korean Immigrants in Small Business: A Comparative Analysis," *Amerasia Journal* 13, no. 1 (1986–1987): 53–71; Illsoo

Kim, *New Urban Immigrants: The Korean Community in New York* (Princeton, N.J.: Princeton University Press, 1981); Eui-Hang Shin and Shin-Kap Han, "Korean Immigrant Businesses in Chicago," *Amerasia Journal* 16, no. 1 (1990): 39–60.

21. Elizabeth Hill, *Coming Clean: The Potential for Toxics Reduction in the Garment Care Industry*, University of California at Los Angeles, Pollution Prevention Education and Research Center, February 1995; a 1994 estimate by the California Fabricare Institute indicated that as many as 55 percent of dry-cleaning facilities in California were Korean owned. A subsequent analysis of dry cleaners in the southern California region identified as many as 50 percent of the cleaners in the region with Korean surnames. CFI's 55 percent estimate was provided to PPERC research associate Hill. The 50 percent estimate comes from the KYCC analysis of the AQMD data base, while the 30–40 percent estimate has been widely used by industry sources. At the national level, dry-cleaning industry sources have used a figure of 30–40 percent as the estimate of Korean-owned cleaning establishments within the United States. "Never before has there been a group with such an impact in our industry," commented one leading industry figure (William Seitz, executive director of NCA-1—National Cleaners Association, at the U.S. Environmental Protection Agency Apparel and Textile Design for Environment Conference, March 30, 1998); see also the comments by Bill Fisher, International Fabricare Institute, at the "Drycleaning Workgroup," U.S. Environmental Protection Agency DfE Garment and Textile Care Strategy Development Partnership, January 12, 1998 (available through International Fabricare Institute, Silver Spring, Maryland).

22. "The language difficulties of the new immigrants" statement was made by Ed Boorstein at the U.S. Environmental Protection Agency Design for Environment Apparel and Textile Industry Conference, March 31, 1998; see also Mary Scalco, "Care Labeling and the Fabric Care Industry" in U.S. Environmental Protection Agency, *Proceedings*, September 1996.

23. Dave Johnston, "Seitz Calls Past 15 Years 'Best and Worst,'" *Drycleaners News* 45, no. 22 (May 1996): 1ff.

24. Hill cites one estimate indicating that only 25 percent of the CFI's membership was Korean. On the other hand, Korean trade associations had far more members in their exclusively Korean associations (e.g., the Los Angeles Korean Dry Cleaners Association had 600 active members compared to only 280 Korean members of the statewide association. Hill, *Coming Clean*, 18–19.

25. R.D. Stewart et al., "Accidental Vapor Exposure to Anesthetic Concentration of a Solvent Containing Tetrachloroethylene," *Industrial Medicine and Surgery*, 30 (1961): 327–330, 1961; R.D. Stewart, "Acute Tetrachloroethylene Intoxication," *JAMA* 208, no. 8 (1969): 1490–1492.

26. While coin-operated facilities were phased out in the United States in the 1970s and 1980s, they remained a significant aspect of the dry-cleaning business in certain European countries such as France where hundreds of such machines continued to be used on a daily basis even during the 1990s, which significantly

compounded the health issues associated with perc use in those countries. Coin-ops even resulted in deaths from acute exposures, such as a two-year-old who died from heavy perc exposures due to double curtains in his bedroom that had been dry cleaned that day in a coin-operated facility. "Coin-Operated Dry Cleaning Machines May Be Responsible for Acute Tetrachloroethylene Poisoning: Report of 26 Cases Including One Death," Robert Garnier et al., *Clinical Toxicology* 34, no. 2 (1996): 191–197; "Preaching Safety to Perc Users," *Chemical Week* 98 (February 26, 1966): 59.

27. King is cited in "Perc Injunction Won by Industry: CPSC Is Loser," *Chemical Marketing Reporter* 214 (September 18, 1978): 5ff.; on the cancer risks, see National Cancer Institute Carcinogenesis Technical Report Series no. 13, "Bioassay of Tetrachloroethylene for Possible Carcinogenicity," DHEW Publication no. (NIH) 77–813, 1977; Elizabeth Bartman Smith, "A Look at Perchloroethylene," *Job, Safety, and Health*, May 1978, 25–28; Aaron Blair, Pierre Decoufle, and Dan Grauman, "Causes of Death among Laundry and Dry Cleaning Workers," *American Journal of Public Health* 69, no. 5 (May 1979): 508–511; on the response of the perc manufacturers, Dow had in fact long argued that there was "no medical demonstration of permanent damage to the liver or any other organ by exposure to perchloroethylene." *Chemical Week*, February 26, 1966, 59; see also "Perc Study Enters the Final Stages; Industry Is Doubtful," *Chemical Marketing Reporter* 212 (July 25, 1977): 4ff.; "1,1,-Trichlor, Perc No Hazard to Man, Claims Dow Study," *Chemical Marketing Reporter* 208 (August 4, 1975): 3ff.

28. International Agency for Research on Cancer, "Tetrachloroethylene: IARC Mongraphs on the Evaluation of Carcinogenic Risks to Humans," vol. 63, *Dry Cleaning, Some Chlorinated Solvents and Other Industrial Chemicals*; Agency for Toxic Substances and Disease Registery, *Toxicological Profile for Teterachloroethylene* (Washington, D.C.: U.S. Department of Health and Human Services, ATSDR, October 1992); see also Kimberly Thompson, "Cleaning Up Dry Cleaners," in *The Greening of Industry: A Risk Management Approach*, edited by John Graham and Jennifer Kassalow Hartwell (Cambridge: distributed by Harvard University Press, 1997).

29. The "one industry figure estimated," is from Stan Golumb, "A Flashback to the 'Early Days,'" *National Clothesline* 39, no. 8 (May 1999): 60; the 60 percent figure is discussed in California Air Resources Board, *Survey Database of California Dry Cleaners* (Sacramento, Cal.; CARB, 1992); see also South Coast Air Quality Management District, *Staff Report to Propose Adoption of Rule 1421: Control of Perchloroethylene Emissions from Dry Cleaning Systems and Repeal Rule 1102.1: Perchloroethylene Dry Cleaning Systems* (Diamond Bar, Cal.: SCAQMD, December 1994).

30. The lead chemical industry organization around the perc debates, the Halogenated Solvents Industry Alliance (HSIA), warned EPA administrator Lee Thomas in a 1987 letter that designating perc as a "probable human carcinogen" would open the door to new and/or more stringent regulations under the

Safe Drinking Water Act, Section 112 of the Clean Air Act, the OSHA Hazard Communication Standard, the reporting requirements under Superfund, and the liability of perc users and manufacturers. The letter concluded with an explicit threat to take legal action if EPA were to base any regulatory decision on the use of such a classification. Letter from the Halogenated Solvents Industry Alliance to U.S. EPA administrator Lee Thomas, June 1, 1987, cited in Elizabeth Drye, "Perchloroethylene," in *Harnessing Science for Environmental Regulation,* edited by John Graham (New York: Praeger, 1991), 115–116; the Fisher statement is in "Perc Designation Called Harmful to Chemical's Users," *Chemical Marketing Reporter* 232 (August 3, 1987): 9ff.; see also "Perchloroethylene Assayed," *Chemical Marketing Reporter* 231 (March 16, 1987): 5.

31. Elizabeth Hill, *Coming Clean,* 35.

32. *Clean Air Act Amendments of 1990,* Pub. L. no. 101–549, 104 Stat. 2399 (1990); see also U.S. Environmental Protection Agency, *Technical Background Document to Support Rulemaking Pursuant to the Clean Air Act—Section 112 (g). Ranking of Pollutants with Respect to Hazard to Human Health,* Emissions Standards Division, Office of Air Quality Planning and Standards (Research Triangle Park, N.C.: U.S. EPA 1994).

33. *National Emission Standards for Hazardous Air Pollutants for Source Categories: Perchloroethylene Dry-Cleaning Facilities—Final Rule,* 40 CFR Parts 963, *Federal Register,* vol. 58, no. 182, September 22, 1993, p. 49354; the text of Section 112 of the Act, "Hazardous Air Pollutants," is available at http://www.epa.gov/oar/caa/caa112.txt

34. U.S. Environmental Protection Agency, *Implementation Strategy for The Clean Air Act Amendments of 1990: Attainment of Air Quality Standards* (Washington, D.C.; U.S. EPA, 1994); see also Katy Wolf and Mike Morris, "P2 and the Clean Air Act Amendments of 1990," *Pollution Prevention Review* 6, no. 3 (Summer 1996): 1–13.

35. Paul Cammer's comments are cited by E.M. Kirschner, "Perchloroethylene Gets Taken to the Cleaners," *Chemical Week,* December 4, 1991, 16; the assumption about any alternatives to PCE other than more effective control technologies is elaborated in Jacobs Engineering Group Inc., *Source Reduction and Recycling of Haolgenated Solvents in the Dry Cleaning Industry,* prepared for the Metropolitan Water District of Southern California and the Environmental Defense Fund (Los Angeles: MWD, 1993). In terms of the cost of the control equipment, the EPA has estimated that the average price to purchase a refrigerated condenser, as required by the NESHAP, would be $6,000 to $8,000, with installation adding another $1,000 to $2,000 to this cost. The EPA also estimated that the average yearly cost for operating a refrigerated condenser would be $460. U.S. Environmental Protection Agency, *New Regulation Controlling Emissions from Dry Cleaners* (Washington, D.C.: U.S. EPA, May 1994), EPA-453-F-94-025.

36. Elaine Murphy, Dow Chemical's *Spot News* publication, cited in "Declining Perc Use Does Not Trouble Dow," *National Clothesline* 38, no. 10 (July 1998), 8.

37. In the period shortly after the Greenpeace chlorine campaign was launched, the Clinton/Gore EPA seemed of two minds about how to proceed. On the one hand, EPA administrator Carol Browner released a hastily developed announcement of EPA's interest in a possible chlorine ban, which immediately generated intense negative feedback from the chemical producers. On the other hand, the EPA's reliance on nonregulatory interpretations of pollution prevention did not suggest that such a comprehensive form of intervention would occur. The chlorine ban concept was in fact never pursued. See "Study of Chlorine and Chlorinated Compounds," President Clinton's Proposal for the *Clean Water Act,* January 28, 1994; Janice Mazurek et al., "Shifting to Prevention: The Limits of Current Policy," in *Reducing Toxics,* 72–73.

38. Personal communication with Ohad Jehassi, July 27, 1998; see also the opening remarks by U.S. EPA official Mary Ellen Weber at the Roundtable, in *Proceedings, International Roundtable on Pollution Prevention and Control in the Drycleaning Industry,* May 27–28, 1992, Falls Church, Virginia, Office of Pollution Prevention and Toxics (Washington, D.C.: US EPA, November 1992), 1–2, EPA/774/R-92/002.

39. See Judy S. Schreiber, "Investigations of Indoor Air Contamination in Residences above Drycleaning Establishments," in *Proceedings,* 53–57; Wendy L. Cohen, "Investigations of Ground-Water Contamination by Perchloroethylene in California's Central Valley," *Proceedings,* 63–66; and Mary Ellen Weber, Opening Remarks, 1; see also Judith Schreiber, "Assessing Residential Exposures to Health," in *Garment and Textile Care Program: An Eye to the Future, 1998 Conference Proceedings,* Design for the Environment (Washington, D.C.: U.S. EPA, June 1998), EPA 744-R-98-006; Judith Schreiber et al., "An Investigation of Indoor Air Contamination in Residences Above Dry Cleaners," *Risk Analysis* 13, no. 3 335–344; personal communication with Jack Weinberg, Greenpeace International, May 31, 1998.

40. Personal communication with Jo Patton, Center for Neighborhood Technology, May 21, 1998; interviews with Ohad Jehassi, Jack Weinberg.

41. Personal communications with Beverly Thorpe, May 31, 1998, and Jack Weinberg; see also William Eyring, *Eco Clean: An Assessment of a Drycleaning Alternative* (Chicago: Center for Neighborhood Technology, July 1992).

42. On the CNT approach to small business, see, for example, *Beyond Recycling: Materials Reprocessing in Chicago's Economy* (Chicago: Center for Neighborhood Technology, 1993).

43. Eyring, *Eco-Clean,* 17.

44. U.S. Environmental Protection Agency, *Multiprocess Wet Cleaning: Cost and Performance Comparison of Conventional Dry Cleaning and An Alternative Process,* Office of Pollution Prevention and Toxics (Washington, D.C.: U.S. EPA, 1993), EPA 744-R-93-004; Jehassi's "promising" comment is cited in Irwin Stambler, "EPA Seeks Pollution-Free Clothes Cleaning Processes," *R&D Magazine* 36 (March 1994): 35.

45. See, for example, "Perc Still Target of Greenpeace," *Drycleaners News* 46, no. 9 (September 1997): 1ff.; see also the Greenpeace report, *Dressed to Kill: The Dangers of Dry Cleaning and the Case for Chlorine-free Alternatives* (Washington, D.C.: Greenpeace International, 1998), available at http://www.greenpeace.org/search.html.

46. Anthony Star and Cindy Vasquez, *Wet Cleaning Equipment Report* (Chicago: Center for Neighborhood Technology, May 1997). A description of the wet-clean process can be found in Pollution Prevention Education and Research Center, *Pollution Prevention in the Garment Care Industry: Assessing the Viability of Professional Wet Cleaning* (Los Angeles: PPERC, December 1997), 1–6 and 7.

47. Personal communication with Jo Patton, May 21, 1998.

48. Davis had learned of PPERC from the Chicago Center for Neighborhood Technology, which had been aware of research undertaken by our center regarding dry-cleaning industry issues. Although based in San Diego, Davis approached our center about participating in research through a small grant to be provided through the South Coast Air Quality Management District, to evaluate alternative technologies in garment care. To participate in the research, a commercial wet-cleaning facility needed to be located in SCAQMD's service area in Los Angeles or Orange County. In terms of the advisory boards: CNT's Advisory Board included several leading, nationally based dry-cleaning trade organizational representatives. PPERC also set up an advisory committee that included the heads of the two primary dry-cleaning organizations in California, the California Cleaners' Association and the Korean Drycleaners Association. The list of PPERC advisory committee members is included in Appendix 1-B of *Pollution Prevention in the Garment Care Industry* (hereafter PPERC Report). During the twelve-month demonstration period, there were twenty separate tours and workshops, including six conducted in Korean for Korean cleaners, arranged in collaboration with the Korean Youth and Community Center. See PPERC Report, 1–15.

49. For the performance evaluation, three sets of forty pairs of identical garments were ordered, in part on the advice of dry cleaners who served on the advisory board for the evaluation and helped select those garments (e.g., silk ties or wool suits). (Garments were selected from mail-order catalogues such as Tweeds, Land's End, and Chadwicks, and also included several donated garments from Banana Republic. See PPERC Report, Appendix 3-J, for the list of garments and their characteristics.) These garments were seen as potentially causing the most difficulties for wet cleaning. Two of the three identical pairs of garments were worn by volunteer wearers over a six-month period, while the third identical garment was held as a control garment. Each month, after having worn the garments, the soiled garments were then brought in for cleaning, one to Cleaner by Nature, and the other identical garment to a dry cleaner. Neither Cleaner by Nature nor the two dry cleaners used for the comparative evaluation were aware that the clothes were part of an evaluation. After cleaning, the garments were

returned to the volunteer wearers who would repeat the cycle. There were six separate "repeat cleanings," each a month apart. After the first and sixth cleanings, an extensive laboratory analysis was undertaken to evaluate a wide range of cleaning performance criteria, such as shrinkage and stretching, whether colors bled, soil and stain removal, and the odor present on the garments. A "customer satisfaction" survey was also designed to provide comparative information about the satisfaction of the real-world customers of Cleaner by Nature and local area dry cleaners. The same criteria for the "laboratory" evaluation—degree of shrinkage or stretching, odor, color run, stain removal, etc.—were used for this "real-world" survey evaluation. PPERC Report, "Methods for Assessing Viability," Section 2.1 and 2.2; additional evaluations (through surveys) were also made to determine possible bias on the part of the volunteers who wore the clothes as well as their satisfaction with the cleaning of each of the garments (parallel to the study's customer satisfaction survey). The survey indicated that the wearers could not effectively distinguish which garment was wet cleaned or dry cleaned, nor were any significant differences in the outcomes of the respective cleaning processes noted. The wearer issues are discussed in PPERC Report, 3–29 to 3–31. The laboratory analysis for the repeat clean test was performed by a team from the Department of Apparel Design and Merchandizing, California State University at Long Beach and led by professors Sue Stanley and Hazel Jackson. For the customer satisfaction survey, the dry-clean customers were identified through a random telephone survey from telephone prefixes within approximately a one-mile radius from the Cleaner by Nature site to establish a parallel customer group. The survey is reproduced in PPERC Report, Appendix 3-S and the methods and results are discussed in pages 3–35 through 3–49.

50. For the financial evaluation, start-up and operating costs were analyzed for both sets of businesses, focusing on the "process dependent" costs that distinguished the two processes. PPERC Report, 4–1 through 4–29.

51. For the environmental evaluation, the environmental impacts associated with each process were analyzed, including those associated with a particular regulatory concern such as air emissions or hazardous waste disposal. Wet-cleaning's water use and discharges into the sewer system were also identified as areas that might reveal potential problems for wet cleaning. The wastewater discharge analysis was performed in conjunction with the City of Los Angeles Bureau of Sanitation through meters set up at the wet-cleaning machine drain. The methods for the environmental evaluation are discussed on pages 5–1 through 5–31.

52. Cleaner by Nature established both an "agency" (drop-site store where customers brought clothes in and picked them up at a small store location in Santa Monica), and "plant" (where the machines were located and the clothes were cleaned) in a warehouse district in Los Angeles. A small delivery business was also established, with the use of a natural-gas-powered van designed to increase the environmental profile of the business. Interestingly, in its first several years of business, Davis estimated that operating a natural-gas van had proved to be less expensive than if she had operated a gas-powered vehicle. In 1998, Davis also opened a second "agency" store in West Los Angeles about a mile and a

half from her first drop store location, and expanded her delivery service, and by 1999 her revenues had more than tripled. Pollution Prevention Education and Research Center, *Supporting Pollution Prevention in the Garment Care Industry: Final Report* (Los Angeles: PPERC, February 2000).

53. For the repeat clean lab evaluation, slight (but not statistically significant) differences between the two processes were identified regarding shrinkage (more shrinkage for wet cleaning) color change (more for wet cleaning) and odor (more chemical smell from dry cleaning). However, other unanticipated issues were also identified, including more stretching in dry cleaning, and more problems for dry cleaning in stain removal. The customer satisfaction random sample survey had greater variation in several categories (color change, the feel of the garment, damage to buttons or decorations) where wet cleaning's performance was identified as superior to dry cleaning. While there were overall similarities (most garments were cleaned successfully, most customers were satisfied) there were some variations between the lab and customer satisfaction evaluations. Moreover, both processes had problems with specific types of garments (for example, those with acetate linings) that could be traced to issues of manufacturing and garment construction rather than the cleaning process itself. While the focus of the performance evaluation had been on wet-cleaning's capacity to clean the range of dry-clean-only garments, it also underlined the performance problem areas for dry cleaning. This included its difficulties with stain removal, odor, or losing buttons or decorations, which had become a constant source of irritation within the dry-cleaning industry itself. This was reflected in negative publicity about poor quality in dry cleaning (for example, special TV news investigative reports) as well as warnings from the dry-cleaning trade organizations about the need to increase the quality of the cleaning in the face of the growing trend toward discount cleaners. The dry-cleaning trade publication, *National Clothesline*, in commenting about the rash of negative publicity from press reports, put it this way: "One may question the basic fairness of these tests, but a simple fact remains: it is more of a challenge to find a 'good drycleaner' than it is for an investigative reporter to find a bad one." "So Much to Learn, So Little Time," *National Clothesline*, February 1999, 1ff. The performance evaluation results are in the PPERC Report, Appendices 3-N, 3-O, and 3-P.

54. For the financial evaluations, see PPERC Report, Appendices 4-A through 4-G.

55. In relation to the environmental evaluation, water use, as anticipated, was greater for wet cleaning than dry cleaning, although dry cleaning, despite its name, did require some use of water in relation to its pollution control equipment. In relation to water quality, monitoring by the city of Los Angeles Bureau of Sanitation indicated that Cleaner by Nature discharges were minimal for each of the heavy metals or volatile organic compounds evaluated and that no permit would be required. A parallel evaluation for the two dry cleaners involved in the comparative study was not undertaken, since the city (like most local and state entities around the country) did not allow any discharge from dry cleaners directly into the sewer system. However, the problem of perc-related

contamination of soil and groundwater has remained a significant problem due to a number of factors. These include illegal discharges (dry cleaners deciding to avoid the costs of hazardous waste disposal by dumping the wastewater down the drain), spills (for example, in transferring the perc from the delivery truck), or other equipment malfunctions. This potential for contamination has remained a concern for water utilities managing local groundwater supplies. A hypothetical wholesale conversion of dry cleaning to wet cleaning in the entire region was also evaluated. According to planners with the regional water agencies, the amount of additional water required for the region was small enough (about 600 acre feet or the equivalent of providing for an additional 1,000 people per year) that the potential for contamination due to perc use remained a far greater concern. The loss of even one small production well due to perc contamination, water officials pointed out, would be greater than the additional water use if every dry cleaner became a wet cleaner. Energy use was also broadly equivalent, involving a small trade-off between greater electricity use for dry cleaning and greater natural gas use for wet cleaning. Personal communications with Lisa Anderson, June 11, 1997, and Warren Teitz, June 30, 1997, Metropolitan Water District; PPERC Report, 5–9; 5–10.

56. One editorial commentary in a dry-cleaner publication put it this way about the shift in concern during the 1990s: "As the decade started, *DCN* [*Drycleaner News*] devoted most of its news space to topics about cleaning and pressing; as it ends, the focus has turned to matters environmental." Dave Johnston, "The Curses of Interesting Times," *Drycleaners News* 48, no. 5 (May 1999): 4.

57. Jo Patton and William Eyring, *Alternative Clothes Cleaning Demonstration Shop Final Report*, Center for Neighborhood Technology (Chicago: CNT, September 1996); personal communication with Sylvia Ewing-Hoover, Center for Neighborhood Technology, October 14, 1997; see also the parallel evaluation by Environment Canada, "Final Report for the Green Clean Project," October 1995.

58. On the range of nonwet cleaning alternatives see William E. Fisher, "The Outlook for the Industry," *National Clothesline*, 40, no. 7, April 2000, 22–24; "At the NCA-I Show, Reflections on the Past, Projections into the Future," *National Clothesline* 40, no. 2 (November 1999): 1ff. *Drycleaners News*, "Change, Choices and Challenges," March 1999, 2; "Choices—and Questions—for the Future," *National Clothesline* 39, no. 11 (August 1999): 1ff.; see also the comments by Mary Ellen Webber at the Design for the Environment Conference in *Garment and Textile Care Program: An Eye to the Future, 1998 Conference Proceedings*, Design for the Environment (Washington, D.C.: U.S. EPA, June 1998), EPA 744-R-98-006. Among the alternatives, CO_2 received the greatest attention, despite its limited commercialization. With the hope of quickly establishing CO_2 as an alternative to both perc and wet cleaning, two major R&D initiatives to develop CO_2 as a perc substitute had been launched during the mid-1990s. One group was centered in the West and included researchers from the Lawrence Laboratory at Los Alamos as well as the environmental technologies research wing of Hughes/Raytheon. The second group was centered in North

Carolina, bringing together researchers from Research Triangle Park and several investors and promoters, including a former Dow Chemical official. The high price tag for a CO_2 machine (as much as $150,000 per unit) ultimately represented its major barrier. Whether the process could work effectively—that is, clean the full range of garments that would otherwise be dry cleaned—also remained to be demonstrated. See, Environmental Law Institute, *Barriers to Environmental Technology Innovation and Use* (Washington, D.C.: January 1998); presentation by Sid Chao, Pollution Prevention Forum Series, UCLA, May 12, 1997; "Micell Shows Off Its First Working CO_2 Installation," *Drycleaner News*, March 1999, 1ff.; "Micell Invents Effective and Environmentally Safe Dry Cleaning Alternative," (http:/www.micell.com/micareNEW.htm); Richard Kinsman, "'Commercializing' Dry Wash: What Does It Mean?" *American Drycleaner* 65, no. 3 (June 1998): 74–84. Kinsman worked with Global Technologies, a company affiliated with one of the two main CO_2 groups. Prior to that, he had replaced Ohad Jehassi at EPA's Design for the Environment program. On CO_2's application to other cleaning processes, see David Jackson and Barry Carver, "It Looks Like Snow . . . Using Solid-State CO_2 in Critical Cleaning," *Precision Cleaning* 8, no. 5 (May 1999): 16–29; David Leviten, Kristi Thorndike, and Jason Omenn, "P2 Technology Review: Supercritical CO_2 Cleaning," *Pollution Prevention Review* 6, no. 4 (Autumn 1996): 35–42. Aside from CO_2, a number of industry figures began to take a look again at petroleum, which had maintained a small (10–15 percent) share of the market subsequent to perc's ascendance in the 1960s. New petroleum-based machines were being promoted as less prone to fires, and new petroleum-based solvents, introduced by Exxon, were also being promoted as more effective than previous petroleum products. However, the environmental hazards associated with petroleum use, including hazardous waste concerns, had not been evaluated and were likely to constitute a significant constraint on its position as "alternative," let alone a "pollution prevention alternative." See Liz Church, "The Industry Witnesses a Rebirth of Petroleum," *Western Cleaner and Launderer* 39, no. 2 (November 1999): 1ff.; "Regulatory Overview for Petroleum Drycleaning," International Fabricare Institute Bulleting—Legislative and Regulatory Overview, no. 17, *Fabricare Resources: The Fabricare Journal of Technical Information*, September 1997.

59. One strategy to address the high cost problem was through tax relief. The Micell group for one helped develop legislation along these lines, introduced by its local congressman (North Carolina Democrat David Price). This legislation, the *Drycleaning Environmental Tax Credit Act of 1999*, would amend Internal Revenue Code 1986 to allow an income tax credit for dry-cleaning machinery that reduced the amounts of hazardous substances (defined as perc-free and petroleum-free) to be utilized. But even the additional tax benefits were still not seen as sufficient in overcoming the cost barrier. See "Bill Benefits CO_2 Solvent," *Drycleaner News*, May 1999, 22.

60. U.S. Environmental Protection Agency, *Cleaner Technologies Substitute Assessment for Professional Fabricare Processes*, Office of Pollution Prevention

and Toxics (Washington, D.C.: U.S. EPA, June 1998), EPA 744-B-98-001, p. ES-2; "CTSA Manages to Upset Nearly Everyone," *Drycleaners News* 47, no. 10 (October 1998): 14–19; "EPA's CTSA Report—Helpful Guide or Hot Potato?" *American Drycleaner* 65, no. 7 (October 1998): 44–52. Soon after the Design for the Environment conference, a possible reassessment of the alternatives issue began to take place among some dry-cleaner groups. See, for example, the argument by IFI executive director William E. Fisher, "A Frank Assessment: Cleaning Technologies and Environmental Issues for the U.S. Cleaning Industry," presented at the International Drycleaners Congress convention, Munich, Germany, May 5, 1998, reprinted in *National Clothesline* 38, no. 9 (June 1998): 1ff.

61. A series of court cases and administrative rulings in the mid- and late 1980s underscored the problem of liability for realtors and lenders under Superfund. As a consequence, a number of realtors and lenders decided to reevaluate their policies toward dry cleaners. For example, banks and realtors began to require that any property transaction involving a commercial site with a cleaner trigger a site assessment evaluation. This might include a questionnaire about the practices of the dry cleaner as well as an initial soil sampling to determine if any possible contamination problems were associated with the site. If this "phase one" assessment indicated any possible problem, that would lead to a more in-depth (and expensive) phase-two assessment that could involve soil gas surveys, more extensive soil sampling, and/or groundwater sampling. The costs for those evaluations alone might cost as much as $5,000–$10,000 and would often be borne by the cleaners themselves. Once any contamination was identified, a far more expensive process would begin to unfold. On the court rulings on liability, see, for example, *United States v. Northernaire Plating Co.*, 670 F. Supp 742 (W.D. Mich. 1987) (where a landowner could be held liable even if the tenant was the sole cause of the contamination); *U.S. v. A&N Cleaners and Launderers Inc.* 788 F. Supp 1317 (S.D. NY 1992) (where an owner or operator cannot contract away its responsibility for CERCLA liability); and *Arthur Spitzer et al.*, California State Water Resources Control Board, Order No. WQ 89-8 (1989) (where hazardous waste cleanup responsibilities could be extended to owners who leased or purchased previously polluted land). See also Craig Tranby, "CERCLA Issues Surrounding Landlord vs. Tenant Liability for Dry Cleaning Activities on Commercial Properties," paper prepared for the Pollution Prevention Education and Research Center (Los Angeles: PPERC, December 1996); personal communications with Evan Henry and Tom Beeler, Bank of America, June 5, 1996; Barbara Russell, American Properties, October 7, 1996; David Simon, May 21, 1996; Thomas H. Clarke Jr., "Environmental Liability of Shopping Center Owners for Past and Future Contamination from Dry Cleaners," *Shopping Center Legal Update* 15, no. 3 (Winter 1996): 2–5, International Council of Shopping Centers; "The Problem with PERC," *Environmental Manager* 7, no. 5 (December 1995); Robert Gottlieb, "Real Estate Trend: Why Do They Shy from Dry?' *Skylines*, BOMA International, October 1996, 26–28.

62. Congressman Barton had signed on during the 104th Congress as the main sponsor of legislation (the *Small Business Remediation Act*, or H.R. 1711) that had been introduced to change Superfund cleanup standards by significantly lowering the risk threshold. The Barton Bill had been written by the lobbyist/ law firm for dry-cleaning interests, the Washington, D.C. law firm, Baise, Miller & Freer, which was itself a classic Beltway insider firm. Baise, Miller & Freer's senior partner, Gary Baise, had been an assistant administrator in the Nixon administration EPA when that agency was first organized in 1970. See also "Sensitive to the Issue: Gary Baise," in *National Clothesline* 38, no. 9 (June 1998): 4; "Capitol Hill March a Success," *American Drycleaner* 65, no. 8 (November 1998): 6; the "perc stigma" concept is discussed in a paper by John L. Payne presented at the 1999 International Drycleaning Convention and reprinted in *National Clothesline* 40, no. 2 (November 1999): 22; the Barton Bill ultimately reinforced the feeling of vulnerability. Without the support of the financial and real estate groups, and the failure of the chemical suppliers to join the lobbying, the dry-cleaning groups were unable to move the legislation beyond the committee process. Perhaps most disturbing for the dry cleaners was the cautious and ultimately antagonistic position of the realtors themselves. For example, the testimony at the Barton Hearings of John Ayres, an environmental consultant, who was also the chair of the Toxic Waste Task Force of the International Council of Shopping Centers (ICSC), was indicative of this problem. Ayres, vice president of an engineering and construction management firm, failed to support dry-cleaner contentions that perc was not hazardous. Testimony of John Ayres in *Environmental Compliance Problems Facing Dry Cleaners, Hearing before the Subcommittee on Oversight and Investigations of the Committee on Commerce*, House of Representatives, 104th Cong., 2nd sess., September 13, 1996, serial no. 104–105, 18–20, 28–29; interviews with John Ayres, May 2, 1997; on the chemical industry position, see "Dow's Doubts: Perc Producers Are Lukewarm to Barton Bill, Dow Says in Newsletter," *Spot News*, Dow Chemical, Summer 1997, available at http://www.pond.com/-hhorning/sfund/bartdow.html; see also the remarks by Barney Deden ("GLADCA Hosts Deden to Speak on Barton Bill") in *Western Cleaner and Launderer* 38, no. 9 (June 1998): 1ff.

63. *Pollution Prevention in the Garment Industry*, Appendix 6-A and 6-B; on state legislation see *Drycleaner News*, May 1997, 1ff.; the Oregon liability law, sponsored by the state dry-cleaning association, is discussed in Terry L. Obteshka, "P2 in Dry Cleaning," *Pollution Prevention Review* 7, no. 4 (Autumn 1997): 63–64; and also, "Oregon Drycleaner Advisory Committee Discusses Funding," *National Clothesline*, September 1998, 38, no. 12, 61; see also "More States Legislate Protection for Drycleaners," an article circulated by Dow Chemical's own news service and reprinted in the dry-cleaner trade publication, *Western Cleaner and Launderer* 38, no. 8 (May 1998): 24ff.

64. The Fisher statement is in "IFI CEO William Fisher Addresses IDC in Toronto," *Western Cleaner and Launderer*, July 1999, 1ff.; On dry-cleaner

vulnerability see "Capitol Hill March a Success," *American Drycleaner* 65, no. 8 (November 1998).

65. "When the Call for Help Comes, Answer," *National Clothesline* 39, no. 8 (May 1999): 2.

66. "MASS Program Honored by EPA," *Drycleaner News* 47, no. 11 (November 1998): 1ff.; Dave Johnston, "Bay State Focus Put on Results," *Drycleaners News* 46, no. 9 (September 1997): 1ff.

67. Lack of compliance in Massachusetts included problems of inaccurate information or lack of reporting as well as more direct emission-related violations such as a missing door gasket on a perc machine that could increase emissions considerably. One facility, in fact, was due to be shut down for its particular compliance failure. The problem of lack of compliance in all these states has been further compounded by the limited number of inspections. According to an EPA review, only 26 percent of the dry cleaners in their database had been inspected, with the average inspections occurring every seven to eight years. U.S. Environmental Protection Agency Enforcement and Compliance Assurance, *Profile of the Dry Cleaning Industry* (Washington, D.C.: U.S. EPA, September 1995), 63, EPA-310-R-95-001; "MASS Program Honored by EPA"; the southern California 95 percent noncompliance figure was based on 340 inspections of dry cleaners between April 1999 and June 1999, Edwin Pupka, South Coast Air Quality Management District, senior manager, Stationary Source Compliance, July 1, 1999; see also "An Evaluation of the Bay Area Air Quality Management District's Air Pollution Control Program" (Sacramento: California Air Resources Board, Compliance Division Staff, March 1998); "Cleaners Are Fined by EPA Inspectors for Clean Air Violations, Perhaps Signaling No More Mr. Nice Guy," *Drycleaners News* 39, no. 6 (March 1999): 1ff.

68. While the FTC draft rule was receiving comments, the environmentally oriented clothing company Patagonia had decided it wanted to provide a wet clean label on two new garments it was introducing that would have otherwise required a dry-clean-only label for the fabrics involved. The question for Patagonia was how to proceed, given that the garment care label rule change was still only at the draft stage. Patagonia had approached the issue of the test protocol for a wet-clean label by going directly to a commercial wet cleaner and then evaluating whether the garment was successfully cleaned, a type of procedure often used by manufacturers in developing their dry-clean label. However, because of the delays in the FTC rule change, Patagonia discovered that it had to include for its new garments a dry-clean label as well, even as it prepared to alert its customers that the company preferred this new, pollution prevention alternative. Federal Trade Commission, *Trade Regulation Rule on Care Labeling of Textile Wearing Apparel and Certain Piece Goods*, 16 CFR Part 423; Connie Vecellio, "Changing the Care Labeling Rule to Fit Changing Times," in *Garment and Textile Care Program, Conference Proceedings*, 105–109; presentation by Marilyn Regazzi, Patagonia, Clean Products Network Workshop, Los Angeles, February 26, 1999.

69. Manfred Wentz, "Textile Care Technology Spectra and Care Labeling Issues," *American Drycleaner*, November 1996, 88–96; Stan Golumb, "Stan Golumb: His Life and Times," *National Clothesline* 39, no. 6, March 1999, 64; for Wentz's position in the industry see, for example, "Conference Examines Development in Clothes Cleaning Technology," *National Clothesline* 38, no. 2 (November 1997): 1ff.; "Wentz to Replace Pulley as IDC Executive," *National Clothesline*, September 1998, 38, no. 12, 36.

70. The Halogenated Solvents Industry Alliance was formed in 1980 by the chlorinated solvents industry "to meet the growing challenges of government regulation," as described by the group. HSIA companies included manufacturers and distributors of perc, methylene chloride, TCE, and other halogenated compounds. See "HSIA: Halogenated Solvents Industry Alliance, Inc.," *Western Cleaner and Launderer*, July 1999, 14ff.; HSIA informationals have also been used, often without attribution, in dry-cleaner publications. See, for example, "Dirty Drinking Water Goes to the Movies," *National Clothesline*, February 1999, 6; see also presentation by Steve Risotto, HSIA, at the "Drycleaning Industry Outlook Roundtable," 1996 California Fabricare Association annual meeting, Long Beach, California, August 9, 1996; Halogenated Solvents Industry Alliance, "The Safe Handling of Perchloroethylene Drycleaning Solvents Beyond Regulatory Compliance," Washington D.C.: October 1999.

71. The former owners of Pilgrim Cleaners were seeking a monetary award of $12 million due to cleanup costs from Dow Chemical and other perc suppliers and distributors, including R.R. Street. *Pilgrim Enterprises et al. v. Kleen-Rite, Inc. et al.*, No. 95–54786, 215ᵗʰ Judicial Court of Texas.

72. Kenny Slatten, "Don't Point Fingers; Take Responsibility," *National Clothesline*, October 1998, 39, no. 1, 18; "Jury Finds R.R. Street Not Liable in Cleanup Suit," *National Clothesline* 38, no. 11 (August 1998): 26; "R.R. Street Exonerated, But Pilgrim Contamination Suit Leaves No Winners," *American Drycleaner* 65, no. 6 (September 1998): 6; "Despite a Jury Verdict to the Contrary, Judge Holds Street Liable in Houston Case," *National Clothesline* 38, no. 2 (November 1998): 1ff.; "Pilgrim Cleanup Suit Jury Verdict Reversed by Judge," *American Drycleaner* 65, no. 8 (November 1998): 10–14; "Notwithstanding Jury Verdict in Street's Favor—Judge Makes Award to Pilgrim—Which Street Will Appeal," *Western Cleaner and Launderer*, November 1998, 4ff. According to the November 1998 *Western Cleaner and Launderer* article, the implications of the ruling for manufacturer or supplier responsibility was similar to the "extended producer responsibility" concept. "The possibilities associated with this even within our own drycleaning industry are mind numbing and endless," the article exclaimed, "particularly if one is a manufacturer, distributor, drycleaning trade association, or any other service provider to the industry." (p. 10)

73. By 1999, however, the Cleaner by Nature success story began to be recognized in different industry forums. A special issue on wet cleaning in one of the trade publications that highlighted Davis's success story followed her prominent

role as a featured speaker in the biannual industry "Clean Show," the most prominent of the dry-cleaner gatherings. "Wetcleaning: Making Waves in the Industry," *American Drycleaner*, September 1999, 76–90.

74. "Community Enterprises," Fifth Avenue Committee, Brooklyn, New York, 1998; see also Jed Emerson and Fay Twersky, editors, *New Social Entrepreneurs: The Success, Challenge and Lessons of Non-Profit Enterprise Creation* (San Francisco: The Roberts Foundation, 1996); David Scheie, "Promoting Job Opportunity: Strategies for Community-Based Organizations," *Shelterforce: The Journal of Affordable Housing Strategies* 89 (September–October 1996): 9–11.

75. Personal communication with Aaron Schiffman, Fifth Avenue Committee, December 2, 1998.

76. Wages were established at $7–$9.85 per hour, which were slightly higher than wage rates at neighborhood dry cleaners. Personal communication with Aaron Schiffman.

77. As another example of the community enterprise model, in 1997, a Philadelphia-based environmental/public interest coalition, the Women's Health and Environmental Justice Network, began to explore the idea of developing a toxic-free or pollution prevention enterprise as part of its efforts to address toxic and public health issues. The group decided to focus on establishing a nonprofit wet-cleaning operation but was encouraged by other community development groups to establish a for-profit operation, albeit one with environmental and social goals. As a result, the group decided to explore a public-private partnership, where a for-profit business could be formally associated with a public interest organization. The group also engaged students at the Wharton School of Management to undertake a feasibility study of establishing a wet-clean operation in the Philadelphia area and to help identify how such a model could evolve. Personal communication with Priscilla Rosenwald, January 14, 1999. On Eco-Mat's troubles, see the Ecomat Prospectus, filed with the U.S. Securities and Exchange Commission, December 9, 1996, describing the losses accrued by the company and its franchising strategies, 18, 28–30.

78. Regina Freer, "From Conflict to Convergence: Interracial Relations in the Liquor Store Controversy in South Central Los Angeles" (Ph.D. diss., University of Michigan, 1999).

79. *KYCC*, a brochure of the Korean Youth and Community Center (Los Angeles: KYCC, 1999); personal communication with Bong Hwan Kim, September 10, 1995.

80. Regina Freer, "From Conflict to Convergence"; personal communication with Deborah Ahn, August 9, 1995.

81. Jenni Cho, "The Role of Non-Profits as Entrepreneurs: Issues for KYCC's Venture in Wet Cleaning," UCLA Department of Urban Planning, June 1996; Korean Youth and Community Center, "Wet Clean Outreach and Education," a proposal to the Urban Resources Partnership, 1996; "Dry Cleaning's Most Up-to-Date Technology Helps to Protect the Environment," *Korea Daily Times*,

October 12, 1996; see also "Cleaner Assistance Program" (Los Angeles: Korean Youth and Community Center, 1997).

82. Personal communication with Joe Whang, May 21, 1998. Our center undertook a financial and "conversion" analysis of Whang's transition to wet cleaning and documented the issues associated with the transition. Peter Sinsheimer, Jenni Cho, and Robert Gottlieb, *Switching to Pollution Prevention: A Performance and Financial Evaluation of Cypress Plaza Cleaners*, Pollution Prevention Education and Research Center (Los Angeles: Urban and Environmental Policy Institute, 1999).

83. Whang's switch had been compounded by the fact that his location was not optimal—in a shopping mall that had lost its anchor tenant and thereby experienced a substantial reduction in mall activity—and that, unlike Davis, he did not have any marketing experience. His 10 percent growth rate after the first year, including a 97 percent retention rate of his earlier dry-clean customer base, also demonstrated, aside from a successful startup business like Cleaner by Nature, the potential for a successful transition. See the Cypress Plaza Evaluation, Pollution Prevention Education and Research Center, 1999.

84. A survey of the cleaners attending tours at Cypress Plaza Cleaners and the implications of the survey results for a pollution prevention transition is discussed in PPERC's Phase 2, *Pollution Prevention in the Garment Care Industry* final report to the South Coast Air Quality Management District, February 2000.

85. One set of issues to be explored might involve the quality of garment construction associated with the global nature of the production system. Another aspect of the "community of interests" approach could then include the issues of overseas production (e.g., labor or "sweatshop" conditions in garment manufacture).

86. Comments by Joe Whang, Cypress Plaza Cleaners, at the Professional Wet Cleaners Forum, Occidental College, May 10, 1998.

4 Janitors and Justice: Industry Restructuring, Chemical Exposures, and Redefining Work

1. Presentation by Mike Brown, Patagonia, Pollution Prevention Forum Series, UCLA, April 17, 1996; personal Communication with Dorothy Clemen, American Janitorial Service, August 15, 1995.

2. Personal communication with Mike Brown, August 1, 1995.

3. "A Study of Ergonomic Hazards Faced by Janitors," Labor Occupational Safety and Health Program, Center for Occupational and Environmental Health and Institute of Industrial Relations, UCLA, 1993, 3–4. The study surveyed 103 janitorial workers from Local 399 of the Service Employees International Union; see also Marketdata Enterprises, Inc., Full Service Market Research and Consulting, *The U.S. Commercial & Residential Cleaning Services Industry* (Tampa, Fla.: Marketdata Enterprises, January 1999).

4. Janitors working for Campus Maintenance at UCLA, for example, cleaned an average of 28,000 to 30,000 square feet each night in 1995, up from 16,000 square feet in 1983. The Research Department of the Service Employees International Union (SEIU) Local 399 states that janitors in Los Angeles clean an average of 25,000 square feet per shift. The 1994 Cleaning Management Magazine Statistical Survey of contract cleaners also identified a range of cleaning averages per employee per hour, depending on the size of the contract firm. These varied from 2,500 square feet per hour per employee for firms that earned less than $100,000, to 3,559 square feet for firms earning greater than $1 million. The average size, according to this survey, was 3,312 square feet per hour per employee. See, "Janitorial Cleaning Services at UCLA," a paper prepared for the Pollution Prevention Education and Research Center, Department of Urban Planning, UCLA, 1995; the shift to part-time and temporary labor also represented an effort to reduce benefit payments. See Heather E. McDonald, "Creative Hiring Options Cut Employer Costs," *Cleaning Maintenance Management*, June 1999; see also "A Study of Ergonomic Hazards Faced by Janitors," Labor Occupational Safety and Health Program, 10; "1994 Contract Cleaner Statistical Survey," *CM: Cleaning Management Magazine* 31, no. 12 (December 1994): 6; "1995 In-House Survey," *CM: Cleaning Management Magazine* 32, no. 4 (April 1995): 19; personal communication with Jono Shaffer, Justice for Janitors Organizing Committee, Service Employees International Union, Local 399, Los Angeles, March 10, 1994.

5. One 1995 trade publication survey identified an average starting wage of $6.79/hour for in-house workers, while contractors paid their janitors an average starting wage of $5.64/hour. The differences between overall average wages of in-house and contract janitorial workers is even more striking: $8.22/hour for in-house workers and $6.62/hour for contract workers. See "1995 In-House Survey," 19.

6. The estimated size of the janitorial workforce has varied in terms of the source of the data. An accurate portrait of the janitorial workforce also depends on how janitorial work is defined, over what statistical region data it is taken, and what characteristics of the workforce can be identified. For example, a review of the number of janitorial and building maintenance workers provided from two key data sources, the *Occupational Compensation Survey* and the *County Business Patterns*, indicates differences both in the actual numbers and the trends in employment. U.S. Department of Labor, *County Business Patterns*, Los Angeles County, 1991; Census Bureau of the United States, *County Business Patterns, United States*, Table 1b, 1991; see also Executive Office of the President, Office of Management and Budget, *Standard Industrial Classification Manual*, 1987; "What Is Justice for Janitors?" The Justice for Janitors Campaign brochure, Service Employees International Union, Washington, D.C., provides yet another type of estimate. Personal communication with Jono Shaffer, SEIU, June 9, 1994.

7. While a number of these trends reflect the dominant shift toward contracting and away from in-house operations, especially those involving public facilities,

it should be noted that in-house services still remain a significant segment of the industry. For example, according to a 1995 survey of in-house operations, 84 percent of those surveyed reported that their operation has always been in-house, while only 5 percent indicated a shift to contracting arrangements for services that had been previously in-house. Ultimately, the same issues—worker empowerment, stakeholder participation, the need for safer cleaners—came into play for both in-house and contract janitorial operations. See, "1995 In-House Survey," 19.

8. Marketdata Enterprises, Inc., Full Service Market Research and Consulting, *The Commercial & Office Building Cleaning Industry* (Valley Stream, New York: August 1991), 7–11.

9. Ray Gross's quote is from a BHI Corporation press release ("BHI Corporation Announces OneSource: New Name for ISS Facilities Division"), January 25, 1999, available at www.pcnewswire.net; BHI Corporation, subsequently Carlisle Holdings Limited, had annual sales of more than $1 billion in 1999, with One-Source its largest single operation. Background on ISS history is available from the ISS 1997 financial report (www.iss.dk/news/press_73.htm).

10. American Building Maintenance Industries information is available at http://www.cbvcp/com/abm/.

11. Lower wages also coincided with a reduction or elimination of benefits in many cases, some of which have been associated with the increase in the use of part-time labor, particularly in private as opposed to government or public office buildings. Even some of the largest contractors (e.g., those with gross sales greater than $1 million) fail to pay benefits to their workforce. According to one industry survey, only 44 percent of the largest companies paid benefits, with the industrywide average at just 23 percent of all contracting companies paying benefits. Payroll costs, as a portion of sales, also decreased through the 1980s, although labor costs as a percentage of cleaning operating costs remained by far the largest cost category (73 percent for in-house services, 50.9 percent for contractors). See "1999–2000 Contract Cleaner Statistical Survey," *Cleaning and Maintenance Management*, 37, no. 1, January 2000, 7; "1998 Contract Cleaner Statistical Survey," *Cleaning and Maintenance Management*, 35, no. 12, December 1998; "Minimum Pain for Minimum Wage Hike?" *Cleaning and Maintenance Management*, 33, no. 5, May 1996, 10; see also "1996 In-House Survey," *Cleaning and Maintenance Management* 33, no. 4 (April 1996): 2, 8; Paul Schimek, "From the Basement to the Boardroom: Los Angeles Should Work for Everyone," report for Justice for Janitors, SEIU Local 399, August 1, 1989, 5; on the living wage issues, see John Seeley, "The Bare Minimum," *L.A. Weekly*, September 20, 1996, 11ff.; Grace Lee, "Wage Wait: City's Living Wage Has Yet to Reach Most Workers," *L.A. Weekly*, July 18, 1997, 13–14.

12. Cited in Richard Mines and Jeffrey Avina, "Immigrants and Labor Standards: The Case of California Janitors," in *U.S.-Mexico Relations: Labor Market Interdependence*, edited by Jorge Bustamante, Clark W. Reynolds, and Raul A. Hinojosa-Ojeda (Palo Alto: Stanford University Press, 1992), 431.

13. *BOMA Experience and Exchange Report* (Washington, D.C.: BOMA International, 1999, 1988, 1982).

14. Bureau of Labor Statistics Area Wage Survey, Los Angeles-Long Beach SMSA, 1977–1988; Sonia Nazario, "For This Union, It's War," *Los Angeles Times*, August 19, 1993; Jon Shaffer interview, July 13, 1994.

15. Marketdata Enterprises, Inc., *The Commercial & Office Building Cleaning Industry*, 8–11.

16. Downsizing pressures and wage restructuring have also been experienced by public sector and in-house operations, which in turn has led to a greater need to justify costs and performance in competitive terms with private sector counterparts (contract cleaners). Despite their higher wage scale and more permanent workforce, the trends for public sector and in-house operations have also pointed toward lower wages and greater pressures on "productivity" (more square feet cleaned per amount of labor required). Moreover, privatization, in the form of contracted services, also remains an option strongly encouraged by those elected officials who argue that public operations need to be run more as a "business." Personal communications with Tony Royster, City of Los Angeles, General Services Department, July 10, 1996, and October 28, 1996; for the industry's 1995 figures see, "The Cleaning Industry: Profile and Projections," *Services Magazine*, Building Service Contractors Association International, information available at http://www.bscai.org/serv/clean.htm

17. Even as the trend toward integration has occurred, the number of small, part-time, and "mom and pop" establishments has also grown. Yet due to the squeeze on labor costs and the increased amount of square footage cleaned per employee, the average number of employees per establishment has been declining overall. For example, the 1991 *County Business Patterns* lists 1,051 building maintenance service establishments for Los Angeles County; of these, 787, or about 75 percent have fewer than ten employees. See U.S Department of Labor, *County Business Patterns*, Los Angeles County, 1991, table 2—Counties—Employees, Payroll, and Establishments, by Industry; for the discussion on integration, see John R. Johnson, "Cleaning Up with Janitorial Supplies," *Industrial Distribution* 85 (February 1996): 31–32; at the same time, customer dissatisfaction with contract cleaning companies remains fairly high. Average customer loss rates can run as high as 45 percent per year for janitorial businesses. Two possible reasons for this are the expressed lack of concern about "professionalism" in the "mom and pop" establishments that dominate the industry, and an overall low standard of quality by contractors. See Marketdata Enterprises, Inc., *The Commercial & Office Building Cleaning Industry*, 12–13, 9–10; see also Cleaning Management Institute 1993 Contract Cleaner Survey, 4–7; Victoria Fraza, "Brushing Up an Old Line," *Industrial Distribution* 85 (October 1996): 45–46ff. for ABM background, see American Building Maintenance Industries, http://www.cbvcp/com/abm/; see also Ellen Paris, "Clean-up Job," *Forbes* 142, no. 7 (October 3, 1988): 44.

18. In one survey the expenditures for chemicals averaged nearly twice as high as the next largest cost category (vehicle operations) and four times as high as employee training. See Table C: Average Annual Expenditures of Contract Janitorial Service Companies, Cleaning Management Institute 1993 Contract Cleaner Statistical Survey, 5; see also "Cleaning Up with Janitorial Supplies," *Industrial Distribution*, February 1996, 31; Fraza, "Brushing Up an Old Line," 45–49. Trends also indicate that cleaning expenses have declined as a portion of total operating expenses and as a portion of commercial rental income, primarily due to decreasing janitorial wages and increasing rental rates. However, rental income, total operating costs, and total cleaning costs have been decreasing in real terms, with cleaning expenses averaging slightly more than 15 percent of operating expenses while maintaining a relatively constant portion of both operating expenses and rental income. Labor costs, on the other hand, have been both more substantial and variable. For example, in the states of California, Washington, and Oregon, labor costs account for about half of all the operating costs for contract cleaning companies. See Table B: Median Rent Income, Operating Expenses, and Cleaning Expenses in 1985 Dollars (inflation adjusted) per Square Foot, *BOMA Experience and Exchange Report, 1986–1991* (Washington, D.C.: Building Owners Management Association, 1992); Paul Schimek, report for Justice for Janitors, 14. See also *BOMA Experience and Exchange Report*, BOMA International, 1981, 1984, 1987; Cleaning Management Institute, *1993 Contract Cleaner Statistical Survey*, 6.

19. The "long and hard work" quote is from Suellen Hoy, *Chasing Dirt: The American Pursuit of Cleanliness* (New York: Oxford University Press, 1995), 97; on the 1891 garden caretaking function, see Henry Lincoln Clapp, "School Gardens," *Education* 21 (1901): 611; the "janitor must know the correct chemical" is from the 1947–1948 *Annual Report of the Los Angeles County Janitors Department* (Los Angeles: Building Services Department, 1948), 3; in its comments about gender, the report also states (p. 4) that the work of janitoresses had to be confined to ladies restrooms and lavoratories but they could also be used in "cases of illness and in assisting others, and in other demonstrations of courtesy to the county employees and the public."

20. By the end of the decade, Green Seal did begin to explore a standard for cleaning products for commercial buildings. For the household cleaner standard, see Green Seal, Inc., *Summary Background Report and Proposed Environmental Standard for General Purpose Household Cleaners*, Standard GS-8-1992; Rita Rousseau, "Clean and Green," *R&I*, April 1, 1997, 96.

21. The EarthWorks Group, *50 Simple Things You Can Do to Save the Earth* (Berkeley, Cal.: Earthworks Press, 1989); Joel Makower with John Elkington and Julia Hailes, *The Green Consumer* (New York: Penguin Books, 1993); see also Dan Stein, *Dan's Practical Guide to Least Toxic Home Pest Control* (Seattle: Washington Toxics Coalition, 1991); *Safeguard Your Home from Harmful Products*, New York City Department of Sanitation, 1996, and letter from Elizabeth

Whelan, American Council on Science and Health, to Mayor Rudolph Giuliani, July 25, 1996, challenging the department's publication for "diverting attention from the real health risks—risks associated with personal lifestyle."

22. Michael Price, "Clean Air Keeps You Out of Court: Poor Indoor Air Quality Could Cost You Plenty," *Cleaning Maintenance Management* 33, no. 7 (July 1996): 30–36; "The Sick Building Syndrome," *Services Magazine*, Building Service Contractors Association International, available at http://www.bscai.org/serv/sick.htm; Frank A. Lewis, "Clean and Green—The Ecological Janitor," *Enviros: The Healthy Building Newsletter* 3, no. 12 (December 1993); Claudia H. Deutsch, "It May Be Time for a Silver Lining in Air Pollution," *New York Times*, November 30, 1997; see also Nicholas A. Ashford and Claudia S. Miller, *Chemical Exposures: Low Levels and High Stakes*, 2d ed. (New York: Van Nostrand Reinhold, 1998).

23. William R. Griffin, "What's New with Cleaning Chemicals," *Maintenance Supplies* 42, no. 6 (June 1997): 22–33.

24. The PPERC study evaluated twenty-five of the most common chemical ingredients among the twenty-two cleaning compounds used in the four sites. Virginia Leenkneckt, "Pollution Prevention in the Workplace: Glycol Ethers in Janitorial Supplies," UCLA School of Public Health, Pollution Prevention Education and Research Center, June 1997; in relation to the health risk information on glycol ethers, see *Proceedings of the International Symposium on Health Hazards of Glycol Ethers*, Pont A Mousson, France, April 19–21, 1994, edited by Andre Cicolella, Bryan Hardin, and Gunnar Johanson, *Occupational Hygiene—Risk Management of Occupational Hazards*, 1996; see also B.I. Ghanayem, "An overview of the hematotoxicity of ethylene glycol ethers," in *Occupational Hygiene, Vol 2*, 253–268; K.S. Hammond, "Tiered Exposure-Assessment Strategy in the Semi-Conductor Health Study," *American Journal of Industrial Medicine* 28 (1995): 661–680; F.P. Gijsenbergh, M. Jenco, H. Veulemans, D. Groesenken, R. Verberckmoes, H.H. Delooz, "Acute butyl-glycol intoxication: a case report," *Human Toxicology* 8 (1989): 243–245; M. Hours, B. Dananche, E. Caillat-Vallet, J. Fervotte, J. Philippe, O. Boiron, J. Fabry, "Glycol ethers and myeloid acute leukemia: a multicenter case control study," *Occupational Hygiene, Vol. 2*, 405–410; D.B. McGregor (1996), "A review of some properties of ethylene glycol ethers relevant to their carcinogenic evaluation," *Occupational Hygiene, Vol. 2*, 213–235.

25. The no-obvious-return-on-investment quote is from Alan Plass, "Budgeting for Compliance," *Maintenance Supplies* 43, no. 2 (February 1998): 58; see also Larry Harvey, "Clean-up Isn't Easy Anymore: Traditional Chlorinated Solvents Are Banned for Production Cleaning Applications," *Industrial Distribution*, August 1995, 94.

26. 29CFR Part 1910.1200.

27. *Occupational Safety and Health Act*, Section 6 (b) (7).

28. See *Reauthorization of Superfund: Hearings before the Subcommittee on Water Resources of the House of Representatives*, Committee on Energy and

Commerce, 99th Congress, 1st sess., 1635 March 26–28, May 1, July 24–25, 1985.

29. At the time of its passage, reporting facilities under the Hazard Communication Standard represented 14 million employees in 300,000 manufacturing establishments, but not initially the janitorial cleaning services sector. See "OSHA 'Hazard Determination and Communication' Regulations," (Kutztown, Penn.: Transportation Skills Programs Inc., 1985).

30. On TRI's influence on industry activity, see Andrew Wood, "10 Years After Bhopal: Charting a Decade of Change," *Chemical Week*, December 7, 1994, 25–30; *Reducing Toxics*, 131–139.

31. Paul Kolp et al., "Comprehensibility of Material Data Safety Sheets," *American Journal of Industrial Medicine* 23 (1993): 135–141; Paul W. Kolp, Phillip L. Williams, and Rupert Burtan, "Assessment of the Accuracy of Material Safety Data Sheets," *American Industrial Hygiene Association Journal* 56 (February 1995): 178–183; Anne Dantz, "Imperfect MSDS Better Than None," *Cleaning and Maintenance Management*, July 1995, 8.

32. Steven Ashkin and Barbara Sattler, "Selecting Environmentally Friendly Cleaning Products," Workshop Session, Cleaning and Maintenance Management, "Satellite Conference '96," June 11, 1996; Ed Feldman, "Managing in a Downsized Environment," Workshop Session, June 11, 1996.

33. To sort out these conflicting signals regarding environmentally preferable cleaning products, our Pollution Prevention Center, in conjunction with the School of Social Ecology at University of California at Irvine, surveyed readers of *Maintenance Supplies*, the suppliers and distributors trade publication. The survey results underlined the ambivalence about green products. While there was interest in the use of environmentally preferable cleaners, "performance"—that is, how quickly and efficiently the cleaning tasks could be performed—was assumed to be a problem area for such products. Moreover, there was little consensus about environmental characteristics or "attributes" of such products, nor were such products being used widely among those sampled. While a few companies had established "green lists" or promoted products as "environment friendly," there was little focus on the specific ingredients in products, such as glycol ethers, nor any desire to create a standard that would identify what constituted environmental "preferability." As a self-selected sample, survey respondents were not representative as such of suppliers and distributors within the janitorial services industry. However, the survey generated the magazine's single largest response rate where no incentives were provided as inducement to fill out the survey questionnaire. Robert Gottlieb, Pollution Prevention Education and Research Center, "Clean and Green Revisited," *Maintenance Supplies* 41, no. 4 (April 1996), with Andrea Brown, Elaine Vaughan, and Jennifer Fishman; An example of a company's "green list" is the ISS (International Service System, Inc.) (now One Source) "Green List Manual" (vol. 3, December 1, 1993), which was identified by the company as a "value-added service" that could provide "an extra selling tool over our competition," 4.

34. "Buyers Green Demands Challenge Suppliers," *Chemical Week*, August 23, 1995, 43; see also, Bob Preuss, "Environmental Cleaning and Purchasing," *Cleaning and Maintenance Management*, 37, no. 4, April 2000, 46–50.

35. Fran McPoland, "The Role of the Federal Environmental Executive," in *Pollution Prevention 1997: A National Progress Report*, Office of Pollution Prevention and Toxics (Washington, D.C.: U.S. EPA, June 1997), EPA 742-R-97-00.

36. *Executive Order 12873—Federal Acquisition, Recycling and Waste Prevention*, Washington, D.C., October 20, 1993, Section 201, Section 503; Sean Cummings, "Rocky Road to a Federal 'Green' Giant," *Cleaning Maintenance and Management*, September 1997.

37. *Commercial Cleaning Supplies*, U.S. General Services Administration Federal Supply Service, February 1996; *Environmentally Preferable Purchasing Program: Cleaning Products Pilot Project*, EPA 742-R-97-002, February 1997; *Final Report: East Philadelphia Field Office Pilot Study on Cleaning Systems*, prepared by Westat Inc. for the Chemical Management Division, Office of Pollution Prevention and Toxics, U.S. Environmental Protection Agency, June 1994, AR-130-01; personal communication with Conrad Fleissner, U.S. EPA, April 9, 1996; Linda Steubesand, "The GSA Report: Establishing a Criteria for 'Environmentally Preferable Products,'" *Maintenance Supplies Magazine* 39, no. 4 (April 1994): 50ff.

38. The EPA/GSA initiative resulted in a 1996 publication that laid out seven environmental attributes and was made available to federal agencies that would be purchasing cleaning products. These seven attributes included skin irritation, food chain exposure, air pollution potential, fragrances added, dyes added, packaging, and degree of exposure to the concentrate. The GSA then evaluated specific products purchased by federal agencies through the ranking system—for example, skin irritation identified as negligible, slight, moderate, or strong. The staff involved with the Cleaning Products Pilot Project had originally hoped to use an EPA-approved list of cleaning products, but "it was clear from the beginning that such a list was outside the authority of GSA and EPA," as an EPA history of the project later put it. See, U.S. Environmental Protection Agency, *History of the Cleaning Products Pilot Project*, Office of Pollution Prevention and Toxics, available at http://www.epa.gov/opptintr/epp/cleaners/select/history .htm; for an analysis of how CBI provisions in TSCA have continued to present significant obstacles in the use of TSCA as a pollution prevention regulatory tool, see Janice Mazurek, Robert Gottlieb, and Julie Roque, "Shifting to Prevention: The Limits of Current Policy," in *Reducing Toxics: A New Approach to Policy and Industrial Decisionmaking* (Washington, D.C.: Island Press, 1995), 64–66.

39. Personal communication with Brian Johnson, April 3, 1996.

40. Commonwealth of Massachusetts, Operational Services Division, FRF#GRO04 for Cleaning Products, Environmentally Preferable, nd; *Chemical Week*, August 23, 1995.

41. "Nontoxic Cleaning Products: Pilot Project Report," Saint Paul Neighborhood Energy Consortium, Saint Paul, 1997; "Procuring Green Cleaners: Minnesota's Experience," Minnesota Office of Environmental Assistance, State of Minnesota, March 1998. For city programs see, for example, Environmental Health Coalition, *Toxic Turnaround: A Step by Step Guide to Reducing Pollution for Local Governments* (San Diego: Environmental Health Coalition, August 1997); "Municipal Pollution Prevention Project for the City of Chula Vista," San Diego, Environmental Health Coalition, February 14, 1995; for an NGO perspective on government procurement, see Philip Dickey's comments on "Government Procurement of 'Environmentally Preferable' Products," Washington Toxics Coalition, presented at the U.S. EPA Pollution Prevention Conference, Woods Hole, June 15, 1994.

42. American Society for Testing and Materials, *Innovation by Consensus: ASTM's First Century*, http://www.astm.org/ANNIVER/consensus.htm

43. Stephen P. Ashkin, "Seven Principles of Environmentally Preferable Products," presented at the Cleaning Management Institute, Washington, D.C., June 6, 1996, and San Francisco, June 10, 1996; see also Stephen P. Ashkin, "Clean Meets Green: EPA Promotes 'Environmentally 'Preferred' Products," *Cleaning and Maintenance Management*, October 1996.

44. Personal communication with Steve Ashkin, September 9, 1999; personal communication with Brian Johnson, August 16, 1996; notes from ASTM meetings provided by PPERC's participant, Andrea Brown.

45. See American Society for Testing and Materials, *Standard Guide on Stewardship for the Cleaning of Commercial and Institutional Buildings*, E1971–98, developed by ASTM Subcommittee E50.03 (West Conshohocken, Pa.: American Society for Testing and Materials, 1999). The language changes referred to are in ASTM E50.06.06, Draft Document, *Standard Guide on Stewardship for the Cleaning of Commercial and Institutional Buildings* Section 7.1.3.3, September 1996, and Proposed Revisions to the Draft Document, nd.

46. Sabrina Gates and Dan Carl, "Toxics Reduction for Santa Monica's Internal Operations," Pollution Prevention Education and Research Center, UCLA Graduate School of Architecture and Urban Planning, June 1992.

47. City of Santa Monica Environmental Programs Division, 1994 Sustainable City Program Document, see http://www.ci.santa-monica.ca.us/environment/policy; see also *Sustainable City Progress Report Update*, City of Santa Monica, Task Force on the Environment, October 1994.

48. The evaluation of Santa Monica's products are identified in Virginie Leenknecht's "MSDS Review," in *Pollution Prevention in the Workplace*, Pollution Prevention Education and Research Center, 2–3.

49. Sibylle Fahrenkamp and Angèle Ferré, "Project Evaluation of Pilot Project: Natural, Non-Toxic Cleaning Products," City of Santa Monica, S.A.F.E Consulting for the Earth, 1994.

50. After the six-month evaluation, the city staff moved to formalize the product substitution process based on a set of pass/fail criteria, including compliance with California's limits on volatile organic compounds and Santa Monica's own ordinance on CFC products. Products that successfully met the pass/fail criteria were scored in three categories: human health, environmental health, and corporate environmental responsiveness. Nineteen criteria were established, including the toxicity of product ingredients, flammability, VOC percentage, biodegradability, and the recycled content/recyclability of the packaging of the products. The nineteen criteria were then weighted (e.g., greater weight was given for health hazard criteria than for corporate environmental responsiveness). See Brian J. Johnson and Deborah O. Raphael, *City of Santa Monica Toxics Use Reduction Program* (Santa Monica, Cal.: Environmental Programs, 1994); Andrea Gardner, "Safer Cleaning Products for Custodial Workers: An Evaluation of a Local Toxics Use Reduction Program," UCLA School of Public Policy and Social Research, June 1995.

51. City of Santa Monica, Environmental Programs, "Notice Inviting Bids for Environmentally Preferable Cleaning Products," September 1994. After it closed, 19 manufacturers and distributors had submitted bids for 227 different products in 19 product categories.

52. Robert Gottlieb and Andrea Brown, "Janitorial Cleaning Products: It's Not Just What Gets Used, But How It Gets Used That Counts," *Pollution Prevention Review*, Winter 1998.

53. Author's notes, City of Santa Monica custodial staff focus group session, September 13, 1997. Despite their acceptance if not enthusiasm about the change, the janitors also identified problems with the new products. During the training/stakeholder process, the janitors had raised issues of safety equipment, proper dilution ratios, adequate ventilation, and appropriate disposal of cleaning products, critical factors for the work environment. A couple of the products were judged to be poor performers, and, in one case, the old floor stripper was reintroduced until the new cycle of bids could identify a preferable, less hazardous product. The janitors had discovered that the floor wax and strippers needed to be compatible, since the less toxic strippers could not remove the previous layers of wax. Nevertheless, at the conclusion of the first round of product use, both janitors and city staff were enormously pleased, even as more changes were identified that needed to be designed into the process.

54. Personal communication with Debbie Raphael, February, 1, 1999; "1998 Specifications for Cleaning Products," City of Santa Monica, 1998.

55. The U.S. EPA issued a document describing the Santa Monica experience as a model for incorporating environmental preferability into purchasing decisions, U.S. Environmental Protection Agency, *The City of Santa Monica's Environmental Purchasing: A Case Study*, Office of Pollution Prevention and Toxics (Washington, D.C.: U.S. EPA, March 1998), EPA742-R-98-001.

56. Author's notes, City of Santa Monica, Environmental Programs, award presentation, August 27, 1996.

57. Cited in John Howley, "Justice for Janitors: The Challenge of Organizing in Contract Services," *Labor Research Review* 15 (1990): 64; one contractor argued that his ascendance in the early 1980s was explicitly related to shifting from a higher paid unionized workforce to a low-wage contract labor force that "paid pennies above the minimum wage." "They wanted to get rid of—I don't want to use 'get rid of'—*phase out* the union cleaners," the executive commented, based on the ability of the contractor to come in as the low bidder. Cited in Michael Winerip, "The Blue-Collar Millionaire," *New York Times Magazine*, June 7, 1998, 72–74; see also "Industry Report: Report to the International Executive Board" (Washington, D.C.: Service Employees International Union, June 1995).

58. Lydia A. Savage, "Geographies of Organizing: Justice for Janitors in Los Angeles," in *Organizing the Landscape: Geographical Perspectives on Labor Unionism*, edited by Andrew Herod (Minneapolis: University of Minnesota Press, 1998), 225–252; Richard W. Hurd and William Rouse, "Progressive Union Organizing: The SEIU Justice for Janitors Campaign," *Review of Radical Political Economics* 21, no. 3 (Fall 1989): 70–75; "Taking It to the Streets: Justice for Janitors Causes a Dust-Up. But Are They Heroes or Hooligans," Mary Ann French, *Washington Post*, April 14, 1995; see also "Traffic Disruption Campaign by 'Justice for Janitors,'" *Hearings before the District of Columbia Subcommittee of the Committee on Government Reform and Oversight*, House of Representatives, 104th Cong., 1st sess., October 6, 1995; John Cohen, "J4J Will Get You if You Don't Watch Out," *Forbes* 23 (October 1989): 390.

59. Roger Waldinger et al., "Justice for Janitors: Organizing in Difficult Times," *Dissent*, Winter 1997, 31–44; Richard Mines and Jeffrey Avina, "Immigrants and Labor Standards: The Case of California Janitors," in *U.S.-Mexico Relations: Labor Market Interdependence*, edited by Jorge Bustamante, Clark Reynolds, and Raul Hinojosa Ojeda (Stanford: Stanford University Press, 1992), 42–448; Roger Waldinger et al., "Helots No More: A Case Study of the Justice for Janitors Campaign in Los Angeles" (Los Angeles: UCLA School of Public Policy, 1996).

60. Harold Meyerson and Joseph Trevino, "How the Janitors Changed Los Angeles," *L.A. Weekly*, April 21–27, 2000, 13–14; Nancy Cleeland, "Janitors' Quest Complicated by Shifting Nature of the Job," *Los Angeles Times*, April 7, 2000; Nancy Cleeland, "Organization, Commitment Power Strike," *Los Angeles Times*, April 13, 2000; Robin Wood, "Justice for Janitors Campaign Makes Inroads," *Cleaning and Maintenance Management*, October 1998; Sean Cummings, "And Justice for All? National Union's 'Justice for Janitors' Campaign Changes the Face of Private Contract Cleaning," *Cleaning and Maintenance Management*, December 1996.

61. Jed Emerson and Fay Twersky, editors, *New Social Entrepreneurs: The Success, Challenge and Lessons of Non-Profit Enterprise Creation* (San Francisco: The Roberts Foundation, 1996); Jed Emerson, "What Is a Social Entrepreneur," available on Pueblo Nuevo Web site, http://www.pueblonuevo.org;

Didacus Ramos and Nicole McAllister, "Pueblo Nuevo History," section of "BOS Drop-Off Reuse Center Business Plan," prepared for the Integrated Solid Waste Management Office, City of Los Angeles, 1996.

62. Susan Adams, "God Is His Business Planner," *Forbes*, July 27, 1998, 90–91; personal communications with Philip Lance, July 19, 1996; August 10, 1999.

63. Lance interview, August 10, 1999.

5   Global, Local, and Food Insecure: The Restructuring of the Food System

1. Eulalio Castellanos et al., Research Group on the Los Angeles Economy, *The Widening Divide: Income Inequality and Poverty in Los Angeles* (Los Angeles: UCLA Graduate School of Architecture and Urban Planning, 1989); Cynthia Pansing, Hali Rederer, and David Yale, "A Community at Risk: The Environmental Quality of Life in East Los Angeles," Graduate School of Architecture and Urban Planning, 1989; see also Edward W. Soja, Allan D. Heskin, and Marco Cenzatti, "Los Angeles: Through the Kaleidoscope of Urban Restructuring" UCLA Graduate School of Architecture and Urban Planning, 1985.

2. See, for example, *RLA Grocery Store Market Potential Study* (Los Angeles: Rebuild LA, October 1995).

3. Linda Ashman et al., *Seeds of Change: Strategies for Food Security for the Inner City* (Los Angeles: Interfaith Hunger Coalition, 1993).

4. Robert Chambers, *Sustainable Livelihoods: Environment and Development; Putting Rural People First*, IDS Discussion Bulletin No. 240 (Brighton, England: Institute for Development Studies, 1988); see also Amartya Sen, "The Political Economy of Hunger: On Reasoning and Participation," paper presented at the Conference on "Overcoming Global Hunger," Washington, D.C., the World Bank, December 1, 1993.

5. Harriet Friedmann, "After Midas's Feast: Alternative Food Regimes for the Future," in *Food for the Future: Conditions and Contradictions of Sustainability*, edited by Patricia Allen (New York: John Wiley & Sons, 1993), 213–233; see also Harriet Friedmann, "The Political Economy of Food: A Global Crisis," *New Left Review* 197 (1997) 29–57; See also Robert Gottlieb and Andrew Fisher, "First Feed the Face: Environmental Justice and Community Food Security, "*Antipode* 28, no. 2 (1996): 193–203.

6. Ashman, *Seeds of Change*, 167–173.

7. One recent study argued that while comparative prices in low-income and middle-income supermarkets might be slightly higher in low-income communities, the poor tend to focus on maximizing purchases in terms of price and thus may not pay more overall. However, those comparisons need to be put in context—the poor ultimately have to pay a far higher percentage of their overall income for food than middle-income consumers—more than three times the per-

centage according to the *Seeds of Change* study (161–165). See also Philip R. Kaufman et al., *Do the Poor Pay More for Food: Item Selection and Price Differences Affect Low-Income Household Food Costs*, USDA Agricultural Economic Report no. 759, Washington, D.C., November 1997; *The Poor Pay More: Food Shopping in Hartford* (Hartford, Corn.: Citizen's Research Education Network, February 1984); Judith Bell and Bonnie M. Burlin, "In Urban Areas: Many of the Poor Still Pay More for Food," *Journal of Public Policy and Marketing* 12, no. 2 (Fall 1993): 268–270; James M. McDonald and Paul E. Nelson Jr., "Do the Poor Still Pay More? Food Price Variations in Large Metropolitan Areas," *Journal of Urban Economics* 30 (1991): 344–359.

8. Charlotte Neumann et al., *Prevalence of Hunger and Malnutrition Among Los Angeles Elementary School Children*, UCLA School of Public Health, A Report for the Los Angeles Unified School District, 2000.

9. The concept of "transit dependent" has been defined as an area where car ownership is less than 80 percent of the population in a given census tract. See Robert Gottlieb and Andrew Fisher, "Food-Related Transportation Strategies in Low-Income and Transit-Dependent Communities," University of California Transportation Center, Working Paper no. 957, Berkeley, 1996.

10. The "just how far the scientific management point of view" quote is from Edwin Nourse, "The Apparent Trend of Recent Economic Changes in Agriculture," *Annals of the American Academy* 144 (May 1930): 46–47.

11. The "diminishing income for farm workers" quote is from Paul Taylor's testimony before a congressional subcommittee and is presented in Richard Lowitt, *The New Deal and the West* (Bloomington: Indiana University Press, 1984), 184; Carey McWilliams, *Factories in the Field: The Story of Migratory Labor in California* (Santa Barbara: Peregrine Publishers, 1971); see also Steven Stoll, *The Fruits of Natural Advantage: Making the Industrial Countryside in California* (Berkeley: University of California Press, 1998).

12. The National Research Council report, for example, pointed out that while the total acres under cultivation between 1964 and 1982 remained relatively constant, the total pounds of pesticide active ingredients increased 170 percent. Board on Agriculture, National Research Council, *Alternative Agriculture: Committee on the Role of Alternative Farming Methods in Modern Production Agriculture* (Washington, D.C.: National Academy Press, 1989), 54, 85, 44; the increase in pesticide use fortyfold is identified in U.S. Department of Agriculture (USDA) Study Team on Organic Farming, *Report and Recommendations on Organic Farming* (Washington, D.C.: USDA, July 1980), 62.

13. "If we define *agriculture* broadly to include the provision of inputs and the distribution of outputs," sociologist Frederick Buttel argued, "*farming* (or food raising) has become a small component of agriculture." Frederick H. Buttel, "Social Relations and the Growth of Modern Agriculture," in *Agroecology*, edited by C. Ronald Carroll, John Vandermeer, and Peter Rosset (New York: McGraw-Hill, 1990), 117; see also Laurian J. Unneveher, "Suburban Consumers and Exurban Farmers: The Changing Political Economy of Food Policy,"

American Journal of Agricultural Economics 75 (December 1993): 1140–1144; Stanley R. Johnson and Sheila A. Martin, "Industrial Policy for Agriculture and Rural Development," in *Industrial Policy for Agriculture in the Global Economy*, edited by Stanley R. Johnson and Sheila A. Martin (Ames, Iowa: Card Publications/Iowa State University, 1993), 6; the shift toward manfactured and processed food items is described by Bruce W. Marion, *The Organization and Performance of the U.S. Food System* (Lexington, Mass.: Lexington Books, 1986), 205; the end of farming in the U.S. scenario is described by University of California, Davis, economist Steven Blank in *The End of Agriculture in the American Portfolio* (Westport: Quorum Books, 1998).

14. The quote is from Frederick H. Buttel's "Social Relations and the Growth of Modern Agriculture" essay, 118; see also Martin Kreisberg, "Miracle Seeds and Market Economics," *Columbia Journal of World Business*, March/April 1969; Frances Moore Lappe, *Diet for a Small Planet* (New York: Ballantine Books, 1975); Dan Morgan, *Merchants of Grain* (New York: Viking Press, 1979).

15. The term "agribusiness" was increasingly used during the 1950s and 1960s to describe how agriculture had become intertwined with its various suppliers, such as pesticide manufacturers, and reflected the totality "of all operations performed in connection with the handling, storage, processing, and distribution of farm commodities." This quote is from a 1956 article in the *Harvard Business Review* by former USDA Assistant Secretary John Davis who introduced the term "agribusiness." Davis also argued that new forms of vertical integration associated with the growing of food needed to supplant the role of government in influencing growing decisions. "In the dynamic era ahead," Davis wrote, "the term 'farm problem' will become a misnomer; farm problems will be recognized as business problems and vice versa. More precisely, farm problems will be 'agribusiness problems. Therefore, we must solve them through the agribusiness approach." John H. Davis, "From Agriculture to *Agribusiness*," *Harvard Business Review*, January–February 1956, 109, 115; the concept of the "agribusiness market economy" is described in H.F. Breimeyer, "The Three Economies of Agriculture," *Journal of Farm Economics* 44 (August 1962): 679–689; see also James Rhodes, "Industrialization of Agriculture: Discussion," *American Journal of Agricultural Economics* 75 (December 1993): 1137–1139; Michael Boehlje, "Industrialization of Agriculture: What are the Implications?," *Choices*, 11, no. 1, 1996, 30–33; one illustration of the reach of agribusiness as the domain of the input providers and middle players is how the success of the large vegetable growers became increasingly dependent on their ability to deal with financial institutions, processors, and food distributors. "The increasing scale of vegetable farming," economist Phil Leveen wrote in the late 1970s, "derives not from their more efficient use of resources over the smaller farm, but because of their ability to relate to the needs of the increasingly concentrated set of institutions called agribusiness." E. Phillip LeVeen, "The Prospects for Small-Scale Farming in an Industrial Society: A Critical Appraisal of *Small is Beautiful*," in *Appropriate Visions: Technology, the Environment and the Individual*, edited by Richard C.

Dorf and Yvonne L. Hunter (San Francisco: Boyd and Fraser Publishing Co., 1978), 118.

16. The term "middlemen" was used more often by farm organizations in the early part of the twentieth century in referring to the various input providers. However, my use of the term "middle players" seeks to broaden the category to include what William Heffernan has described as the range of players and stages within a food system rather than a concept that focuses on influences in relation to food growing. Put another way, the middle players occupy the space between food grown and food consumed. As part of this analysis, however, I've separated the food retail sector from other middle players due to its direct impact on key community food security considerations such as access that are also addressed in this book. See William D. Heffernan, "Agriculture and Monopoly Capital," *Monthly Review* 50, no. 3 (July–August 1998): 46–59.

17. A.V. Krebs, *The Corporate Reapers: The Book of Agribusiness* (Washington, D.C.: Essential Books, 1992), 308–310; Marion, *The Organization and Performance of the U.S. Food System*, 126–152; see also *Technology, Public Policy, and the Changing Structure of Public Policy*, U.S. Congress, Office of Technology Assessment (Washington, D.C.: OTA, March 1986), OTA-F-285, p. 99.

18. William D. Heffernan, "Concentration of Agricultural Markets," Department of Rural Sociology, University of Missouri, October 1997.

19. William D. Heffernan, "Constraints in the U.S. Poultry Industry," *Research in Rural Sociology and Development* 1 (1984): 238–239; Ruth Simon, "Will the Turkey Fly This Time?" *Forbes*, May 18, 1987, 210–212; Steven Greenhouse, "Priest vs. 'Big Chicken' in Fight for Labor Rights," *New York Times*, October 6, 1999.

20. Jimmy M. Skaggs, *Prime Cut: Livestock Raising and Meatpacking in the United States, 1607–1983* (College Station: Texas A&M University Press, 1986), 184–187.

21. Marion, *The Organization and Performance of the U.S. Food System*, 134; see also Michael Boehlje, "Industrialization of Agriculture: What Are the Implications?" *Choices* 11, First Quarter, 1996, 30–33; C. Hurt, "Industrialization in the Pork Industry," *Choices* 9, Fourth Quarter, 1994, 9–13; Stephanie Simon, "Heartland Riven over a Corporate Hog," *Los Angeles Times*, March 2, 1999; Richard N.L. Andrews, *Managing the Environment, Managing Ourselves* (New Haven: Yale University Press, 1999), 308; Krebs, *The Corporate Reapers*, 378–380; Jane Rissler, "Bright Lights, Pig City," *Nucleus*, Spring 1999, 8–9; Nancy Thompson, "Spotlight on Pork III," Center for Rural Affairs, Walthill, Nebraska, 1998; Peter T. Kilborn, "Storm Highlights Flaws in Farm Law in North Carolina," *New York Times*, October 17, 1999.

22. "The Five Large Packers in Produce and Grocery," in *Report of the Federal Trade Commission on the Meat Packing Industry* (Washington, D.C.: GPO, 1920), 29.

23. Richard J. Arnould, "Changing Patterns of Concentration in American Meat Packing, 1880–1963," *Business History Review* 45, no. 1 (Spring 1971): 18–34.

24. Krebs, *The Corporate Reapers*, 370; Skaggs, *Prime Cut*, 178–182.

25. The ConAgra official (Charles Monfort) is quoted in Bill Jackson, "Monfort + Soviets = Trade," *Greeley Tribune*, September 29, 1991, in Carol Andreas, "Monfort's Disposable Meatpackers," *Covert Action 50* (Fall 1994): 62; see also A.V. Krebs, "America's New 'Centrally Planned' Food Economy," *Prairie Journal*, Summer 1992, 6; Bruce Marion is cited in Lewie Anderson, *Return to the Jungle: An Examination of Concentration of Power in Meat Packing* (Gaithersburg, Md.: United Food and Commercial Workers, 1989) and is quoted in Carol Andreas, *Meatpackers and Beef Barons: Company Town in a Global Economy* (Niwot, Colo.: University Press of Colorado, 1994), 134; see also Lourdes Gouveia, "Global Strategies and Local Linkages: The Case of the U.S. Meatpacking Industry," in *From Columbus to ConAgra: The Globalization of Agriculture and Food*, edited by Alessandro Bonanno et al. (Lawrence, Kansas: University Press of Kansas, 1994), 125–148.

26. A.V. Krebs describes one example of the workforce shift when the Armour Company, whose roots went back to Upton Sinclair's novel, *The Jungle*, was sold by its corporate parent, the Greyhound Corporation, to ConAgra, one of the big three in beef that consolidated its position in the 1980s. Prior to the sale, Greyhound had shut down thirteen of the Armour plants that had employed 1,500 workers with an average wage of $10/hour. ConAgra subsequently reopened twelve of those plants, but only employed 900 workers, with a starting pay of $6/hr. A.V. Krebs, *Heading Toward the Last Roundup: The Big Three's Prime Cut* (Washington, D.C.: Corporate Agribusiness Project, 1990), 52; see also Kathleen Stanley, "Industrial and Labor Market Transformation in the U.S. Meatpacking Industry," in *The Global Restructuring of Agro-Food Systems*, edited by Philip McMichael (Ithaca: Cornell University Press, 1994), 129–144; Carol Andreas, *Meatpackers and Beef Barons*, 3; Nancy Cleeland, "For Meatpackers, Walkout Was Step Forward and Back," *Los Angeles Times*, July 9, 1999; Marla Cone, "State Dairy Farms Try to Clean Up Their Act," *Los Angeles Times*, April 28, 1998.

27. Several analysts have pointed out that the 1840s Irish Potato Famine reflected an earlier loss of diversity—only two varieties were grown by the Irish, thus making the product vulnerable to late blight disease, which in turn triggered the famine. See Cary Fowler and Pat Mooney, *Shattering: Food, Politics, and the Loss of Genetic Diversity* (Tucson: University of Arizona Press, 1990); Paul Raeburn, *The Last Harvest: The Genetic Gamble That Threatens to Destroy American Agriculture* (New York: Simon & Schuster, 1995); Joan Dye Gussow, "Dietary Guidelines for Sustainability—12 Years Later," keynote address, Society for Nutrition Education Annual Meeting, Albuquerque, New Mexico, July 19, 1998; the National Research Council report, *Genetic Vulnerability of Major Crops*, was published in 1972, Washington, D.C., National Research Council (U.S.), Committee on Genetic Vulnerability of Major Crops, 1; see also Susan

George in *Ill Fares the Land: Essays on Food, Hunger and Power* (Washington, D.C.: Institute for Policy Studies, 1984).

28. The McDonald's example is also noteworthy in its influence in further centralizing and globalizing the food processing sector as well. As McDonald's expanded and globalized in the 1960s and 1970s, it shifted from as many as 175 local produce suppliers to one primary supplier, the Simplot Company, which also reorganized its own activities to emphasize the manufacture of frozen french fries. Simplot in turn became the largest potato processor in the United States, while frozen french fries, which represented just 2 percent of the market in the late 1950s expanded to as much as 25 percent of the market by the 1980s, with McDonald's alone accounting for about one-fourth of total frozen french fry production. At the same time, the fast-food giant also began to change both the growing and processing of food through its supplier contracts in countries where it had also established a presence. See John Love, *McDonald's: Behind the Arches* (New York: Bantam Books, 1995), 324, 332–333; 441.

29. William D. Heffernan's presidential address at the 51st meeting of the Rural Sociological Society, "Confidence and Courage in the Next Fifty Years," *Rural Sociology* 54 no. 2 (Summer 1989): 149–168.

30. "The Fabulous Market for Food," *Fortune*, October 1953, 271.

31. Marion, *The Organization and Performance of the U.S. Food System*, 250, 274–275; Cathleen Ferraro, " 'Slotting Fee' Irks Produce Suppliers: Ralphs Wants Payment for Use of Shelf Space," *Sacramento Bee*, February 5, 2000; Robin Fields and Melinda Fulmer, "Markets' Shelf Fees Put Squeeze on Small Firms," *Los Angeles Times*, January 29, 2000.

32. This "relentless pursuit of convenience food items," as the Breakfast Mates episode indicated, had an almost surreal quality to it. The *New York Times*, in its coverage of the introduction of this product, noted that the amount of time "saved" meant an average of thirteen seconds rather than fourteen seconds for the traditional cereal and milk pouring method. This was calculated in terms of the amount of time required to get a serving bowl and spoon, take out the cereal, pour it and the milk, and then begin eating. Since the additional packaging required for Breakfast Mates increased the unit price per serving, the *Times* noted that the one second savings represented a cost of $1 more per second. The *Times*, however, didn't seek to evaluate either the environmental or nutritional impacts of the new method. Dirk Johnson, "Got-Milk Cereals Get Mixed Reviews," *New York Times*, January 17, 1999.

33. By 1997, food and beverage sales for Philip Morris accounted for more than $31 billion, positioning it only second to the Nestlé Company among the world's top food and beverage companies. Seymour Cooke Food Research International, cited by Rural Advancement Foundation International, *RAFI Communique*, March/April 1999; see also Richard Kluger, *Ashes to Ashes: America's Hundred-Year Cigarette War, the Public Health, and the Unabashed Triumph of Philip Morris* (New York: Alfred A. Knopf, 1996); the energy intensive nature of "value-added" processed foods is described in Dr. Martin Okos et al., *Energy*

Usage in the Food Industry (Washington, D.C.: American Council for an Energy Efficient Economy, October 1998); Joan Gussow's comment about the "excessive processing of food" is from her essay, "Food Security in the United States: A Nutritionist's View," in *Food Security in the United States*, edited by Lawrence Busch and William B. Lacy (Boulder: Westview Press, 1984), 225.

34. Russell C. Parker, *Concentration, Integration, and Diversification in the Grocery Retailing Industry* (Washington, D.C.: Bureau of Economics and Trade Commission, 1986), 23–24; W.P. Hedden, *How Great Cities Are Fed* (Boston: D.C. Heath & Co., 1929), 192.

35. Marion, *The Organization and Performance of the U.S. Food System*, 294; also, Bruce Marion, "Concentration-Relationship in Food Retailing," in *Concentration and Price*, edited by Leonard W. Weiss (Cambridge: MIT Press, 1989), 183–194.

36. Alden C. Manchester, "The Transformation of U.S. Food Marketing," in *Food and Agricultural Markets: The Quiet Revolution*, edited by Lyle P. Schertz and Lynn M. Daft, Economic Research Service (Washington, D.C.: U.S. Department of Agriculture, National Planning Association, 1994), 7–18; Robert Gottlieb and Andy Fisher, *Homeward Bound: Food Related Transportation Strategies in Low Income and Transit Dependent Communities* (Los Angeles: Community Food Security Coalition, 1996).

37. Youngbin Lee Yim, *Spatial Trips and Spatial Distribution of Food Stores*, University of California Transportation Center. Working Paper no. 125. Berkeley, CA: UCTC; 1993.

38. Mark Dohan, "An Analysis of Supermarkets in Los Angeles County" (Los Angeles: Pollution Prevention Education and Research Project and Occidental Community Food Security Project, 1994); Kami Pothukuchi, "Supermarket Development in the Inner-city: A Research Report," Detroit: Wayne State University, 1999; *Food Review* 18, no. 2 (1996): 4; "Briefly Noted," *Nutrition Week* 27, no. 29 (1996): 7; Youngbin Lee Yim, "The Relationship Between Transportation Services and Urban Activities: The Food Retail Distribution Case" (Ph.D. diss., University of California at Berkeley, 1990).

39. By the 1990s, supermarket firms were under pressure to reinvest in the inner cities. While acknowledging that a potentially lucrative market might be available, supermarket executives still complained that the land simply wasn't available to build the parking lots required for the 60,000 square feet and more formats that were now dominant. But parking requirements had no direct relationship to the percentage of automobile ownership in the community or the shorter distances from homes to store. Thus, the problem for the inner-city communities became compounded: first, abandonment by the supermarkets in the 1960s, 1970s, and 1980s. Then, renewed interest by the markets to reenter the inner city in the 1990s, but without any systematic initiative or strategic direction from either the chains or their trade organization to overcome the barriers that they themselves had helped to construct. The decline of inner-city supermarkets was also reinforced by the lack of diversity within the management struc-

ture—and therefore the decision making and orientation of those markets to identify with and reflect the needs of low-income communities. See Stuart Silverstein, "In Supermarkets' Executive Department, A Lack of Variety," *Los Angeles Times*, May 2, 1999; *No Place to Shop: Challenges and Opportunities Facing the Development of Supermarkets in Urban America* (Washington, D.C.: Public Voice for Food and Health Policy, February 1996); "Supermarket Initiatives in Underserved Communities" (Washington, D.C.: Food Marketing Institute, 1996); *Homeward Bound*, 21–35.

40. Similar to the changes in food retail, the restaurant business also evolved in low-income communities where fast-food outlets began to displace sit-down restaurants. One neighborhood food system mapping exercise undertaken by the L.A.-based Community Coalition Against Substance Abuse of a two-square-mile area in South Central Los Angeles identified fifty-three restaurants, fifty-two of which were fast-food outlets and only one sit-down restaurant. Presentation of Survey Findings, Project GROW, Community Coalition, Los Angeles, September 12, 1998.

41. Some of the merger activities even assumed a global dimension. One of the major players in the U.S. retail business was the Netherlands-based Ahold chain. This global entity, with $28 billion in sales and a presence in four continents, had initially purchased several chains in the eastern United States and then began acquiring new chains in other regions of the country as well. James Flanigan, "Grocery Mergers No Bargain for the Southland," *Los Angeles Times*, October 21, 1998; Melinda Fullmer, "Using Size to Their Advantage," *Los Angeles Times*, November 10, 1999, and Melinda Fullmer, "Wal-Mart Food Fight Leaves Big Rivals Standing," *Los Angeles Times*, November 27, 1999; Roberta Cook, "The Changing Structure of Produce Marketing," *Small Farm News*, September–October 1996, 5–7; in terms of the private label, the Kroger chain, for example, included more than 5,000 private-label items in twenty-seven plants, including bakeries, dairies, and beverage facilities, see James F. Peltz, "What Might Be in Store under Kroger," *Los Angeles Times*, October 21, 1998.

42. Janet Poppendieck, *Breadlines Knee Deep in Wheat: Food Assistance in the Great Depression* (New Brunswick: N.J.: Rutgers University Press, 1986).

43. Murray R. Benedict, *Farm Policies of the United States, 1790–1950: A Study of Their Origins and Development* (New York: Twentieth Century Fund, 1953); Wayne D. Rasmussen and Glady L. Baker, *Price Support and Adjustment Programs from 1933 through 1978: A Short History*, Agriculture Information Bulletin No. 424 (Washingto, D.C.: USDA, February 1979).

44. Greg Mitchell, *Campaign of the Century: Upton Sinclair's Race for Governor of California and the Birth of Media Politics* (New York: Random House, 1992); Miriam Buck Arozena, "Utopia 1934: A Personal Reminiscence," Los Angeles, 1999.

45. Ardith L. Maney, *Still Hungry After All These Years: Food Assistance Policy from Kennedy to Reagan* (Greenwood Press: Westport, Conn., 1989), 15–16;

Poppendieck cites Jerome Frank's comment about how American Farm Bureau Federation President Ed O'Neal characterized USDA's role regarding these issues during the New Deal: "What in the hell do we have to do with the consumer? This is the Department of Agriculture." *Breadlines Knee Deep in Wheat*, 226.

46. Michael Harrington, *The Other America: Poverty in the United States* (New York: Macmillan, 1962); Nick Kotz, *Let Them Eat Promises: The Politics of Hunger in the United States* (Englewood Cliffs, N.J.: Prentice Hall, 1969); Maney cites a memo from an economist on the staff of the Council of Economic Advisors who argued in 1963 that food assistance had to be a "politically acceptable program [that] must avoid completely any use of the term 'inequality' or the term 'redistribution of income or wealth.'" *Still Hungry After All These Years*, 48.

47. Along these lines, in 1967, the Field Foundation convened a team of doctors to study and document the widespread nature of hunger and malnutrition; a decade later, in 1977, the team revisited the same sites that had been studied earlier and pointed to a significant decrease in the amount of hunger that could be identified. The Field Foundation findings, as one of the clinicians involved later put it, served to demonstrate that "the hunger problem had been virtually eliminated." Physician Task Force on Hunger in America, *Hunger in America: The Growing Epidemic* (Middletown, Conn.: Wesleyan University Press; 1985), 3; see also Larry Brown, "Hunger in America," *Annual Public Health Reviews*, 1988.

48. The U.S. Conference of Mayors national survey is in their report, *Human Services in FY 82* (Washington, D.C.: U.S. Conference of Mayors, 1982); A study by a team of Cornell University researchers cited one woman who described hunger as "when there is nothing absolutely in the house." But hunger, she also argued, "is when you have to eat the same thing all week long and you have no variation from it and you know sooner or later you're going to run out of that too, because it's only gonna go so far." *Journal of Nutrition Education* 1992 Supplement; on the social safety net concept, see the interview with David Stockman in the *Washington Post*, December 4, 1983.

49. Janet Poppendieck points out that when the TEFAP name was changed from Temporary Emergency Food Assistance Program to The Emergency Food Assistance Program in 1990, the same acronym was kept but the "increasingly obsolete and embarrassing" term "temporary" was dropped. Janet Poppendieck, *Sweet Charity? Emergency Food and the End of Entitlement* (New York: Viking Penguin, 1998), 141, 88–89; on the background and development of the emergency food programs in the early 1980s, personal communication with Dennis Stewart, former director Commodity Programs, U.S. Department of Agriculture, August 16, 1996; see also James K. Morris, "The Federal Government's Golden Hoard," *Progressive*, October 1981, 17–18; Kathryn Porter Bishop, *Soup Lines and Food Baskets: A Survey of Increased Participation in Emergency Food Programs*, Washington, D.C.: Center for Budget and Policy Priorities, May 1983;

J. Larry Brown and H.F. Pizer, *Living Hungry in America* (New York: Macmillan, 1987); Janet Poppendieck, "Dilemmas of Emergency Food: A Guide for the Perplexed," *Agriculture and Human Values,* 1994, 69–76; *Eliminating Hunger: Food Security Policy for the 1990s* (Washington, D.C: Urban Institute, October 1989); U.S. Conference of Mayors, *Hunger in American Cities* (Washington, D.C.: U.S. Conference of Mayors, June 1983).

50. On the increase in demand, see for example *1999 Status Report on Hunger and Homelessness in America's Cities* (Washington, D.C.: U.S. Conference of Mayors, 1999); in terms of the issue of supplies, the Los Angeles Regional Food Bank, as one example, saw its supplies for distribution drop in one year by as much as one-third, from 33.8 million pounds in 1994 to 22.6 million pounds in 1995, related in part to the development of a more active "seconds" market in damaged supermarket goods, and thus a decrease in supermarket donations. The creation of this secondary market (e.g., brokers who buy salvaged goods and manufacturer rejects and then sell them at flea markets, discount stores, and even in Eastern Europe) in fact took off significantly during the early 1990s. This in turn extended the problem of increased demand and decreased supply. Liz Spayd, "Food Banks Are Going Hungry," *Washington Post,* April 17, 1994; the L.A. Regional Food Bank figures were provided by Doris Bloch, executive director, Los Angeles Regional Food Bank, personal communication, August 16, 1996.

51. For example, the Seeds of Change study noted that low-income residents in South Central Los Angeles spent far more of their income on food purchases (36 percent) than comparable middle-income community members (12 percent). Meanwhile the prices paid by a family of three for a minimal food basket based on the government's Thrifty Food Plan were $285 higher per year than the prices paid for the same items in a comparable food market located in a middle-income area. Other studies also reached similar conclusions about the inability of the emergency food system to meet increased demand. Seeds of Change (161–165).

52. A survey by the Bread for the World Institute, for example, indicated that two-thirds of the food banks surveyed had no budget for lobbying and only four of seventy-one had more than 2 percent of their budget earmarked for that type of public role. Although many of the emergency food providers indicated that they'd like to develop an advocacy role, the constraints in funding and institutional support as well as donor sources such as Philip Morris limited their choices. The rise of an anti-hunger/emergency food discourse has also had direct political consequence, as Poppendieck points out. "Perhaps emergency food, with its ubiquitous collection mechanisms for food and money, its high visibility, its sporting events and canned good drives and checkout counter options, has reassured the *public* that no one will starve, and thus given Congress and the president tacit permission, if not enthusiastic support to dismantle the federal guarantee of minimum support for those in need and the newly empowered state governments room to experiment with a variety of time limits, family caps, and other reductions." Janet Poppendieck, *Sweet Charity,* 299 and 137; on the

reluctance to criticize donor sources such as Philip Morris, see Emily Green, "L.A.'s First Lady of Hunger Relief," *Los Angeles Times*, March 22, 2000; the Bread for the World survey is in its annual report, Bread for the World Institute, "Hunger 1994: Transforming the Politics of Hunger, Fourth Annual Report on the State of World Hunger," Silver Spring, Md., 1993, cited in *Sweet Charity*, 271; the tension between a public as opposed to charitable approach to hunger, poverty, and food security was present in the Progressive Era as well, distinguishing between advocates for social change and those focused on individual need. See John Spargo, *The Bitter Cry of the Children* (Chicago, Quadrangle Books, 1965), 61–65.

53. Gussow, "Food Security in the United States," 215.

54. The "sophisticated processing" quote is cited in William Serrin's article, "Let them Eat Junk," *Saturday Review*, February 2, 1980, 18; the "consumer hot button" quote is from Walter Kiechel III, "The Food Giants Struggle to Stay in Step with Consumers," *Fortune*, September 11, 1978, 50–56; see also Warren Belasco's important study, *Appetite for Change: How the Counterculture Took on the Food Industry* (Ithaca: Cornell University Press, 1993), that chronicled the alternative food movements of the 1960s and 1970s as well as the penetration of fast-food culture during this same period.

55. Along these lines concerning the fast-food undermining of local diets, Canadian sociologist Ester Reiter has written, in her analysis of the fast-food workplace, *Making Fast Food: From the Frying Pan into the Fryer* (Montreal: McGill-Queen's University Press, 1996), "Since the late 1960s, the map of Canada, once full of regional distinctions, has been transformed. Fast food restaurants of a handful of chains have colonized the suburbs from coast to coast, obliterating local differences in taste and style." P. 47; the "eateries the same everywhere" quote is from Gregory Hall, "The Psychology of Fast Food Happiness," in *Ronald Revisited: The World of Ronald McDonald*, edited by Marshall Fishwick (Bowling Green, Ohio: Bowling Green University Popular Press, 1983), 83.

56. Love, *McDonald's: Behind the Arches*, 11–12; David Gerald Orr, "The Ethnography of Big Mac," in *Ronald Revisited*, 61; Conrad P. Kottak, "Rituals at McDonald's" in *Ronald Revisited*, 55.

57. When McDonald's landed the franchise at the Arkansas high school, one of the several academic defenders of fast food praised the decision because of the "excellent concentration of high quality protein" in the McDonald's hamburger and the "source of vitamins in the french fries." Cited in "Big Mac Goes to School," *Newsweek*, October 4, 1976, 85.

58. The Newsweek article, "The Roadside Gourmet: Pop Goes the Food," is by Joseph Morgenstern, September 25, 1972, 76; see also Robert L. Emerson, *Fast Food: The Endless Shakeout* (New York: Lebhar-Friedman Books, 1982), 13; "Big Mac Strikes Back," *Time*, April 13, 1989, 58–60; "The Fast-Food Stars: Three Strategies for Fast Growth," *Business Week*, July 11, 1977; "The Fast

Food Furor," *Time*, April 21, 1975, 48–51; James E. Peltz, "Jack in the Box Springs Back," *Los Angeles Times*, October 23, 1999.

59. Theodore Levitt, "Production-line Approach to Service," *Harvard Business Review* 50, no. 5 (September–October 1972): 44; Bruce Lohof, "Hamburger Stand Industrialization and the Fast-Food Phenomenon," in *Ronald Revisited*, 16; the Kroc interview is in "Appealing to a Mass Market: Conversation with Ray Kroc," *Nation's Business*, July 1974, 74; see also Robin Leidner, *Fast Food, Fast Talk: Service Work and the Routinization of Everyday Life* (Berkeley: University of California Press, 1993); Love, *McDonald's: Behind the Arches*.

60. Whalen made this remark at a meeting of the National Soft Drink Association, cited in John Hess, "Harvard's Sugar-Pushing Nutritionist," *Saturday Review*, August 1978, 12.

61. Marion, *The Organization and Performance of the U.S. Food System*, 275; "Kids Head for Wrong Potatoes, Study Finds," *Los Angeles Times*, September 6, 1999.

62. James Gerstenzang, "3 U.S. Industries Leading the Way in Global Economy," *Los Angeles Times*, January 27, 1995.

63. Robert L. Emerson, *The New Economics of Fast Food* (New York: Van Nostrand Reinhold, 1990); Craig R. Whitney, "Protestors Just Say No to 'McDo,' Jospin Glad," *New York Times*, September 15, 1999; Suzanne Daley, "French See a Hero in War on 'McDomination,' " *New York Times*, October 12, 1999; John-Thor Dahlburg, "To Many French, Ugly American Is McDonald's," *Los Angeles Times*, April 22, 2000.

64. The Elizabeth Whelan quote can be found in her article, "The New Skinny on Snack Foods," in *Priorities for Long Life and Good Health* 8, no. 1 (1996), on the web site of Whelan's organization, the American Council on Science and Health (www.acsh.org/publications/olestra/index/html); the FDA approval is discussed in FDA's own press release announcing the approval, "FDA Approves Fat Substitute, Olestra," January 24, 1996, available at www.pharminfo.com/drugpr/olestra_pr.html; see also "Crunch Time for a Fake Fat: Will America Take the Bait?" Glenn Collins, *New York Times*, August 24, 1997.

65. The numbers on growth of GM crops can be found in Clive James, "Global Review of Commercialized Transgenic Crops," *ISAAA Brief No. 8*, The International Service for the Acquisition of Agri-biotech Applications (ISAAA), Ithaca, New York, 1999; Ismail Serageldin, "Biotechnology and Food Security in the 21st Century," *Science*, July 16, 1999, 387–389; Economic Research Service, U.S. Department of Agriculture, "Genetically Engineered Crops for Pest Management," at http://www.econ.ag.gov/whatsnew/issues/biotech.

66. The discussion of gene giants and food industry "clusters" is in William D. Heffernan, *Consolidation in the Food and Agriculture System*, A Report to the National Farmers Union, February 1999.

67. "The Suicide Seeds," Jeffrey Kluger, *Time*, February 1, 1999, 44–45.

68. Monsanto in fact in 1997 re-identified itself as a "life sciences" rather than chemical company. The Monsanto statement is on their Web page at http://www.monsanto.com/ag/_asp/monsanto.asp; the "feed the world" argument has been often used by Monsanto and other biotech companies, particularly in criticizing the low-input strategies of sustainable agriculture. "I can assure you that people won't be fed if minimal agricultural inputs are used, if we try to return to some kind of cute 1890s farm practices," Monsanto's Richard Mahoney said in a 1990 interview. "Somehow people think we can dreamily return to those practices. There's no new technology being proposed in these minimum-agriculture schemes. They don't work." "Taking the Initiative on the Environment: Richard Mahoney of Monsanto," *Institutional Investor* 24, no. 16 (December 1990): 42; see also Jennifer A. Thomson and Zhang-Liang Chen, "Biotech Isn't a Luxury in Some Nations," *Los Angeles Times*, April 10, 2000; the "substantial equivalence" controversy is discussed in Erik Millstone, Eric Brunner, and Sue Mayer, "Beyond 'Substantial Equivalence,'" *Nature* 401 (October 7, 1999): 525–526; see also Economic Research Service USDA, "Value-Enhanced Crops: Biotechnology's Next Stage," *Agricultural Outlook*, March 1999; Martina McGloughlin, "Without Biotechnology, We'll Starve," *Los Angeles Times*, November 1, 1999.

69. See "Table 1: Commercialized Biotechnologies in the U.S.," Gerald Middendorf et al., "New Agricultural Biotechnologies: The Struggle for Democratic Choice," *Monthly Review*, July–August 1998, 89.

70. Goodman and Redclift argued, in anticipating the emergence of the gene giants, that the multinationals like Monsanto recognized during the 1980s the logic of combining genetic research with conventional plant breeding capacity and international marketing networks. "The genetic code," they predicted in 1991, "can now be manipulated and nature refashioned according to then logic of the marketplace." David Goodman and Michael Redclift, *Refashioning Nature: Food, Ecology, and Culture* (London and New York: Routledge, 1991), 167; see also Brian Halwell, "The Emperor's New Crops," *Worldwatch*, July–August 1999, 21–29.

71. Jim Hightower, "Hard Tomatoes, Hard Times: The Failure of the Land Grant College Complex," reprinted in *Radical Agriculture*, edited by Richard Merrill (New York: New York University Press, 1976), 87–110; the FlavrSavr tomato experienced a series of problems, since the tomatoes tended to be soft and bruised and had trouble being marketed as "fresh produce." R. King, "Low-Tech Woe Slows Calgene's Super Tomato," *Wall Street Journal*, April 11, 1996, cited in Luke Anderson, *Genetic Engineering, Food, and Our Environment* (White River Junction, Vt.: Chelsea Green Publishing Company, 1999), 44–45.

72. In fact, the cross-pollination issue had already surfaced by 1998, when a Wisconsin organic food company was subject to nearly $150,000 in losses on its European sales after tests in Switzerland detected genetically engineered Bt corn in its tortilla chips, a problem the company feared could be traced to the

cross pollination from a Novartis Bt corn that had affected the corn grown by one of its organic farmers. Martha Groves, "Groups to Sue EPA Over Risks of Using Bt Insecticide," *Los Angeles Times*, February 18, 1999; Michael Pollan's quote is from his article, "Playing God in the Garden," *New York Times Magazine*, October 25, 1998, 48; see also Deepak Saxena, Saul Flores, and G. Stotsky, "Insecticidal Toxin in Root Exudates from *Bt* Corn," *Nature* 402 (December 2, 1999): 480.

73. Monsanto had in fact sold, by 1998, over 100 million doses of rBGH, at $5.80 a dose, at a huge markup for itself. Martha Groves, "Canada Rejects Hormone that Boosts Cows' Milk Output," *Los Angeles Times*, January 15, 1999.

74. "Milk Controversy Spills Into Canada," *Rachel's Environment and Health Weekly*, no. 621 (October 22, 1998); Wayne Roberts, Rod MacRae, and Lori Stahlbrand, *Real Food For a Change* (Canada: Random House of Canada, 1999), 126–130.

75. Maria Margaronis, "The Politics of Food," *The Nation* 269, no. 22 (December 27, 1999): 11–16; see, for example, the Web site of the "Students for Responsible Research" (http://www.cnr.berkeley.edu/~reynolds/srr/Default.htm) for the debates over bio-tech funding at the University of California at Berkeley.

76. Henry I. Miller, "A Rational Approach to Labeling Biotech-Derived Foods," *Science* 284 (May 28, 1999): 1471–1472; Mare Lappé and Tom Hayden, "Label Genetically Altered Food," *Los Angeles Times*, April 9, 2000. Ronnie Cummins and Ben Lilliston, "The Rise and Fall of 'Franken-food,'" *Earth Island Journal* 14, no. 4 (Winter 1999–2000): 30–31; Anita Manning, "Genetics' Growing Bounty Reaps Fears for Future," *USA Today*, June 29, 1999; "Biotech: The Pendulum Swings Back," *Rachel's Environment and Health Weekly*, no. 649 (May 6, 1999); *Making Consumers Sovereign: How to Change Food Information Systems so Food Shoppers Are the Informed Consumers Governments and Businesses Say They Should Be*, Toronto Food Policy Council, Paper Series #9, September 1998, 28; as opposition in Europe mounted, even Tony Blair, in a commentary in *The Independent*, declared that "there's no doubt that there is potential for harm, both in terms of human safety, and in the diversity of our environment, from GM foods." Cited in Brian Halwell, "Politically Modified Foods," *Worldwatch*, 13, no. 3, May–June 2, 2000.

77. Brewster Kneen, "Restructuring Food for Corporate Profit: The Corporate Genetics of Cargill and Monsanto," *Agriculture and Human Values* 16, no. 2 (June 1999): 164; M. Arax and J. Brokaw, "No Way Around Roundup: Monsanto's Bio-engineered Seeds Are Designed to Rrequire More of the Company's Herbicide," *Texas InfiNet*, January 30, 1997; Jean-Pierre Berlan and Richard C. Lewontin, "Operation Terminator: Cashing in on Life," *Le Monde Diplomatique*, December 1998.

78. The Shapiro quote is cited in David Barboza, "After Deal of 2 Giants, Shares Plunge," *New York Times*, December 21, 1999; see also Barnaby J. Feder, "Monsanto Says It Won't Market Infertile Seeds," *New York Times*, October 5, 1999;

on the European and international opposition, see George Gaskell et al., "Worlds Apart?: The Reception of Genetically Modified Foods in Europe and the U.S." *Science* 285 (July 16, 1999): 384–386; Craig R. Whitney, "Europe Loses its Appetite for High-Tech Food," *New York Times*, June 27, 1999; John Lanchester, "Mad Coke Disease," *New York Times*, July 4, 1999; Paul Jacobs, "Protest May Mow Down Trend to Alter Crops," *Los Angeles Times*, October 5, 1999; Sonni Efron, "Japanese Choke on American Biofood," *Los Angeles Times*, March 14, 1999; John-Thor Dahlburg, "Amid Food Scares, Europe Seeks Better Way," *Los Angeles Times*, June 18, 1999.

79. On the Seattle events, see, for example, Ronnie Cummins's commentary, "After Seattle: Has Global Opposition Killed Biotech?" in *BioDemocracy News* no. 23 (December 1999); on the opposition and turmoil regarding Monsanto and the biotech investments of the gene giants, see Lucette Lagnado, "Raising the Anti: For Those Fighting Biotech Crops, Santa Came Early," *Wall Street Journal*, December 14, 1999; Scott Kilman, "Monsanto's Biotech Spud is being Pulled from the Fryer at Fast-Food Chain," *Wall Street Journal*, April 28, 2000; John Stauber, "Food Fight Comes to America," *The Nation* 269, no. 22 (December 27, 1999): 18–19.

80. Peter Vanderwicken, "P&G's Secret Ingredient," *Fortune*, July 1974, 74–79ff.; Jim Hightower, *Eat Your Heart Out: Food Profiteering in America* (New York: Crown Publishers, 1975), 55; Harriet Friedmann, "Distance and Durability: Shaky Foundations of the World Food Economy," *Third World Quarterly* 13, no. 2 (1992): 371–383; *Paradox of Plenty*, 197; Friedmann's conceptual argument about distance (and durability)—where the constituents of a food product come from, how food is processed or constructed, and where and how food is consumed (a version of the food product life cycle)—also has, in life-cycle terms, powerful energy implications that can serve as a proxy for the environmental consequences of these food system changes. For example, a German study of strawberry yogurt located in a Stuttgart supermarket identified that a truckload of 150 grams of the yogurt ultimately traveled 1,005 kilometers. The strawberries came from Poland, the yogurt from northern Germany, corn and wheat flour from the Netherlands, jam from West Germany and sugar beet from East Germany. Only the milk and the glass jar were from Stuttgart itself. To purchase one truckload of the yogurt that had been delivered, the study estimated that 10,000 liters of diesel fuel were required. Cited in Tim Lang, "Dietary Impact of the Globalization of Food Trade," *IFG News* 3 (Summer 1998): 10–11; see also Richard Douthwaite, *Short Circuit: Strengthening Local Economics for Security in an Unstable World*, Dublin, Ireland: The Lilliput Press, 1996, 282.

81. Harriet Friedmann, "After Midas's Feast," 227–229; the foodshed concept was first introduced in the 1920s as signifying the "homogeneity of a producing territory," W.P. Hedden, *How Great Cities Are Fed*, 20; Boston: D.C. Heath and Company, 1929; see also Jack Kloppenburg Jr., John Hendrickson, and G.W. Stevenson, "Coming in to the Foodshed," in *Rooted in the Land: Essays on Community and Place*, edited by William Vitek and Wes Jackson (New Haven: Yale

University Press, 1996), 113–123; on seasonal diets, see Jennifer Wilkins, "Seasonal and Local Diets: Consumers' Role in Achieving a Sustainable Food System," *Research in Rural Sociology and Development* 6 (1995): 149–166.

6 The Politics of Food: Agendas and Movements for Change

1. The idea for the Chicago meeting had first been explored by myself, Andy Fisher of the *Seeds of Change* group, and Mark Winne of the Hartford Food System at Richard Riordan's Pantry restaurant in downtown Los Angeles. Winne, whose Hartford Food System was then the most established of the community food security groups, was particularly interested in making community food security more visible among the various food movements. See Andy Fisher, "Community Food Security: A Food Systems Approach to the 1995 Farm Bill and Beyond: A Policy Options Paper," presented to the Working Meeting on Community Food Security, Chicago, August 25, 1994.

2. Jeffrey A. Zinn and A. Berry Carr, "The 1985 Farm Act: Hitting a Moving Target," *Forum for Applied Research and Public Policy*, Summer 1988, 17–18; Ken Cook, "Pinch Me. I Must Be Dreaming," *Journal of Soil and Water Conservation*, March–April 1986; Center for Resource Economics, *1990 Farm Bill: Environmental and Consumer Provisions, Volume II: Detailed Summary* (Washington, D.C.: Island Press, 1991).

3. Garth Youngberg, Neill Scaller, and Kathleen Merrigan, "The Sustainable Agriculture Policy Agenda in the United States: Politics and Prospects," in *Food for the Future: Conditions and Contradictions of Sustainability*, edited by Patricia Allen (New York: John Wiley & Sons, 1993); David Ostendorf and Dixon Terry, "Toward a Democratic Community of Communities: Creating a New Future with Agriculture and Rural America," in *Environmental Justice: Issues, Policies, and Solutions*, edited by Bunyan Bryant (Washington, D.C.: Island Press, 1995), 157–158; Barbara Meister, "Analysis of Policy Options for Promoting Sustainable Rural Development in the 1995 Farm Bill" (master's thesis, JFK School of Government, Harvard University, 1994).

4. Valerie B. Straus, "The Farm Crisis of the 1980s and the Neopopulist Political Response," UCLA Graduate School of Architecture and Urban Planning, 1993; Andrew Fisher and Robert Gottlieb, "Community Food Security: Policies for a More Sustainable Food System in the Context of the 1995 Farm Bill and Beyond," Working Paper no. 11, Lewis Center for Regional Policy Studies, UCLA School of Public Policy, May 1995; Patricia Allen and Carolyn Sachs, "Sustainable Agriculture in the United States: Engagements, Silences, and Possibilities for Transformation," in *Food for the Future*, 1993.

5. On the Hartford Food System see Dawn Biehler, Melissa Sepos, and Mark Winne, *The Hartford Food System: A Guide to Developing Community Food Programs* (Hartford, Conn.: Hartford Food System, 1999).

6. Author's notes, Proceedings of the Community Food Security Coalition meeting, Philadelphia, February 9, 1996.

7. *The Community Food Security Empowerment Act* (Los Angeles: Community Food Security Coalition, January 1995); Mark Winne, *Food Security Planning: Toward a Federal Policy* (Hartford, Conn.: Hartford Food System, 1994); a number of the policy recommendations in the Community Food Security Coalition proposed legislation were subsequently incorporated into the USDA Community Food Security Initiative Action Plan released in August 1999 in advance of the USDA Community Food Security Summit in October 1999. See also the *USDA's Community Food Security Initiative Roll Call of Commitments* (Washington, D.C.: USDA, October 1999).

8. In just a few decades, the 1970 USDA publication had noted, a farmer had gone from providing his own inputs to buying "prodigious amounts of his production needs—fertilizers, formula feed, hybrid seeds, insecticides, herbicides, [and] tractor fuel" as well as such customer services as pesticide spraying. Such a change, USDA suggested, was both welcome and inevitable. *1970 USDA Yearbook* (Washington, D.C.: USDA, 1971); the "fairly burst with pride" quote is from "Emptying the Cornucopia," Catherine Lerza, *Food for People, Not for Profit: A Source Book on the Food Crisis*, edited by Catherine Lerza and Michael Jacobson (New York: Ballantine Books, 1975), 46; Hightower's quote is from his essay "Hard Tomatoes, Hard Times: The Failure of the Land Grant College Complex" (Washington, D.C.: Agriculture Accountability Project, 1972); see also Jim Hightower, "The Case for the Family Farmer," *Washington Monthly*, September 1973, reprinted in Lerza, *Food for People, Not for Profit*, 35–44.

9. The organic quote is from Jerome Goldstein, "Organic Force," in *Radical Agriculture*, edited by Richard Merrill (New York: New York University Press, 1976), 213; the "hip farmers" comment is used in a description of five rural communal farms in northeast Ontario that is described in the first issue of *Mother Earth News*, "Morning Glory Farm," January 1970.

10. The number of urban gardeners alone was estimated to be as many as 25 million in 1971, with some attracted to the chemical-free gardening strategies advocated by the Rodale publications. The "simpler, realer one-to-one" quote is from Gurney Norman, "The Organic Gardening Books," *The Last Whole Earth Catalogue*, 1971, San Francisco: Portola Institute, 1971, 50; Robert Rodale, "J.I. Rodale's Greatest Contribution," *Organic Gardening and Farming*, September 1971, 28–29; see also J.I. Rodale, "Why I Started Organic Gardening," *Organic Gardening and Farming*, September 1971, 35–39; "The New Consumers: Food Habits and the Basis of Choice," Tom R. Watkins, in *Sustainable Food Systems*, edited by Dietrich Knorr (Westport, Conn.: AVI Publishing Co., 1983).

11. Ostendorf and Terry in their "Toward a Democratic Community of Communities" essay (p. 170) cite North Dakota's state banks and state grain elevators and southern farm cooperatives as historical examples of this process of "creative disengagement."

12. Paul Barnett, *Stopping the Pesticide Treadmill* (Davis, Cal.: California Agrarian Action Project, 1981); see also *New Directions in Farm, Land, and Food Policies: A Time for State and Local Action*, 2d ed., The Agriculture Project, edited by Joe Belden et al. (Washington, D.C.: Conference on Alternative State and Local Policies, 1981), 13–78.

13. "Policy Implications of Alternative Agriculture," Mark E. Rushefsky, *Policy Studies Journal* 8, no. 5 (Spring 1980): 774.

14. See, for example, *Planting the Future: Developing an Agriculture That Sustains Land and Community*, edited by Elizabeth Bird, Gordon Bultena, and John Gardner, Center for Rural Affairs (Ames, Iowa: Iowa State University Press, 1995).

15. USDA Organic Food Regulations, proposed rule, published in the Federal Register, December 16, 1997, available at http://www.ams.usda.gov/nop/rule.htm.

16. See, for example, Ben Lilliston and Ronnie Cummins, "Organic vs. 'Organic': The Corruption of a Label," *The Ecologist* 28, no. 4 (July–August 1998).

17. Debi Barker, "U.S. National Organic Standards—the Battle Is Not Yet Over," *IFG News* 3, International Forum on Globalization (Summer 1998): 13–14; Comments on the rule are available at the USDA Web site (http://www.ams.usda.gov/nop/view.htm). Melinda Fulmer, "USDA Sets New Rules for Organic Foods," *Los Angeles Times*, March 8, 2000; "USDA Sets National Standards on Organics Food Labeling," *Food Service Director*, 13, no. 4, April 15, 2000, 15.

18. The "not a niche market anymore" quote is cited by Barnaby J. Feder, "Organic Farming, Seeking the Mainstream," *New York Times*, April 19, 2000; "More Buyers Asking: Got Milk Without Chemicals?" Kate Murphy, *New York Times*, August 1, 1999; "Organic Milk Pours into Mainstream," Melinda Fulmer, *Los Angeles Times*, July 24, 1999; Martha Groves, "USDA Allows 'Organic' Labels on Meat," *Los Angeles Times*, January 15, 1999; "California Organic Production Sales" in *Small Farm News*, Spring 1998, 6. In Europe, the late 1990's upsurge in protest against genetically modified foods and concerns over food safety that first erupted around the mad-cow disease episode in England produced increasing interest in organic production methods. Some European Union officials in fact predicted that organic products could grow from 1 percent to 10 percent of the market in just six years. Betsy Lordan, "Europe's Organic Growers Know the Good Life," *Monterey County Herald*, March 5, 1999.

19. "Organics on the Brink: Bonanza or Boondoggle?" Desmond Jolly, *Small Farm News*, Summer 1998, 1ff.; "Can an Organic Twinkie be Certified?" Joan Dye Gussow, in *For All Generations: Making World Agriculture More Sustainable*, edited by J. Patrick Madden and Scott Chaplowe, Glendale, CA: World Sustainable Agriculture Association, 1997, 143–153.

20. Gussow extended this argument in identifying the need for what she and Kate Clancy characterized as the pursuit of "sustainable diets," based on a consumption pattern cognizant of the way the food was grown in a manner that was "not wasteful of such finite resources as top soil, water, and fossil energy." "Dietary Guidelines for Sustainability," Joan Gussow and Katherine Clancy, *Journal of Nutrition Education* 18, no. 1 (1986): 2; see also Joan Gussow, "Can a Community Have a Food System?" in *Community Food Systems: Sustaining Farms and People in the Emerging Economy: Conference Proceedings* (Davis: University of California Sustainable Agriculture Research and Education Program, September 1997), 3–15; Fred Kirschenbaum, "On Becoming Lovers of the Soil," in *For All Generations*, 101–114.

21. See Jack Kittredge, "Community-Supported Agriculture: Rediscovering Community," in *Rooted in the Land: Essays on Community and Place*, edited by Willim Vitek and Wes Jackson (New Haven: Yale University Press, 1996), 253–260; Trauger Groh and Steven McFadden, *Farms of Tomorrow: Community Supported Farms, Farm Supported Communities* (Kimberton, Pa.: Biodynamic Farming and Gardening Association, 1990), and Trauger Groh and Steven McFadden, *Farms of Tomorrow Revisited* (Kimberton, Pa.: Biodynamic Farming and Gardening Association, 1997); R. Kelvin, *Community Supported Agriculture on the Urban Fringe: Case Study and Survey* (Kutztown, Pa.: Rodale Institute Research Center, 1994).

22. CSAs in the United States can be distinguished in part by the degree of participation in the actual farming activities. One type of CSA has been based on a type of "community farm" model (including some that have required members to undertake farm work as part of their share payment). At the other end has been the "subscription" model where the farm crew does all the farm and delivery work and members simply pay their fee and receive a weekly box. By the late 1990s most CSAs tended to develop as "subscription"-type programs. Elizabeth Henderson, "Community Supported Agriculture in North America," *Why* nos. 31–32 (Spring/Summer/Fall 1999): 14–16.

23. The "by entering contracts" quote is from the Prairieland CSA fact sheet, "What Is Community Supported Agriculture," on the Prairieland web site at http://www.prairienet.org/pcsa/factsht1.htm; the Community Supported Agriculture of North America group at the University of Massachusetts Extension estimates that by 1999 there were more than 1,000 CSA farms in the United States and Canada; see http://www.umass.edu/umext/csa/about.html.

24. See Michelle Mascarenhas, *Just Food: Organic Farming in Santa Cruz County* (Los Angeles: Occidental Community Food Security Project, 1997). Mascarenhas defines sustainable agriculture, in the context of her discussion of farmworkers and organic farming, as "agriculture that is environmentally responsible, economically viable, and socially just," p. 5; see also Patricia Allen, "Sustainable Agriculture at the Crossroads," *Capitalism, Nature and Socialism* 2, no. 3 (1991): 20–28.

25. The European allotments were primarily identified as the securing of individual rights to the land and served in part to undermine the securing of "common rights" or the broader community role to the land characteristic of the peasant movements such as the Diggers. The land-use association of gardens constituting an individual amenity as opposed to community value has continued to serve as a powerful constraint in the development of community food strategies in urban, particularly low-income communities. See Thomas J. Bassett, "Reaping on the Margins: A Century of Community Gardening in America," *Landscape* 25, no. 2 (1981): 1–8; Lewis Mumford, *The City in History: Its Origins, Its Transformations, and its Prospects* (New York: Harcourt, Brace & World, 1961), 527.

26. "Pingree's potato patches" were named after Detroit mayor Hazen S. Pingree who initiated a community garden program on city-owned unused vacant lots designed for the unemployed to develop their own food source, with potatoes the main staple. Though designed as a municipal program, it was more directly associated with the concept of "self-help," or "settlement work." It did this by providing "members of Coxey's Army the privilege to cultivate the vacant lots in our cities and have what they produce as a reward for their labor, [so] we can stop this tramping instantly," as Pingree put it. It was also seen as a potential inducement for those "who have failed to win a place in the city to seek work in the country," given the association of the potato patches as a form of farming or food production. See R.F. Powell, "Vacant Lot Gardens Vs. Vagrancy," *Charities* 13 (1904): 25–28; Frederick W. Speirs, Samuel McCune Lindsay, and Franklin B. Kirkbride, "Vacant-Lot Cultivation," *Charities Review* 8, no. 3 (1898): 74.

27. During World War I, more than one and a half million students were recruited to become part of the United States School Garden Army and ultimately converted 60,000 acres to productive land. The appeal to patriotism as a motivation failed to insure the program's survival, despite appeals to the post World War I cause of "fighting Bolshevism" as the "rallying cry of the Victory Gardeners." Joachim Wolschke-Bulmahn, "From the War Garden to the Victory Garden: Political Aspects of Garden Culture in the United States During World War I," *Landscape Journal* 11, no. 1 (1992): 57.

28. At its peak shortly before the U.S. entry into World War I, as many as three quarters of urban school district superintendents claimed in a Board of Education survey to be encouraging some form of gardening. Brian Trelstad, "Little Machines in Their Gardens: A History of School Gardens in America, 1891–1920," *Landscape Journal* 16, no. 2 (Fall 1997): 168; see also Henry Parsons, *Children's Gardens for Pleasure, Health, and Education* (New York: Sturgis & Walton Co., 1910), cited in Bassett, p. 3; M. Louise Greene, *Among School Gardens*, Russell Sage Foundation (New York: Charities Publication Committee, 1910); Louise Klein Miller, *Children's Gardens for School and Home: A Manual of Cooperative Gardening* (New York: D. Appleton & Co., 1904), 2; Henry L. Clapp, "School Gardens," *Education* 21 (June 1901): 611–617.

29. Florence Finch Kelly, "An Undertow to the Land: Successful Efforts to Make Possible a Flow of the City Population Countryward," *Craftsman* 11 (1906): 307–308.

30. Joanna C. Colcord and Mary Johnston, *Community Programs for Subsistence Gardens* (New York: Russell Sage Foundation, 1933), 11–12; Bassett, 2.

31. On the People's Park events, see Warren Belasco's discussion in *Appetite for Change*, 21. The *Berkeley Tribe* described People's Park as "the beginning of the Revolutionary Ecology Movement," cited in *Sources: An Anthology of Contemporrary Materials Useful for Preserving Personal Sanity While Braving the Great Technological Wilderness*, edited by Theodore Roszak (New York: Harper & Row, 1972), 393.

32. One 1977 USDA nationwide food survey indicated that as many as 50 percent of all suburban residents gardened and grew food. In urban areas, community gardening for physical activity as well as for food production and community activity grew more modestly, with one 1982 Gallup poll estimating that more than 10,000 community gardens were being utilized by more than 3 million urban gardeners. See Mark Francis, Lisa Cashdan, and Lynn Paxson, *Community Open Spaces: Greening Neighborhoods Through Community Action and Land Conservation* (Washington, D.C.: Island Press, 1984).

33. Cited in William Olkowski and Helga Olkowski, "Urban Agriculture: A Strategy for Transition to a Solar Society," in *Resettling America: Energy, Ecology and Community*, edited by Gary J. Coates (Andover, Mass.: Brick House Publishing, 1981), 339.

34. In 1995 a conference was held that extended the community garden/urban agriculture concept to include animal farming in the city for the "nutritional value, economic benefits, and healing presence of animals." Though raising animals remained highly restrictive and bound by obvious land use constraints, nevertheless a small but lively expansion of the "chicken in the backyard" strategy often associated with Third World countries emerged in several cities in the United States. On animal farming in the city, see *Can Mrs. O'Leary's Cow Come Home: Explorations in Urban Animal Agriculture* (Chicago: Heifer Project International, February 1996); "Urban Farms: How Green Is My Barrio," Ann Scott Tyson, *Christian Science Monitor*, December 1, 1996; on "worm farms," see "Worming In," Debra Bendis, *Christian Century*, January 21, 1998, 46; on the Nebraska National Guard role, see communication by Tom Kerr, Food Circles Networking Project, posted on the "comfood" list serve, July 14, 1999; on the rooftop garden movement and other inner-city and urban gardening movements see Patricia Hynes, *A Patch of Eden: America's Inner-City Gardeners* (White River Junction, Vt.: Chelsea Green Publishing Company, 1996); Ruth H. Landman, *Creating Community in the City: Cooperatives and Community Gardens in Washington, D.C.* (Westport, Conn.: Bergin and Garvey, 1993); Carolina A. Ojeda-Kimbrough, "Investing in Gardens: A Critical Assessment of Community Gardening Programs in Los Angeles," UCLA Department of Urban Planning, 1996.

35. On the Esperanza garden, see Matea Gold, "Development Uproots Urban Gardeners," *Los Angeles Times*, January 25, 1999; on the New York City situation, see Dennis Duggan, "Folks Seeing Red Over Losing Green," *Newsday*, January 19, 1999; Josh Getlin, "N.Y.'s Slices of Heaven in Trouble Despite Deal," *Los Angeles Times*, May 18, 1999; "Midler, Groups Team Up to Save N.Y. Gardens," *Salt Lake Tribune*, May 14, 1999; "Community Garden Preservation Alert" available at http://www.preserve.org/communitygardens. htm#background.

36. Gladys L. Baker et al., *Century of Service: The First 100 Years of the United States Department of Agriculture*, United States Department of Agriculture, Economic Research Service, Agricultural History Branch (Washington, D.C.: USDA, 1963), 313; Harvey Levenstein, *Paradox of Plenty: A Social History of Eating in Modern America* (New York: Oxford University Press, 1993), 88; on urban agriculture strategies in the Third World, see Scott G. Chaplowe, "Sustainable Prospects in Urban Agriculture," in *For All Generations*, 70–100; Jac Smit and Joe Nasr, "Urban Agriculture for Sustainable Cities: Using Wastes and Idle Land and Water Bodies as Resources," *Environment and Urbanization* 4, no. 2 (1992): 141–152; I. Wade, "Food and Self-Reliance in Third World Cities," Food-Energy Nexus Programme (Tokyo: United Nations University, 1987); Tara Garnett, "Farming the City: The Potential of Urban Agriculture," *The Ecologist* 26, no. 6 (November/December 1996).

37. Eric Bailey, "Gardening Blossoms as Industry in California," *Los Angeles Times*, December 13, 1998.

38. A survey of community garden groups in New York in 1998 indicated that more than 80 percent of the gardens (106 of 128 food-producing gardens surveyed) used organic growing techniques. "Survey Results," in *Vitis Vine*, Summer 1998, 13; on the estimate of the value of community garden produce, see "Community Gardens, Cultivating Communities" (Madison, Wisconsin: City of Madison Advisory Committee on Community Gardens, 1998); see also David Malakoff, "What Good Is Community Greening?" *Community Greening Review* 5 (1995): 4–11.

39. Gail Feenstra, Sharyl McGrew, and David Campbell, *Entrepreneurial Community Gardens: Growing Food, Skills, Jobs and Communities* (Davis, Cal.: University of California, 1999), Agriculture and Natural Resources Publication 21587, 72–77.

40. Laura Lawson and Marcia McNally, "Rethinking Direct Marketing Approaches to Low and Moderate-Income Communities and Urban Market Gardens" (Davis: University of California Sustainable Agriculture Research and Education Program, October 1998).

41. "The Retail Initiative—Models for Success" (New York: Local Initiatives Support Corporation, 1996); James J. O'Connor and Barbara Abell, *Succeessful Supermarkets in Low-Income Inner Cities*, Food and Nutrition Service (Alexandria, Va.: U.S. Department of Agriculture, August 1992); "Joint Venture in the Inner City—Supermarkets General Corporation and New Community

Corporation: A FMI Case Study" (Washington, D.C.: Food Marketing Institute, 1996); Pamela Fairclough and Shelly Herman, *Developing Successful Neighborhood Supermarkets in New York City: A Guide for Community-Based Organizations* (New York: Community Food Resource Center, 1994); testimony of William J. Linder, founder, New Communities Corporation, before the Select Committee on Hunger, U.S. House of Representatives, September 30, 1992; Food Marketing Institute, "Pathmark and New Community Corporation—Joint Venture Helps Revitalize Newark," *FMI Issues Bulletin*, January 1993; *Cities and Supermarkets: Partners in Progress—Case Studies of Successful Collaborative Programs* (Washington, D.C.: National League of Cities and Food Marketing Institute, 1995).

42. Shopping cart loss has been a significant cost to inner-city markets, both in resecuring abandoned carts (where fees are paid to contractors who gather the carts on nearby streets on a per cart basis) and replacement of carts damaged or not found. One inner-city market executive from Food 4 Less estimated, in an interview with one of the *Seeds of Change* group, that shopping cart losses could be $50,000 or more annually at a single market. But the same executive, while acknowledging that the losses were attributable to transportation problems for shoppers, declared no interest in establishing a program such as a supermarket van to increase access for his existing or potential customers. "Shopping cart loss is our transportation program," the executive declared. Cited in Mark Dohan, "An Analysis of Supermarkets in Los Angeles County," Department of Urban Planning, University of California at Los Angeles, 1994; see also Kim Nauer, "Food Flight: The Loss of the Neighborhood Grocer," *Neighborhood Works*, February–March 1994.

43. Personal communication with Gloria Ohland, Surface Transportation Policy Project, September 19, 1998; "Policy Committee Update," *Community Food Security News*, Summer 1998, 8; *Homeward Bound*, 37–54; Gloria Ohland and Cynthia Pansing, "Highway to Hunger," *Community Food Security News*, Spring/Summer 1997, 3–4.

44. The *Progressive Grocer* comment is cited in the article by Stephen Bennett, "The Lure of Local Produce," *Progressive Grocer*, 128.

45. Steve Weinstein, "Khaledi's L.A. Law," *Progressive Grocer*, August 1994, 124–128; Melinda Fulmer, "Independent Stores Bag Neglected Ethnic Niche," *Los Angeles Times*, August 18, 1998, Alecia Vargas, "Harvard MBAs' Chain Riles Chicago's Hispanic Grocers," *Wall Street Journal*, November 6, 1997.

46. Morris L. Sweet, "History of Municipal Markets, *Journal of Housing* 18 (June 1961): 237–247.

47. Leveen, 120; see also John L. Hess, "Return of a Farmers' Market," *Organic Gardening and Farming* 21 (October 1974): 64–67; Carole Turko, "Tips From Other Farmers' Markets," *Organic Gardening and Farming* 23 (June 1976): 111–113.

48. Hal Linstrom and Pete Henderson, "Farmer-to-Consumer Marketing," *National Food Review* (Washington, D.C.: USDA, Fall 1979), 22–23; Michael

Schaaf, "Challenging the Modern U.S. Food System: Notes from the Grassroots," in *Sustainable Food Systems*, edited by Dietrich Knorr (Westport, Conn.: AVI Publishing Co., 1983), 279–301; Robert Sommer, "Farmers' Markets: Myths and Realities," *California Agriculture*, February 1979, 12–13.

49. Presentation by Harry Brown-Hiegel, "History of Farmers' Markets in Southern California," in the minutes from the Southland Farmers' Market Association Retreat, November 17, 1996; at the same time, the successful development of programs such as the Farmers' Market nutrition program, which provided low-income women in the Women, Infants, and Children federal food assistance program with coupons to shop in a farmers' market, further indicated the potential to extend the food in the city strategy to include a low-income base of support. See "Background to the Gardena Farmers' Market," memo from Marion Kalb, director, Southland Farmers' Market Association, June 4, 1996; information on the Women, Infants, and Children Farmers' Market Nutrition Program is available at (http://www.attra.org/guide/marketing/wicfmnp.html).

50. The USDA survey results were presented by USDA Agriculture and Marketing Service official Arthur Burns at a 1999 workshop in Santa Fe on "Farmers' Market Development and Minority Participation," whose proceedings are available at http://www.ams.usda.gov/directmarketing/santafe.htm; the USDA also provides, in its *National Directory of Farmers' Markets* (USDA, 1998), a listing of farmers' markets in each state, and information about their substantial growth in the mid- and late 1990s.

51. In a commentary for the trade publication for California food retailers, farmers' markets, in contrast to supermarkets, brought people together once a week to "create a community—a gathering of villagers." "It arouses a loyalty from participants," the commentary noted, "that's expressed in dollars changing hands." "Lessons from the Midvale Farmers' Market," *California Grocer*, May 1995; see also Mary Ann McGrath, John F. Sherry, and Deborah D. Heisley, "An Ethnographic Study of an Urban Periodic Marketplace: Lessons from the Midvale Farmers' Market," *Journal of Retailing* 69, no. 3 (Fall 1993): 280–319; Carin Rubenstein, "Alienation in Supermarkets," *Psychology Today*, July 1981. Rubenstein noted that a study of shopping patterns of 900 supermarket shoppers and 248 farmers' market shoppers found that only 16 percent of the supermarket shoppers were with other adults in contrast with 75 percent of the farmers' market shoppers, and that only one in ten of the supermarket shoppers chatted socially with other customers, while two-thirds of the farmers' market customers did so.

52. In a 1978 article detailing the reemergence of urban markets, urban planner Padraic Burke noted that the informality of markets, with their "unstructured social environment" were capable of transforming "a normally mundane activity into a different urban experience." Burke also identified the urban market as "one of the few places in the city where economic and social distinctions tend to fade." Though the rise of the middle-class farmers' markets in the late 1990s undercut some of that egalitarian experience, the existence of the "bridge"

markets (such as the Pico market described in the introduction in this book) points to the possibility of creating truly cross-class, cross-cultural urban places. Padraic Burke, "Reviving the Urban Market: 'Don't Fix It Up Too Much,'" *Nation's Cities*, February 1978, 11; see also Andy Fisher, "Hot Peppers and Parking Lot Peaches: Evaluating Farmers' Markets in Low-Income Communities," report prepared for the University of California Sustainable Agriculture Research and Education Program (Los Angeles: Community Food Security Coalition, January 1999).

53. Harvey A. Levenstein, *Revolution at the Table: The Transformation of the American Diet* (New York: Oxford University Press, 1988), 50, 105.

54. The National School Lunch program's mission was defined as safeguarding "the health and well-being of the Nation's children" by "promoting health" and "preventing disease," while also emphasizing the link between childhood and adult dietary patterns. "What children eat helps determine not only how healthy they are as children, but how healthy they will be as adults," according to the program's framing document (59 FR 30225); The surgeon general's comments are cited in *Paradox of Plenty*, 65; see also *The National School Lunch Program*, Washington D.C.: USDA, Food and Nutrition Service, FNS-78, revised December 1982; M.L. Wilson, "Nutrition and Defense," *Journal of the American Dietetic Association* 17 (January 1947): 12–20.

55. While the National School Lunch program had become available by the 1960s to more than 18 million children at three-quarters of all the schools in the country, only about half of those schools actually participated. These schools included a large number of students who would have qualified for free or reduced lunches. According to the USDA administrator in charge of the program, reasons for nonparticipation were often financial, with cafeteria managers assuming they could serve cheaper meals or ultimately identify ways to make the school meal "a profitable proposition." See testimony of Howard P. Davis, deputy administrator, consumer food programs, U.S. Department of Agriculture, in *National School Lunch Act, Hearing Before the Select Subcommittee on Education of the Committee on Education and Labor*, House of Representatives, 89th Cong., 2d sess., July 21, 1966.

56. *School Lunch Program: Cafeteria Managers' Views on Food Wasted by Students*, U.S. General Accounting Office, Report to the Committee on Economic and Educational Opportunities, House of Representatives (Washington, D.C.: U.S. GAO, July 1996); A 1962 text on the school lunch program began the chapter on menu planning by focusing on "what foods may be purchased by USDA for distribution." Marion L. Cronan, *The School Lunch* (Peoria, Illinois: Chas. A. Bennett Company, 1962). On the 1962 orange juice episode, see Ardith L. Maney, *Still Hungry After All These Years*, 37; on the beef and salmon sales issue, see "Be on the Lookout for More Beef on 1999 Menus," *FoodService Director* 12, no. 1 (January 15, 1999): 20, and "Bonus Salmon Coming," *FoodService Director*, September 15, 1999, 70.

57. Advocates like the Food Research and Action Committee (FRAC) had received significant media attention by identifying visible examples of what the Reagan lunch meal would have looked like: e.g., french fries, ketchup, meat and soy patty, one slice of white bread, and a reduced portion of milk. It was in this period that Reagan's memorable comment about ketchup qualifying as a vegetable further exacerbated the concern that any nutritional component of the school lunch had basically been undermined. Maney, 134–135. Judy Jones Putnam and Michael G. Van Dress, "The Changing Food Mix in the Nation's Schools," *National Food Review*, 18, Spring 1982, 16–20.

58. "This epidemic in the U.S." quote is cited by Gerg Critser, "Let Them Eat Fat: The Heavy Truths about American Obesity," *Harper's*, March 2000, 42; The obesity issues are discussed in Ali H. Mokdad et al., "The Spread of the Obesity Epidemic in the United States, 1991–1998," *JAMA* 282, no. 16 (October 27, 1999): 1519–1522, and David S. Ludwig et al., "Dietary Fiber, Weight Gain, and Cardiovascular Disease Risk Factors in Young Adults," *JAMA* 282, no. 16 (October 27, 1999): 1539–1546; the problem of lack of engagement could be seen, for example, in a 1995 conference on "Healthy Eating for Children." The conference brought together policy people, academics, and the food industry to talk about "the nutritional well-being of America's children." In identifying what strategies ought to be pursued (e.g., 53 percent wanted to establish a consensus on appropriate messages for children), only 1 percent said they wanted to work on recognizing and involving children in programs. This was also true despite the fact that teachers were spending 11 to 15 hours per year directly teaching "nutrition," but often with little connection to *what the children were actually eating in the school.* See Rebecca M. Mullis, Anita Owen, and Lin Blaskovich, "National Action Conference on Healthy Eating for Children: A Policy Dialogue," *Journal of Nutrition Education* 27, no. 5 (September/October 1995): 224; "Position of ADA, SNE, and ASFA: School-based Nutrition Programs and Services," *Journal of the American Dietetic Association* 95, no. 3 (March 1995): 367; see also "A Review of School Food Service Research," Leslie A. Lytle, Steven H. Kelder, and M. Patricia Snyder, *School Food Service Research Review* 17, no. 1 (1993): 7–14; Eileen Kennedy and Jeanne Goldberg, "What Are American Children Eating: Implications for Public Policy," *Nutrition Reviews* 53, no. 5 (May 1995): 114.

59. One interesting outcome of an exclusive contract arrangement involved a student at Greenbriar High School in Evans, Georgia, who was suspended for wearing a Pepsi shirt on his school's "Coke Day." School officials had launched Coke Day to show regional Coke officials that the school was highlighting Coke in order to win a $500 contest. In another episode, twenty students at Virginia Commonwealth University undertook a sleep-in outside the VCU Vice President's office, protesting the school's contract with McDonald's providing the fast food company a twenty-year monopoly at the Student Commons. See Michelle Mascarenhas and Robert Gottlieb, "Healthy Farms, Healthy Kids," *Community Food Security News*, Fall 1998, 6; Liza Featherstone, "The New Student

Movement," *The Nation*, May 15, 2000; Constance L. Hays, "Lessons," *New York Times*, April 19, 2000.

60. The "exactly what they want" quote is from "Nuggets Get Top Grades," *FoodService Director* 12, no. 7 (July 15, 1999): 94; the Frito-Lay episode is described in Jane Levine, "Creating Consumers (How the Food Industry Delivers Its Products and Messages to Elementary School Students and What Nutrition Professionals Know and Think About It)" (Ph.D. diss., Teachers College, Columbia University, 1998), 70–71.

61. Cited in "Creating Consumers," 72; see also Steven Manning, "Students for Sale: How Corporations Are Buying Their Way into America's Classrooms," *The Nation* 269, no. 9 (September 27, 1999): 11–18.

62. Toward that end, one school in Hershey, Pennsylvania, even got an art school class to develop a brand logo for the school's nacho cheese and chips machine similar to what was sold commercially in order to achieve student acceptance. See "Pizza Concepts Rule Lunchrooms: Menu Branding in Schools," Amanda Chater, *FoodService Director*, March 15, 1999, 12s; the same publication also cites an International Foodservice Manufacturers Association study on the value of consumer branding to food service operators, including schools. Foodservice outlets, the study noted, "find it difficult to create a reputation for quality. Offering popular customer brands can help create such an image." "IFMA Study Identifies Key Brand Values," *FoodService Director*, March 15, 1999, 2s. At the same time, the value of the school cafeteria relationship for the fast-food companies and the brand name companies was also seen as representing an opportunity in terms of *future sales*—the "consumers today" becoming "the buyers of tomorrow," as Vance Packard so effectively described this perspective more than forty years earlier in *The Hidden Persuaders* (New York: Pocket Books, 1958), 158; see also Naomi Klein, *No Logo: Taking Aim at the Brand Bullies*, New York: Picador USA, 1999; Regarding overall trends, see U.S. General Accounting Office, "School Lunch Program: Role and Impacts of Private Food Service Companies" (Washington, D.C.: August 1996), GAO/RCED/96-217.

63. "The Link Between Nutrition and Cognitive Development in Children," Center on Hunger, Poverty and Nutrition Policy, Tufts University, 1998; on the caffeinated products issue, see Helen Cordes, "Generation Wired: Caffeine Is the New Drug of Choice for Kids," *The Nation*, 266, no. 15 (April 27, 1998): 11–16; Melinda Fulmer, "Products Stimulate Debate," *Los Angeles Times*, September 26, 1998.

64. I. Contento et al., "Nutrition Education for School-aged Children," *Journal of Nutrition Education* 27 (November–December 1995), 298; Leslie Lytle and Cheryl Achterberg, "Changing the Diet of America's Children: What Works and Why?" *Journal of Nutrition Education* 27 (September–October 1995): 252; Connie Evers, "The Childhood Obesity Epidemic," *School Foodservice and Nutrition*, March 1999, 21–23; Juneal Smith et al., "Prevalence of Self-Reported Nutrition-Related Health Problems in the Lower Mississippi Delta," *American Journal of Public Health* 89, no. 9 (September 1999): 1418–1421.

65. Andrea Azuma and Robert Gottlieb, *School Gardens in the LAUSD: Barriers and Opportunities—A Survey Assessment*" (Los Angeles: Urban Environmental Policy Institute, 2000).

66. One event related to the appearance issue involved a bag of oranges speckled with markings from thrips (tiny bugs who eat tiny flecks of the peels from oranges facing the sun). The McKinley cafeteria manager had pulled the oranges from the salad bar, assuming that the principal would not only agree that the oranges were inappropriate but that this type of problem warranted ending the pilot. The principal asked whether the children had been eating the oranges that day. They had in fact been popular items, since the thrip markings, an indication of no spray, also indicate sweetness. The principal decided to keep the oranges in the salad bar since it was a good educational experience for the children to see and taste food grown differently. On the McKinley pilot, see Robert Gottlieb, "McKinley Students Lead Lunch Revolution," *The Outlook*, September 16, 1997.

67. Robert Gottlieb, Jennifer Wilkins, and Andrea Azuma, *Farm-School Connections: A New Framework for Nutrition Education and Community Food Security*, Los Angeles: Occidental Community Food Security Project, May 2000.

68. The numbers of children selecting the salad bar option (as opposed to the hot meal options such as pizza or fried tacos) at the Chinatown school averaged a little less than half of all the students eating cafeteria lunches during its first three months (350 of 800 students). These numbers were achieved despite significant logistical problems due to its unexpected popularity, including long lines of students wishing to select the salad bar (some teachers in fact sent their students to the hot meal area, since there were no lines for the prepackaged meal). More than 90 percent of the primarily Chinese and Latino children in the Chinatown school qualified for free or reduced meal programs, which further underlined the popularity of the salad bar *among low-income children*. This had also been a factor in the Santa Monica experience (where the salad bar option had been *more popular* at the schools where there were greater numbers of students on free or reduced meal plans). Occidental Community Food Security Project, *Healthy Kids, Healthy Farms: Final Report to the California Wellness Foundation* (Los Angeles: Occidental College, December 1999).

69. Author's notes, "School-Food Workshop," USDA Food and Nutrition Service and Community Food Security Coalition, Washington, D.C., December 10, 1999; memo from Robert Gottlieb to Shirley Watkins, undersecretary, Food and Nutrition Services, USDA, November 17, 1999; USDA Food and Nutrition Service and the Community Food Security Coalition, "The Farm-School Connection: An Informal Collaboration between the Community Food Security Coalition and USDA" (Washington, D.C. and Venice, Cal.: USDA and CFSC, 2000). Minutes, "Farm-to-School Workshops" sponsored by the Community Alliance with Family Farmers, January 19, 2000 (Ojai, California) and February 11, 2000 (Davis, California).

70. Marilyn Barrett, *Creating Eden: The Garden as Healing Space* (New York: Harper & Row, 1992); Mitchell Hewson, *Horticulture as Therapy* (Guelph, Ontario: Homewood Health Centre, 1994); Diana Balmori and Margaret Morton, *Transitory Gardens, Uprooted Lives* (New Haven: Yale University Press, 1993); the main horticulture therapy organization, the American Horticulture Therapy Association, has information available through its web site at http://www.ahta.org/info.html

71. Hope Mohr, *Gardens for Survivors: A Feasibility Analysis for Developing Healing and Food Security Strategies for Survivors of Domestic Violence*, The Occidental Community Food Security Project (Los Angeles: Urban Environmental Policy Institute, February 1998).

72. Paul Racko, Carol Williams, and Melissa Yates, "Project GROW Takes Root," *The Grapevine*, 1, no. 1, Summer/Fall 1999, 1.

7   Pathways to Change: A Conclusion

1. The 1972 study by the Council of Environmental Quality is cited by Ann L. Riley in *Restoring Streams in Cities* (Washington, D.C.: Island Press, 1998), xx.

2. "Re-envisioning the Los Angeles River: A Program of Community and Ecological Revitalization" (Los Angeles: Urban and Environmental Policy Institute, 1999–2000). Information available at (www.lariver.oxy.edu).

3. Charles F. Wilkinson, *The Eagle Bird: Mapping a New West* (New York: Pantheon Books, 1992).

4. Sanford Lewis, "Citizens as Regulators of Local Polluters and Toxic Users," *New Solutions* 1, no. 1 (1990): 20–21.

5. Christopher Gunn and Hazel Dayton Gunn, *Reclaiming Capital: Democratic Initiatives and Community Development* (Ithaca: Cornell University Press, 1991), 107, 3; see also Cornelia Butler Flora, "Social Capital and Sustainability: Agriculture and Communities in the Great Plains and the Corn Belt," Journal Paper no. J-16309, Department of Sociology, Iowa State University, 1995.

6. Lynn Kaatz Chary, "Pollution Prevention and Income Protection: Fighting with Empty Hands—A Challenge to Labor," in *Work, Health, and Environment: Old Problems, New Solutions*, edited by Charles Levenstein and John Wooding (New York: The Guilford Press, 1997), 453.

7. Jan Mazurek, *Making Microchips: Policy, Globalization, and Economic Restructuring in the Semiconductor Industry* (Cambridge: MIT Press, 1999), 167–197; Janice Mazurek, Sanford Lewis, and Virginia Parks, *Pollution Prevention Strategies and Decision-Making in the Electronics Industry Sector*, a report to the Liberty Hill Foundation (Los Angeles: Pollution Prevention Education and Research Center, 1999).

8. Campaign for Responsible Technology, "What Is Sematech," http://www.svtc.org/svtc/sematech.htm; "The Bargaining Chip: Bulletin of the Electronic

Industry Good Neighbor Campaign," A Joint Collaboration of the Campaign for Responsible Technology and the Southwest Network for Environmental and Economic Justice (San Jose and Albuquerque: CRT, 1997); "Milestones: Silicon Valley Toxics Coalition History 1982–1996," *Silicon Valley Toxics Action* 15, no. 1, 1997, 7; Patricia O'Hara et al., *Design for Environment, Safety, and Health (DFESH): Implementation Strategy for the Semiconductor Industry*, Washington, D.C., Sematech, November 30, 1995, Doc. #: 95103006A-ENG.

9. Maureen Smith, "Perspectives on the U.S. Paper Industry and Sustainable Production," *Journal of Industrial Ecology* 1, no. 3 (Summer 1997): 69–85; and "Roundtable on the Industrial Ecology of Pulp and Paper," *Journal of Industrial Ecology*, Summer 1997, 87–114; the elaboration of the "carbohydrate economy" approach of the Institute for Local Self-Reliance can be found at their Web site, www.carbohydrateeconomy.org.

10. See John P. Kretzmann and John L. McKnight, *Building Communities from the Inside Out: A Path toward Finding and Mobilizing a Community's Assets*, Center for Urban Affairs and Policy Research, Northwestern University (Evanston: Neighborhood Innovations Network, 1993).

11. Sam Bass Warner, "Eco-Urbanism and Past Choices for Urban Living," in *Urban Policy in Twentieth-Century America*, edited by Arnold R. Hirsch and Raymond A. Mohl (New Brunswick: Rutgers University Press, 1993), 226.

12. Frederick H. Buttel, "Rethinking International Environmental Policy in the Late Twentieth Century," in *Environmental Justice: Issues, Policies, and Solutions*, edited by Bunyan Bryant (Washington, D.C.: Island Press, 1995), 191.

Index